# Chiral Recognition in the Gas Phase

# Contents

The Editor ........ vii
The Contributors ........ ix
Introduction ........ xi

**Chapter 1** Valence Photoelectron Circular Dichroism of Gas Phase Enantiomers ........ 1

*Laurent Nahon and Ivan Powis*

**Chapter 2** High-Resolution Microwave Spectroscopy of Chiral Molecular Contact Pairs ........ 27

*Xunchen Liu and Yunjie Xu*

**Chapter 3** Infrared and Raman Detection of Transient Chirality Recognition in the Gas Phase: The Case of Ethanol ........ 39

*Martin A. Suhm*

**Chapter 4** The Role of Deformation Energy of Bifunctional Entities on the Formation of Diastereoisomers ........ 47

*Katia Le Barbu-Debus*

**Chapter 5** Chiral Recognition in Mass Spectrometry, Focusing on FAB Mass Spectrometry ........ 61

*Motohiro Shizuma*

**Chapter 6** Enantioselectivity in Gas-Phase Ion-Molecule Reactions ........ 87

*Maurizio Speranza*

**Chapter 7** Equilibrium Methods for Characterizing Gas Phase Chiral Recognition ........ 133

*David V. Dearden and Nannan Fang*

**Chapter 8** Deoxy Oligonucleotides as Chiral References for the Discrimination of Enantiomeric Amino Acids under Mass Spectrometry ........ 143

*M. Vairamani and Sangeeta Kumari*

**Chapter 9** Evaluating the Enantioselectivity of Asymmetric Catalytic Reactions and Screening Chiral Catalysts by ESI-MS ........ 167

*Hao-Yang Wang and Yin-Long Guo*

**Chapter 10** Solution Phase vs. Gas Phase Chiral Recognition by ESI-MS: A Case Study of Two Chiral Selector Classes ........ 181

*Kevin A. Schug, Aruna B. Wijeratne, Bilal H. Bazzi, and Daniel W. Armstrong*

**Chapter 11** Recognition of Amino Acid Chirality in Polypeptide Ions by MS/MS ........ 205

*Hongqian Yang and Roman A. Zubarev*

**Index** ........ 219

# The Editor

Anne Zehnacker was born in 1962 and is directeur de recherche at the French Center for Scientific Research (CNRS). She began studying chemistry in Strasbourg (France) and earned a PhD in the electronic spectroscopy of aromatic molecules at the Orsay University in 1988. She spent one year in the theoretical chemistry group in CEN Saclay. Dr. Zehnacker is a member of the French Chemical Society and has been an invited scientist at several universities, including Sendai (Japan), Warsaw (Poland), Seoul (South Korea), Melbourne (Australia), Göttingen (Germany), and Toledo (Spain). She serves as a member of the advisory committee of Physical Chemistry Chemical Physics (PCCP). Her work focuses on molecular interactions and photoinduced processes in clusters. She was awarded the CNRS bronze medal in 1992 and the prize from the physical chemistry division of the French Chemical Society in 2003 for her work on chiral recognition in jet-cooled complexes.

# The Contributors

**Daniel W. Armstrong**
Department of Chemistry and Biochemistry
The University of Texas at Arlington
Arlington, Texas

**Bilal H. Bazzi**
Department of Chemistry and Biochemistry
The University of Texas at Arlington
Arlington, Texas

**David V. Dearden**
Department of Chemistry and Biochemistry
Brigham Young University
Provo, Utah

**Nannan Fang**
Department of Chemistry and Biochemistry
Brigham Young University
Provo, Utah

**Yin-Long Guo**
Shanghai Mass Spectrometry Center
Shanghai Institute of Organic Chemistry
People's Republic of China

**Sangeeta Kumari**
Indian Institute of Chemical Technology
Hyderabad, India

**Katia Le Barbu-Debus**
Laboratoire de Photophysique Moléculaire
CNRS Université Paris XI
Orsay, France

**Xunchen Liu**
Department of Chemistry
University of Alberta
Edmonton, Canada

**Laurent Nahon**
Synchrotron Soleil
St. Aubin, France

**Ivan Powis**
School of Chemistry
University of Nottingham
Nottingham, United Kingdom

**Kevin A. Schug**
Department of Chemistry and Biochemistry
The University of Texas at Arlington
Arlington, Texas

**Motohiro Shizuma**
Osaka Municipal Technical Research Institute
Osaka, Japan

**Maurizio Speranza**
Dipartimento di Chimica e Tecnologie del Farmaco
Sapienza-Università di Roma
Rome, Italy

**Martin A. Suhm**
Institut für Physikalische Chemie
Georg-August-Universitat
Göttingen, Germany

**M. Vairamani**
Indian Institute of Chemical Technology
Hyderabad, India

**Hao-Yang Wang**
Shanghai Mass Spectrometry Center
Shanghai Institute of Organic Chemistry
People's Republic of China

**Aruna B. Wijeratne**
Department of Chemistry and Biochemistry
The University of Texas at Arlington
Arlington, Texas

**Yunjie Xu**
Department of Chemistry
University of Alberta
Edmonton, Canada

**Hongqian Yang**
Department of Medical Biochemistry and Biophysics
Karolinska Institutet
Stockholm, Sweden

**Roman A. Zubarev**
Department of Medical Biochemistry and Biophysics
Karolinska Institutet
Stockholm, Sweden

# Introduction

*Anne Zehnacker*

Chirality is pervasive in nature and has long fascinated humans. The first step toward understanding chirality occurred during the nineteenth century, when French physicist J. B. Biot discovered that organic substances in solution, like tartaric acid, can rotate the polarization plane of a linear polarized light. It was only thirty-six years later when L. Pasteur related optical rotation to a molecular property. He observed that a peculiar form of tartaric acid exceptionally present in wine didn't provoke any rotation of linear polarized light. He realized that sodium ammonium salts of this para-tartaric acid, also called racemic acid, actually contained right- and left-handed crystals.[1] The optical rotation was recovered in solutions of pure left- or right-handed crystals sorted out with tweezers. What Pasteur called molecular asymmetry was a property of the molecule itself. It was only in 1873 that this molecular property was postulated to originate from stereochemical factors, when J. H. van't Hoff and J. A. Le Bel proposed the notion of an asymmetric carbon atom. The name *chirality* was proposed by Lord Kelvin, who later gave the following definition, in the Baltimore Lectures on Molecular Dynamics and the Wave Theory of Light: "I call any geometrical figure, or group of points, chiral, and say it has chirality, if its image in a plane mirror, ideally realized, cannot be brought to coincide with itself."[2]

The importance of Pasteur's discovery was not only the existence of molecular asymmetry, but its intimate relation to life chemistry. Indeed, chiral tartaric acid was produced during wine fermentation, a process driven by bacteria. Most of the bricks of life are chiral, and nature has, one does not know how, made a choice between the two enantiomers. From Pasteur's intuition about "dissymmetry of the cosmic forces" to the discovery of parity violation in cesium atoms, physicists and chemists seek to understand chirality. They follow different approaches, however. Physicists search for the consequences of parity violation of weak electron-nucleus interactions, which must result in an absolute energy difference between enantiomers, with spectroscopic consequences.[3] The other physically based method for probing molecular chirality, namely, chiroptical spectroscopy, rests on the interaction between polarized light and chiral molecules. Recent books have been devoted to the principles of optical activity and its application.[4,5] In particular, circular dichroism relies upon differential interaction with a circular polarized light and consists in measuring the difference in absorption by a chiral molecule of a right and a left circular beam. Thought very small, this effect is well documented in the range of both electronic and vibrational transitions. Because of their limited magnitude ($10^{-6}$ to $10^{-2}$ of the absorption), these methods are mostly limited to a condensed phase. More recent is the observation, in the gas phase, of the photoelectron circular dichroism (PECD) spectroscopy effect, which will be described in Chapter 1. Notwithstanding its magnitude (several orders of magnitude greater than conventional absorption), PECD is exquisitely sensitive to stereochemical factors. It therefore has the long-term capability of studying chiral recognition in molecular pairs isolated in the gas phase. Moreover, there are attempts to explain life's homochirality by asymmetric processes induced by astronomical sources of circular polarized light. PECD, as an asymmetric photon-induced process, could play a role in the origin of life's homochirality, which will also be discussed in Chapter 1.

The chemist does not intervene at the atomic scale or at a single molecule, but at molecular interaction. Indeed, all the chemical approaches rest on chiral discrimination, i.e., the difference in behavior of the two enantiomers of a chiral molecule when interacting with a chiral surrounding. This phenomenon plays a key role in life chemistry: it is, for example, well recognized that the biological activity and bioavailability of enantiomers often differ. This is true for the activity of drugs, but also for smell and taste.[6] The enantiomers of carvone have characteristic odors, caraway for (S)-(+)-carvone and spearmint for (R)-(–)-carvone.

The marked enantioselectivity characterizing most of the processes involving the interaction of a chiral ligand, like a drug with enzymes or protein receptors, has been explained in terms of formation of weakly bound contact pairs involving specific interactions.

The economic importance of enantiomerically pure compounds has prompted growing development and application of chiral chromatographic methods, especially in pharmaceutical sciences.[7] Here again, specific interactions between the chiral stationary phase and the enantiomers to separate are invoked.[8] However, the interaction energy at stake is often weak, i.e., of the same order as the thermal energy at room temperature; the contact pairs responsible for chiral recognition are difficult to isolate and study in solution. It is therefore of prime importance to find a means of studying them at the molecular level, in order to cast some light on the molecular interactions responsible for chiral discrimination in isolated molecular pairs. Gas-phase experiments, on either neutral or ionic adducts of chiral molecules, allow studying the intrinsic properties of chiral recognition in solvent-free conditions. Two directions have been explored so far: the structural and the energetic aspects. Structural aspects have been studied mainly by optical spectroscopy in jet-cooled conditions, in neutral complexes.[9] The combination of supersonic expansion and electronic, vibrational, or microwave spectroscopy has led to a flurry of experimental results that, in conjunction with quantum chemical calculations, brings information on the structure of weakly bound complexes of chiral molecules and the nature of the interactions responsible for chiral recognition. Electronic spectroscopy is limited to complexes containing an aromatic chromophore and only brings indirect structural information, even when mass resolved by using resonance-enhanced multiphoton ionization (REMPI) methods.[9–12] More information is brought by vibrational spectroscopy, which can be obtained, on the one hand, by IR-UV double resonance experiments. In this case, however, an aromatic chromophore is necessary as the method is based on the depletion of a fluorescence or ion signal due to IR absorption. More universal is direct Fourier-transformed infrared (FTIR) absorption or Raman scattering from species cooled down in a slit jet,[13] because these techniques do not require the presence of an aromatic chromophore. Clusters of small molecules can be studied, which are amenable to high-level quantum chemistry calculations. By this way, subtle effects like those related to transient chirality or chirality linked to the nonequivalence of lone pairs can be studied. An example of such an approach is given in Chapter 3, which describes how infrared and Raman spectroscopy can be used to evidence transient torsional chirality in ethanol derivatives. The two *gauche* forms of ethanol are transient enantiomers, which interconvert on a timescale of the order of 1 ps. Hydrogen bond formation from a donor to ethanol quenches the stereomutation. Another method for detecting quenching of stereomutation by hydrogen bonding is microwave spectroscopy, as shown in Chapter 2. The individual conformers of transient chiral molecules, like the gauche form of ethanol, are discriminated by formation of a complex with a permanently chiral species, like *R*,*R*-dimethyloxirane. Despite microwave spectroscopy being applicable to any molecular system with permanent dipole moment, it meets difficulties when applied to larger systems. The dimers of 2-butanol, which can be seen as the smallest truly chiral organic molecule, are already challenging in terms of the interpretation of experimental spectra.[14]

Larger molecular systems demand different approaches. The experiments often rest on medium-resolution infrared spectroscopy combined with theoretical methods with decent computational time, like those resting on the density functional theory. In this respect, numerous complexes of an aromatic chiral chromophore with chiral solvating agents have been studied.[10,15] These studies resort to electronic spectroscopy, by either laser-induced fluorescence (LIF) or REMPI. In both cases, the spectroscopy of the $S_0$-$S_1$ transition is the signature of chiral recognition and is necessary for further investigation, either energetic or structural. From an energetic point of view, REMPI experiments allow measuring the binding energy of the diastereomeric complexes, thanks to a two-color dissociative ionization scheme. The appearance threshold of the fragment resulting from photodissociative ionization of the complex, AP, is related to the binding energy of the complex BE and the adiabatic ionization potential of the chromophore $IP_{ad}$ by the simple equation $BE = AP(Rs^+ \rightarrow R^+) - IP_{ad}(R)$.[16,17] From a structural point of view, the knowledge of the $S_0$-$S_1$ spectroscopy is a prerequisite for recording vibrational spectra via double resonance experiments. Besides its sensitivity, this method has the advantage of being isomer selective, as it allows recording separately the spectra of different species, which absorb in the same energy range. It has been applied to complexes of a chiral alcohol chromophore with aminoalcohols, in which a conformation-dependent interaction site ($NH_2$ or OH) has been observed.[18] More recently, it has been used to compare the efficiency of different chiral chromophores in discriminating between the enantiomers of methyl-lactate.[19,20] It has been shown that the interaction that ensures the stability of the complex (strong conventional hydrogen bonds) is not the same as the one that is responsible for chiral recognition. Indeed, minor CH … π or dispersive interactions play a decisive role in chiral recognition,[20] a hypothesis that has been postulated already in complexes of chiral alcohols.[10] Some of these topics have been reviewed already.[9,12,21] An important issue is that supersonic expansions are not a medium in thermodynamic equilibrium. The formation of complexes is mainly governed by kinetic factors, which in turn play an important role in the chiral recognition efficiency. These aspects have been studied in detail thanks to IR-UV double-resonance spectroscopic studies accompanied by quantum chemical calculations, with either *ab initio* or density functional theory (DFT) methods. They will be illustrated with the example of complexes involving molecules with an intramolecular hydrogen bond in Chapter 4.

Mass spectrometry approaches of chiral recognition in ionic complexes have grown in number during the last decades and have been the subject of several review articles.[22–25]

As for neutral species, chiral recognition in ionic systems rests on the formation of complexes involving enantiospecific interactions. These diastereoisomeric adducts are endowed with different stability, which leads to thermodynamic enantioselectivity ($\Delta\Delta G \neq 0$). They also show different activation barrier to reactivity, which results in kinetic enantioselectivity ($\Delta\Delta G^* \neq 0$). All the methods at the basis of mass spectrometry study of chiral recognition rest on one of these properties. They are basically classified in terms of the following three approaches.

The first one rests on the comparison of the relative peak intensity of diastereoisomeric adducts in a single-stage mass spectrometry experiment. As the two adducts must be measured during the same experiment to avoid artifacts, one of the enantiomers of either the reference compound or the analyte must be mass tagged, usually in a substituent remote to the chiral center, so that the mass of the molecule can be correlated with its absolute configuration. A way of doing that rests on the use of isotopically labeled species, so that the corresponding mixture of the diastereoisomeric adducts can be mass resolved. This approach is valid provided that the stereochemical effect is not altered by isotope effects and is described in Chapter 5. This chapter describes how the intensity of the ions produced by fast atomic bombardment (FAB) nicely reflects the composition of the matrix. By means of a temperature-controlled probe, it is therefore possible to deduce the thermodynamic parameters of the enantioselective host-guest complexation. Mass tagging can be also obtained resorting to what are called quasi-enantiomers, i.e., molecules that would be enantiomers if it were not for a minor chemical substitution, supposed to be ineffective in the reaction or complex formation process. This technique, pioneered by Horeau and Nouaille[26] and Guo et al.,[27] is described in Chapter 9. It has been applied successfully for screening asymmetric catalysts based on the detection of the catalytic intermediates rather than that of the products.[28]

The second method rests on ion/molecule reactions, mainly exchange reactions. A chiral analyte incorporated in a complex with a chiral host undergoes an exchange reaction by a nonchiral reference. The kinetics of displacement of the guest by the nonchiral reference is an indication of the difference in interaction energy between the two enantiomers of the analyte with the host. This method is illustrated in Chapter 7 in the example of crown-ether as chiral hosts. The enantioselectivity of oligosaccharides like cyclodextrins or their linear analogues, namely, maltose-based oligomers, toward amino acids and pharmacologically important molecules has been probed by the same method and described in Chapter 6.[29] More recently, more complex macrocycles have been used as chiral hosts.[30] Recognition by calixarenes and resorcinarenes grafted with chiral substituents at their upper or lower rim has proven to depend on the binding site of the chiral guest. Because of the complexity of these guest-host systems, molecular dynamics calculations are necessary for getting information on their molecular structure. As pointed out in Chapter 6, different isomeric forms of the host-guest adduct coexist in the experimental conditions, which can show different enantioselectivity. This points out the fact that the strongest binding sites do not correlate with the strongest enantioselectivity; this observation has also been made for neutral complexes in the gas phase[20] as well as in solution.[31]

The last methods rest on collision-induced dissociation (CID) of trimeric ionic clusters. The widespread kinetic method (KM) pioneered by R. G. Cooks and his group consists of comparing the efficiency of dissociation of the chiral reference (ref*) and the analyte (A) from the trimeric cationic complex, $M^{II}(A_R)(ref^*)_2 - H^+$, for the two enantiomers of the analyte.[32] It has proven to be a very efficient method for enantiomeric excess measurements, in particular when using transition metal cations, which provide multiple interaction for chiral recognition.[25] The chiral recognition ratio (CR) introduced by the group of Che[33,34] is also based on dissociation of diastereoisomeric complexes, but measures the ratio of the intensity of one fragment ion, $M^{II}(A_R)(ref^*) - H^+$, to that of the parent ion, $M^{II}(A_R)(ref^*)_2 - H^+$. These dissociation-based methods also apply to anionic complexes, as illustrated in Chapter 8. The conformational wealth of oligonucleotide sequences used as aptamers prompted their use as stationary phase in chiral chromatography.[35] In this context, Chapter 8 describes the use of small DNA sequences as auxiliaries for discriminating the enantiomers of amino acids.

The experiments described above must satisfy the same conditions as extensively discussed for measurements of binding affinities scales in general,[36–39] like reaching the Bolztmann equilibrium in a collision-free environment and defining an effective temperature, studying systems with a dominant unimolecular dissociation pathway in CID, and understanding the role of flexibility, i.e., entropy in the studied processes. These issues are beyond the scope of this book, but one should notice that in the case of chiral recognition, the different complexes at play only differ in the configuration of one of the subunits, in the case of real or pseudo enantiomers, or in a substituent remote to the chiral center, in the case of isotope-labeled or quasi-enantiomers. The similarity of the systems therefore makes the comparison between them pretty safe.[38] However, a strong limitation is that the ionic clusters must retain their original configuration; racemization is, of course, strongly undesirable. Moreover, the interpretation of thc information obtained by the kinetic method rests on the assumption that no isomeric form of the complex complicates the analysis of the data. This assumption might be wrong in clusters of complex chiral molecules, which can show multiple binding sites, as described in Chapter 4.

The approaches resting on mass spectrometry (MS) only, as described in review articles[40] or in this book, do not reach the outstanding efficiency of HPLC coupled with MS/MS detection, which affords sensitivity down to 0.0025% of the minor enantiomer.[41] However, it makes possible fast and direct analysis of enantiomeric composition down to 0.5% enantiomeric excess.[40] An important issue related to analytical applications is whether the gas-phase results mimic those observed in solution, in the case of electro-sprayed solutions, or in the matrix, in the case of fast atomic bombardment. In some cases, different ionization methods give rise to contrasted differences in the enantioselectivities, a case where FAB was supposed to match the solution results in a more quantitative way.[42] The comparison between the selectivity observed in solution and that in the gas phase has been discussed in Chapter 10. Indeed, comparison between gas-phase properties and what is observed in solution helps us to understand the mechanism responsible for the separation of enantiomers and fathom the relative importance of dispersion vs. purely electrostatic interactions, as those observed in ion exchange mechanisms. In this respect, it brings interesting information on the chiral recognition mechanisms at play when cinchona alkaloid carbamates or antimony (III) tartrates are used as stationary phases for chiral stationary phase chromatography.

Last, it is impossible to describe chiral recognition phenomena without mentioning homochirality of life. R. G. Cooks has proposed that homochirality of life results from three successive processes: chiral selection of a single enantiomer via symmetry breaking, chiral accumulation, and chiral transmission.[43] The observation of an especially stable protonated homochiral serine octamer strengthens the case that serine could play a key role in chiral accumulation and chiral transmission. Moreover, the L-serine cluster can accommodate a D-sugar molecule, which in turn dimerizes within the cluster to form a life-related C6 sugar. The symmetry-breaking step could be achieved either through parity violation effects or via irradiation by a circular polarized light, as proposed in Chapter 1.

Among the promising new methods, which are still under development, one can mention the use of ion mobility, to observe either the difference of mobility between diastereoisomeric adducts or the difference of mobility of enantiomers in a drift gas seeded with a chiral molecule.[44,45]

Coupling laser spectroscopy techniques and ion traps has been applied widely to the study of biologically relevant molecules by infrared multiphoton dissociation (IRMPD). Amino acids[46,47] and peptides[48] have been the subject of particular interest. So far, the only application to adducts of chiral molecules has been limited to protonated serine clusters, with sizes ranging from the dimer to the famous octamer.[49] The vibrational spectra in the range of 3 μ are compatible with the structure proposed for the octamer. However, no attempt for chirality-dependent spectroscopic fingerprint has been made, probably because the width of the spectra obtained at room temperature wipes out the subtle chirality effects. The use of cold ion trap combines the advantages of mass spectrometry and resolved spectra as obtained at low temperature, which makes spectroscopic measurements as precise as in supersonic expansions.[50] Besides the fact that supersonic expansions are meant mainly to study neutral species, the lack of thermodynamic equilibrium is one of the main points differentiating them from ion traps. It would be promising to perform experiments in a temperature-controlled ion trap to assess the relative importance of the most stable adducts and those of lesser binding energy in chiral recognition. Last, comparison of the chiral recognition efficiency in neutral, ionic, protonated, or cationized complexes of the same molecules is made possible thanks to the always broader range of experimental techniques available. This comparison would cast light on molecular and chiral recognition processes in different solvent conditions, or environment related to life chemistry.

Experiments resting on electron capture dissociation (ECD) have been reported very recently and compared to collision-activated dissociation (CAD) results in terms of fragmentation specificity. ECD experiments have proven to be much more sensitive to hydrogen bonding patterns than CAD. For this reasons, they are a precious tool for studying the stereochemistry of peptides, which can profoundly influence their biological activity, as observed in the case of opioids. Indeed, changing the chirality of a single amino acid strongly modifies the pharmaceutical activity of the peptide, which makes peptide stereoisomers interesting candidates for new drugs. The study by ECD of the consequences of changing the chirality of a single amino acid in a peptide, as described in Chapter 11, opens the way to promising, highly stereoselective experiments.

## REFERENCES


1. Pasteur, L. 1861. *Leçons de chimie professées en 1860 par MM. Pasteur, Cahours, Wurtz, Berthelot, Sainte-Claire Devile, Barral et Dumas*. Paris: Hachette.
2. Lord Kelvin. 1904. *Baltimore lecture*. London: Clay.
3. Quack, M. 2002. *Angew. Chem. Int. Ed.* 41:4618.
4. Berova, N., Nakanishi, K., Woody, R. W. 2000. *Circular dichroism. Principle and applications.* New York: Wiley.
5. Barron, L. D. 2004. *Molecular light scattering and optical activity*. Cambridge: Cambridge University Press.
6. Bentley, R. 2006. *Chem. Rev.* 106:4099.
7. Maier, N. M., Franco, P., Lindner, W. 2001. *J. Chromatogr. A* 906:3.
8. Pirkle, W. H., Pochapsky, T. C. 1989. *Chem. Rev.* 89:347.
9. Zehnacker, A., Suhm, M. A. 2008. *Angew. Chem. Int. Ed.* 47:6970.
10. Le Barbu, K., Brenner, V., Millie, P., Lahmani, F., Zehnacker-Rentien, A. 1998. *J. Phys. Chem. A* 102:128.
11. Filippi, A., Giardini, A., Piccirillo, S., Speranza, M. 2000. *Int. J. Mass Spectrom.* 198:137.
12. Speranza, M., Satta, M., Piccirillo, S., Rondino, F., Paladini, A., Giardini, A., Filippi, A., Catone, D. 2005. *Mass Spectrom. Rev.* 24:588.
13. Borho, N., Suhm, M. A. 2004. *Phys. Chem. Chem. Phys.* 6:2885.
14. Hearn, J. P. I., Howard, B. J. 2007. *Mol. Physics* 105:825.
15. Latini, A., Toja, D., Giardini-Guidoni, A., Palleschi, A., Piccirillo, S., Speranza, M. 1999. *Chirality* 11:376.
16. Latini, A., Toja, D., Giardini-Guidoni, A., Piccirillo, S., Speranza, M. 1999. *Angew. Chem. Int. Ed.* 38:815.
17. Le Barbu, K., Zehnacker, A., Lahmani, F., Mons, M., Piuzzi, F., Dimicoli, I. 2001. *Chirality* 13:715.
18. Seurre, N., Sepiol, J., Le Barbu-Debus, K., Lahmani, F., Zehnacker-Rentien, A. 2004. *Phys. Chem. Chem. Phys.* 6:2867.
19. Le Barbu-Debus, K., Broquier, M., Mahjoub, A., Zehnacker-Rentien, A. 2008. *J. Phys. Chem. A* 112:9731.
20. Le Barbu-Debus, K., Broquier, M., Mahjoub, A., Zehnacker-Rentien, A. 2009. *Phys. Chem. Chem. Phys.* 11:7589.
21. Speranza, M. 2004. *Adv. Phys. Organic Chem.* 39:147.
22. Sawada, M. 1997. *Mass Spectrom. Rev.* 16:73.
23. Speranza, M., Gasparrini, F., Botta, B., Villani, C., Subissati, D., Fraschetti, C., Subrizi, F. 2009. *Chirality* 21:69.
24. Schug, K.A., Lindner, W. 2005. *J. Separation Sci.* 28:1932.
25. Tao, W. A., Cooks, R. G. 2003. *Anal. Chem.* 75:25A.
26. Horeau, A., Nouaille, A. 1990. *Tetrahedron Lett.* 31:2707.
27. Guo, J. H., Wu, J. Y., Siuzdak, G., Finn, M. G. 1999. *Angew. Chem. Int. Ed.* 38:1755.
28. Markert, C., Pfaltz, A. 2004. *Angew. Chem. Int. Ed.* 43:2498.

29. Ahn, S., Ramirez, J., Grigorean, G., Lebrilla, C. B. 2001. *J. Am. Soc. Mass Spectrom.* 12:278.
30. Fraschetti, C., Pierini, M., Villani, C., Gasparrini, F., Filippi, A., Speranza, M. 2009. *Collect. Czech. Chem. Commun.* 74:275.
31. Nishijima, M., Pace, T. C. S., Nakamura, A., Mori, T., Wada, T., Bohne, C., Inoue, Y. 2007. *J. Organic Chem.* 72:2707.
32. Tao, W. A., Gozzo, F. C., Cooks, R. G. 2001. *Anal. Chem.* 73:1692.
33. Yao, Z. P., Wan, T. S. M., Kwong, K. P., Che, C. T. 2000. *Anal. Chem.* 72:5383.
34. Yao, Z. P., Wan, T. S. M., Kwong, K. P., Che, C. T. 2000. *Anal. Chem.* 72:5394.
35. Michaud, M., Jourdan, E., Villet, A., Ravel, A., Grosset, C., Peyrin, E. 2003. *J. Am. Chem. Soc.* 125:8672.
36. Cooks, R. G., Wong, P. S. H. 1998. *Accounts Chem. Res.* 31:379.
37. Schalley, C. A. 2001. *Mass Spectrom. Rev.* 20:253.
38. Blair, S. M., Kempen, E. C., Brodbelt, J. S. 1998. *J. Am. Soc. Mass Spectrom.* 9:1049.
39. Dalleska, N. F., Honma, K., Armentrout, P. B. 1993. *J. Am. Chem. Soc.* 115:12125.
40. Young, B. L., Cooks, R. G. 2007. *Int. J. Mass Spectrom.* 267:199.
41. Badaloni, E., Cabri, W., Ciogli, A., Deias, R., Gasparrini, F., Giorgi, F., Vigevani, A., Villani, C. 2007. *Anal. Chem.* 79:6013.
42. Sawada, M., Takai, Y., Yamada, H., Nishida, J., Kaneda, T., Arakawa, R., Okamoto, M., Hirose, K., Tanaka, T., Naemura, K. 1998. *J. Chem. Soc. Perkin Trans.* 2:701.
43. Nanita, S. C., Cooks, R. G. 2006. *Angew. Chem. Int. Ed.* 45:554.
44. Dwivedi, P., Wu, C., Matz, L. M., Clowers, B. H., Siems, W. F., Hill, H. H. 2006. *Anal. Chem.* 78:8200.
45. Mie, A., Jornten-Karlsson, M., Axelsson, B. O., Ray, A., Reimann, C. T. 2007. *Anal. Chem.* 79:2850.
46. Lucas, B., Gregoire, G., Lemaire, J., Maitre, P., Ortega, J. M., Rupenyan, A., Reimann, B., Schermann, J. P., Desfrancois, C. 2004. *Phys. Chem. Chem. Phys.* 6:2659.
47. Wu, R. H., McMahon, T. B. 2008. *Chemphyschem* 9:2826.
48. Correia, C. F., Balaj, P. O., Scuderi, D., Maitre, P., Ohanessian, G. J. 2008. *Am. Chem. Soc.* 130:3359.
49. Kong, X. L., Tsai, I. A., Sabu, S., Han, C. C., Lee, Y. T., Chang, H. C., Tu, S. Y., Kung, A. H., Wu, C. C. 2006. *Angew. Chem. Int. Ed.* 45:4130.
50. Stearns, J. A., Guidi, M., Boyarkin, O. V., Rizzo, T. R. 2007. *J. Chem. Physics* 127:154322.

# 1 Valence Photoelectron Circular Dichroism of Gas Phase Enantiomers

*Laurent Nahon and Ivan Powis*

## CONTENTS

1.1 Introduction ........ 1
1.2 PECD Formalism ........ 2
1.3 Methods for Studying PECD ........ 4
  1.3.1 Experimental Approaches ........ 4
  1.3.2 Theoretical Approaches ........ 5
1.4 The Main Features of PECD: The Showcase of Camphor ........ 7
1.5 PECD as a Probe of Structure ........ 11
  1.5.1 Chiral Scattering of the Photoelectron ........ 11
  1.5.2 Chemical Structure: Fenchone vs. Camphor ........ 15
  1.5.3 Glycidol Conformers ........ 16
1.6 Reflections on PECD ........ 22
  1.6.1 Applications and Developments ........ 22
  1.6.2 A Photophysical Approach Contributing to Life's Homochirality? ........ 23
Acknowledgments ........ 25
References ........ 25

## 1.1 INTRODUCTION

To those broadly familiar with the theory or practice of chiral molecular spectroscopies, one of the more recently investigated chiroptical phenomena, photoelectron circular dichroism[1] (PECD), may nevertheless be surprising when first encountered because from randomly oriented, noninteracting molecular enantiomers it routinely yields asymmetry factors ranging from 0.01 to 0.3. Asymmetries of this magnitude exceed those encountered in more traditional circular dichroism (CD) measurements by several orders of magnitude, and so promise a number of advantages associated with the potentially greater ease of detecting such large effects.

Foremost among these is the ability (indeed, in practice a necessity) to work with dilute gas phase samples. In this manner one both removes restrictions imposed by solvent absorption cutoffs, and so greatly extends the wavelength range available for making measurements, and eliminates the usually indeterminate contributions to the dichroism made by any induced chirality in a solvent shell.[2] The application of standard supersonic molecular beam techniques for sample introduction into the gas phase likewise offers an extended temperature range, extending down to near absolute zero, over which the effects may be investigated, and further offers prospects for the experimentalist to reintroduce, in a controlled manner, molecule–molecule and molecule–solvent interactions through the formation of small cluster species.

Because it is based on the use of photoelectron spectroscopy, a universally applicable spectroscopic tool not suffering from restrictive selection rules, PECD can be applied, in principle, to any chiral species and does not require that there be a specific chromophore present in the molecule, or that the molecule be so tagged or labeled. The asymmetry observed is specific to each individual orbital, and to the emitted electron's energy, and so as photon energies increase above the ionization threshold of more and more orbitals, PECD measurements become ever more structured—even, it will be seen, under conditions where the conventional photoelectron spectrum (PES) does not fully resolve the individual orbital contributions. Although in this chapter we focus on the PECD of valence orbitals, the phenomenon applies equally to the ionization of core levels, and PECD has been recorded with photon energies ranging from ~8 eV in the vacuum ultraviolet (VUV) to over 500 eV in the soft X-ray region.[3,4]

The multidimensional nature (orbital vs. photon energy) of PECD spectroscopy provides data of an immensely rich structure from a given species. Theoretical understanding and modeling of PECD has kept pace with experimental developments in this decade and, at the least, one can be confident of an ability to assign absolute configuration by comparison of PECD experiment and theory. But it has become increasingly clear that PECD strongly probes molecular conformation and chemical substitution even, in larger species, at sites that are somewhat remote from an initially localized ionizing orbital or from a chiral center. It is possible to conceive of PECD as a kind of electron diffraction study performed with electrons generated *in situ* at various locations and with varying energies. The chiral nature of the scattering/diffraction process lends a much enhanced sensitivity to the detailed molecular structure, as compared to the photoelectron measurements that are possible for nonchiral molecules.

In the following sections the basic nature and formalism of PECD will be outlined, as will the experimental and theoretical models that have been applied. A benchmark study of camphor will then be presented, followed by examples where molecular conformation and chemical substitutions are probed. We conclude with a look at future prospects for this recently introduced chiroptical technique.

## 1.2 PECD FORMALISM

Photoelectron circular dichroism arises from a chiral asymmetry in the predicted angular distribution of photoelectrons emitted upon ionization with circularly polarized radiation (CPL). The traditional, well-known form of photoelectron angular distribution is

$$I(\theta) = 1 + \beta P_2(\cos\theta) \tag{1.1}$$

where $\theta$ is the angle between the electron's direction and the linear polarization vector of the light. $P_2$ is the second Legendre polynomial, effectively introducing an anisotropic term varying between $\sin^2\theta$ and $\cos^2\theta$ depending upon the so-called anisotropy parameter, $\beta$.

Although not widely recognized as such, Equation 1.1 is in fact a specific instance of a more general expression first developed by Ritchie[5] that we write as

$$I(\theta) = 1 + b_1^p P_1(\cos\theta) + b_2^p P_2(\cos\theta) \tag{1.2}$$

and which applies to pure polarization states of the ionizing light beam and assumes random orientation of the molecular targets. The parameters $b_1^p$ and $b_2^p$ depend upon the photon polarization, $p$, and the photoionization dynamics of the considered orbital[5,6] and include both angular momentum coupling terms and electric-dipole photoionization matrix elements. However, for linear polarization ($p = 0$) the "additional" chiral parameter, $b_1^0$, will necessarily be zero, so that Equation 1.1 is recovered. It can be seen that $b_2^0$ is then just the familiar $\beta$ anisotropy parameter. In the case of circular polarization states ($p = \pm 1$) $b_1^{\{\pm 1\}}$ may become nonzero, but only in the event that the target molecule is chiral, i.e., lacks any symmetry element*—this is linked to the odd parity of the electric-dipole operator. In this case, the angle $\theta$ is taken to be referenced from the direction of light beam propagation, and the first Legendre polynomial term, $P_1$, now introduces a $\cos\theta$ term to the angular distribution. It is readily appreciated that this term is odd with respect to the inversion $\theta \rightarrow \pi - \theta$ and so must introduce some forward-backward asymmetry to the electron angular distribution.

A detailed examination of the form of the $b_j^p$ parameters reveals that they are governed by certain symmetry relations for the different combinations of light and molecular helicity. In particular:

$$b_2^{\{\pm 1\}} = -\tfrac{1}{2} b_2^0 \;\; (\equiv -\tfrac{1}{2}\beta) \tag{1.3a}$$

$$b_1 s = -b_1^{\{-1\}} \tag{1.3b}$$

Equation 1.3b indicates that the forward-backward asymmetry in the angular distribution will be inverted if the photon helicity (i.e., handedness of the CPL) is reversed; a similar negation of the $b_1$'s, and hence inversion of the angular asymmetry, is expected should the handedness of the molecular enantiomer be exchanged.

* More exactly, a chiral molecule may in fact possess, at most, one twofold ($C_2$) rotational axis.

It is trivial to obtain from Equations 1.2 and 1.3 and properties of the Legendre polynomials an expression for the forward-backward asymmetry for given CPL state and enantiomer:

$$I(0^{\circ}) - I(180^{\circ}) = 2b_1 \tag{1.4}$$

Equivalently, the dichroism (difference between left- and right-CPL ionizations) is

$$g = I_{L(p=+1)}(\theta) - I_{R(p=-1)}(\theta) = 2b_1^{+1}\cos(\theta) \tag{1.5}$$

The asymmetry factor, $g$, in the form of Equation 1.5, should also, like the $b_1$ parameters, be negated for an exchange of molecular enantiomer.

Like other physically based methods for probing molecular chirality, PECD relies upon differential interaction with a chiral environment—frequently, as here, this will be circularly polarized radiation. But while circular dichroism observed on *integral* measurements such as in the total absorption cross section is a well-known phenomenon, it is weak—perhaps 1 in $10^4$ or less—as the effect arises from interference between electric- and higher-order magnetic-dipole and electric-quadrupole terms. In stark contrast, Equation 1.2, expressing a *differential* measurement, is derivable in the pure electric-dipole approximation, and PECD therefore exists without the need to invoke the weaker higher-order radiation-molecule interaction terms. It is for this reason that PECD asymmetries are found to be typically three orders of magnitude greater than in conventional absorption CD.

## 1.3 METHODS FOR STUDYING PECD

### 1.3.1 Experimental Approaches

To our knowledge, all valence shell PECD studies have been carried out by using the synchrotron radiation (SR) as a VUV photon source providing an intense, widely tunable, and highly polarized CPL.

Besides this common use of the SR, different types of setup have been employed. Our collaborative group has been using, for several studies on camphor,[7] fenchone,[8] and glycidol,[9] an electron/ion imaging spectrometer called DELICIOUS, which has been described in detail previously.[10] Briefly, it is based upon the original Eppink and Parker velocity map imaging (VMI) design[11] that provides two very interesting features: (1) a multiplex dual angular and radial capability in the momentum distribution analysis of electrons that is very well suited for studies on PECD, and (2) a relative insensitivity of the momentum determination to the exact ionization point, due to the presence of inhomogeneous focusing fields produced by hollow electrodes. This provides a spectacular de-blurring of the recorded images made with a source of finite dimensions, and therefore gives rise to an improved energy resolution, of particular benefit for SR-based studies for which ionization volumes are typically relatively large.

As compared with the original Eppink and Parker VMI design,[11] the main advance of the DELICIOUS VMI is the insertion of an optimized double-refocusing einzel lens in the drift tube, permitting the collection of up to 18 eV kinetic energy electrons, allowing the study of dynamical variations of the PECD over an extended photon energy range above an ionization threshold, with an optimized energy resolution of 6%.[10]

Liquid or solid samples can be placed in a small reservoir immediately behind the nozzle of a continuous molecular beam source and heated (330–400 K) to raise their vapor pressure. A stream of helium gas at a backing pressure of ~1 bar is passed through the vapor and expands through the nozzle to produce a seeded supersonic molecular beam. After skimming, this beam passes through the source of the spectrometer, where it is intersected by the VUV SR beam. Note that one of the advantages of using such a molecular beam inlet is that the jet expansion conditions can be tuned so that the temperature of the species, and therefore their possible conformer population, can be varied.

A full description of the acquisition methodology and the subsequent data analysis has been given elsewhere.[7] In brief, for a given enantiomer and a selected photon energy, several photoelectron images are first recorded with alternate light helicity to minimize the effect of environmental instabilities. The data are then reduced to a single raw image generated as the total difference between images obtained with left and right CPL. The difference images are then analyzed using the pBasex method[12] to extract the angular distribution as a function of image radial coordinate (electron energy). After normalization and scaling for the measured circular polarization rate of the beamline at a given photon energy, the final PECD curve showing asymmetry vs. electron kinetic energy (or equivalently, ionization energy) is obtained.

Other groups have been studying valence shell PECD on bromocamphor[13] and camphor,[14] methyloxirane,[15] and 3-hydroxy-tetrahydrofuran[16] by using an effusive jet coupled to twin hemispherical electrostatic electron spectrometers, positioned at the so-called magic angle (54.7°) with respect to the photon axis. In this specific direction, $P_2$, the second Legendre polynomial, is zero so that any such measurement is independent of the β parameter (see Equation 1.2). The chiral $b_1^{\{\pm 1\}}$ parameter may then be inferred without determining the full electron angular distribution.

The advantage of our imaging approach, besides its evident multiplex capability over a wide kinetic energy range, is that it provides a direct global vision of the cosine dependence of PECD (see Equation 1.5 and Figure 1.3). Also, the β parameter (see Equations 1.2 and 1.3a) can be obtained simultaneously providing complementary information, especially in the context of an experiment/calculation comparison approach. Conversely, in the case of fast electrons (say, above a few eV kinetic energy) the alternative approach, utilizing dispersive electron energy analyzers, provides potentially higher kinetic energy resolution (typically in the 300–100 meV range) than our imaging technique.

### 1.3.2 Theoretical Approaches

While it is only electric-dipole matrix elements that are required for numerical evaluation of the $b_j^p$ distribution parameters, their calculation nevertheless poses a more challenging problem than analogous, and nowadays routine, bound state excitation problems, due to the continuum nature of the final state in a photoionization process (free photoelectron plus molecular ion core).[17] For the calculation of PECD, this challenge is compounded by two further factors. First, chiral molecules by definition possess no, or at most only one, element of symmetry. They are also larger, typically, than the small-molecule species normally featured in photoionization dynamics calculations; to date PECD calculations have examined between $C_3$ and $C_{10}$ molecules. Consequently, full molecule calculations are necessarily large scale and lack any scope for efficiency savings achieved by exploitation of molecular symmetry. It is further found that calculations for $b_1^{\{\pm1\}}$ parameters converge more slowly than do either cross-section or β parameter calculations, adding to the overall computational effort. The calculation of matrix elements for PECD therefore requires that only computationally efficient methods be considered.

The first reported PECD calculations[6,18] utilized the continuum multiple scattering (CMS-Xα) method.[19] This is a parameterized approach in which a model, self-consistent, neutral molecule potential is constructed after partitioning the molecule into overlapping spherical regions located at each atomic center. In each region the exchange contribution to an effective one-electron potential is represented using the Xα local density approximation.[20] The bound and continuum electron wavefunctions are then expanded in a basis of spherical harmonic functions, truncated at some value, $\ell_{max}$, in each spherical region of the model potential, with the radial terms obtained by direct numerical integration. For the continuum calculation, the self-consistent Xα potential does not have the correct asymptotic form (Coulomb attraction for ion plus electron) but may be adapted to have this form. By choosing a frozen core approximation the matrix elements can finally be obtained as one-electron functions between an initial ionizing orbital and the free photoelectron. A more extensive description of the CMS-Xα methodology applied by the authors can be found elsewhere.[21]

An alternative approach has been applied by Stener, Decleva, and coworkers.[22–25] In their B spline Linear Combination of Atomic Orbitals (LCAO) Density Functional Theory (DFT) method, a regular LCAO basis set is adapted for the continuum by the addition of B spline radial functions in a multicenter expansion. The system is then solved at the DFT Kohn-Sham level, using an appropriate exchange correlation functional—mainly the LB94 potential that has the correct asymptotic Coulomb behavior. This again provides an independent electron, frozen core approximation. Both CMS-Xα and B spline approaches are currently performed in a fixed nuclei approximation, i.e., are unable to unravel a possible vibronic dependence of PECD.

Where a detailed three-way comparison between PECD predictions from these two theoretical models and relevant experiment has been attempted, very reasonable agreement has been obtained.[1,22] Nevertheless, it is clear that the B spline method holds greater long-term potential for development. The parameterization implicit in the available exchange correlation functionals provides a more sophisticated treatment than does that of the Xα local potential model, while avoiding the latter method's somewhat unrealistic partitioning of the molecular potential into spherical regions.

## 1.4 THE MAIN FEATURES OF PECD: THE SHOWCASE OF CAMPHOR

In this section we will present the main characteristics of valence shell PECD effect, illustrating the formal predictions presented in Section 1.2, for the prototypical chiral showcase of camphor: a very rigid, one-conformer, easy-to-handle molecule that has been the subject of several studies.[7,14]

A typical angle-integrated photoelectron spectrum (PES) taken using the DELICIOUS spectrometer at $h\nu = 20$ eV is shown in Figure 1.1 together with a higher-resolution camphor spectrum recorded with a dispersive hemispherical

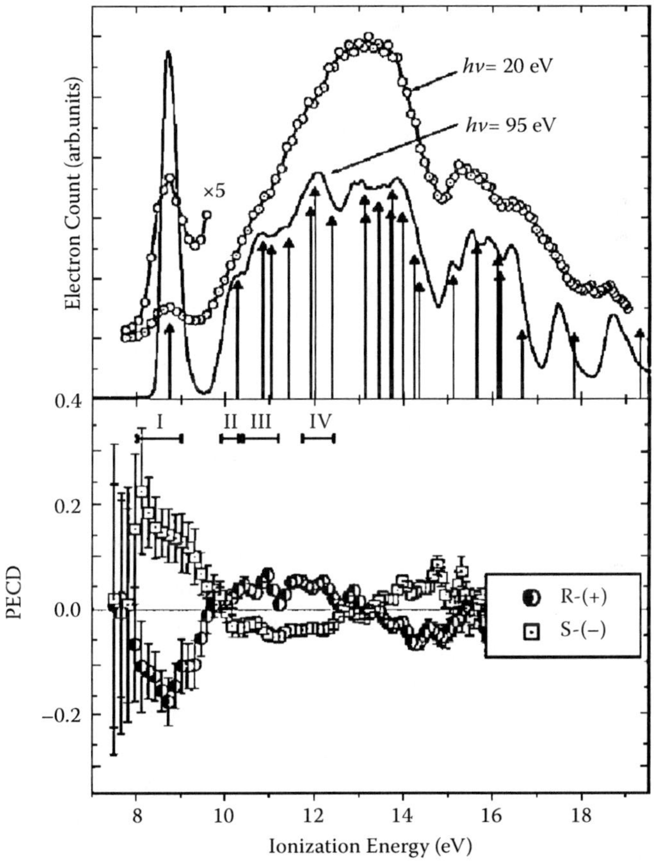


**FIGURE 1.1** Photoionization of camphor recorded with the VMI spectrometer at $h\nu = 20$ eV. Top panel shows an angle-integrated photoelectron spectrum. This is compared with a higher-resolution conventional photoelectron spectrum recorded at $h\nu = 95$ eV, taken from Rennie et al.[26] Calculated vertical ionization energies reported in the same reference are also indicated. Lower panel shows the PECD deduced across this spectrum for both the R and S enantiomers. The regions **I** to **IV** of the photoelectron spectrum used when forming mean PECD values are also indicated here. (Reproduced from L. Nahon, G. A. Garcia, C. J. Harding, E. A. Mikajlo, and I. Powis, *J. Chem. Phys.* 125 (2006): 114309. With permission.)

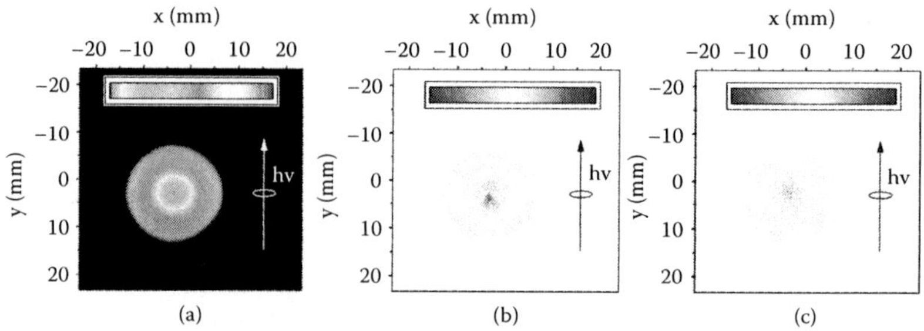


**FIGURE 1.2** (Color Figure 1.2 follows page 46.) (a) Raw photoelectron image of R camphor taken at hν = 10.3 eV. The two rings correspond to the two outermost orbitals that can be ionized. (b) Untreated subtraction between R camphor images recorded with different light helicity (left minus right circularly polarized light). The color scale reveals the forward/backward asymmetry along the propagation direction of the light, with an opposite sign for the two electronic states. (c) Idem with the S enantiomer. The intensity distribution patterns are antisymmetric with respect to the switch of enantiomers, as predicted by theory. (Reproduced from L. Nahon, G. A. Garcia, C. J. Harding, E. A. Mikajlo, and I. Powis, *J. Chem. Phys.* 125 (2006): 114309. With permission.)

electron analyzer at $h\nu$ = 95 eV.[26] Despite the moderate resolution achieved with DELICIOUS-VMI, the outermost orbital, or highest occupied molecular orbital (HOMO), is always fully resolved, and the region of the second band corresponding to the next (HOMO-1) orbital is equally distinct in these spectra.

Figure 1.2a shows a typical detector image of camphor photoelectrons taken at a photon energy of 10.3 eV, while Figure 1.2b and c shows the dichroism at this energy as the difference between such raw images, for the R and S enantiomers of camphor, respectively. At this photon energy, two ionic states can be reached, appearing as a double-ring structure in Figure 1.2a. The outer ring pattern corresponds to ionization of the HOMO, the inner one to ionization from the HOMO-1. Even before the mathematical analysis, looking at the difference images one can clearly notice the asymmetry in intensity, which changes sign along the forward and backward directions of ejection with respect to the light beam's propagation.

Following the full data analysis described in Section III of Nahon et al.,[7] we can then reconstruct from the 2D images shown in Figure 1.2b and c the laboratory frame difference (L-CPL – R-CPL) in electron distribution intensity as a function of the initial kinetic energy (KE) and ejection angle. These are presented in Figure 1.3 as surface/contour plot representations of the angle-resolved photoelectron spectrum for each enantiomer. Such a representation offers an intuitive understanding of the chiral molecule photoionization. Some important features are immediately seen in these images. First, the angular distribution has the predicted cosine dependence about the light propagation direction, and so the asymmetry peaks in the forward-backward direction ($\theta$ = 0 and 180°). Second, as predicted from the properties of the $b_1$ coefficients, this forward-backward asymmetry reverses when the enantiomer is switched. Third, the magnitude, and more strikingly the sign of the asymmetry, changes from the outer to the inner ionic states, clearly indicating a dependence on

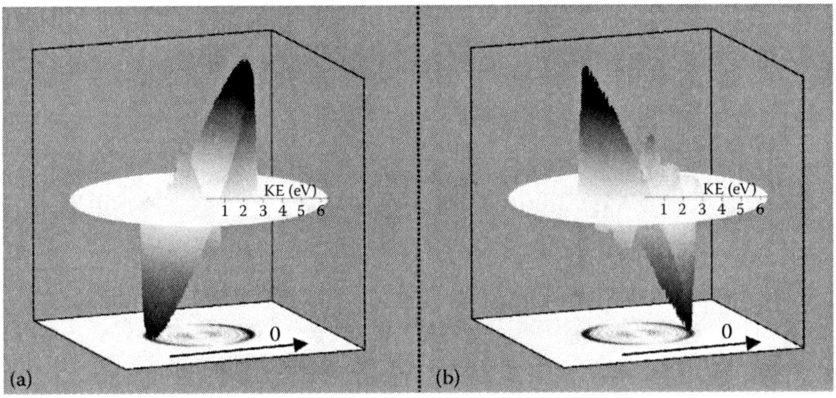


**FIGURE 1.3** Three-dimensional reconstruction of the PECD, weighted by the PES, observed on the R (a) and S (b) enantiomers of camphor photoionized at 10.3 eV. The arrows indicate the CPL photon propagation axis. (Reproduced from L. Nahon, G. A. Garcia, C. J. Harding, E. A. Mikajlo, and I. Powis, *J. Chem. Phys.* 125 (2006): 114309. With permission.)

the initial orbital being ionized. In this sense, PECD appears to be (at least partly) an initial state effect.

The mathematical analysis of such difference images allows quantitative values for the dichroism to be extracted. Figure 1.1 includes an example of such quantitative PECD data as a function of ionization energy for $h\nu$ = 20 eV. In this example, the absolute magnitude of the observed chiral asymmetry peaks at around 20% of the mean intensity, in the region of the HOMO. This asymmetry ($2b_1$) is many orders of magnitude greater than that usually associated with circular dichroism in total photon absorption (typically in the $10^{-4}$ range), with an intensity especially surprising considering that we are dealing with randomly oriented targets. This is because, as stated in Section 1.2, as a differential measurement, PECD occurs in the pure electric-dipole (E1) approximation. Note that at the lower photon energy of 10.3 eV, the PECD absolute magnitude peaks at 7% for the HOMO and 5% for the HOMO-1. Therefore, the magnitude of the effect is quite similar for both orbitals, despite their very different nature: the HOMO, based upon the O lone pair, is very localized on the carbonyl group, while the HOMO-1 is much more diffuse over the whole molecule skeleton.

In Figure 1.1, one can note how the R enantiomer data exactly mirror the S enantiomer data across the zero baseline. The excellence of this mirroring corroborates theoretical expectations derived from the electric-dipole approximation, since it would not be observed if some nondipolar effect would contribute to PECD. It also indicates something of the reproducibility of otherwise identical quantities, and thus the quality of the recorded data. As at 10.3 eV, the $h\nu$ = 20 eV PECD has an opposite sign for the HOMO and the HOMO-1. Furthermore, it can be observed that the PECD recrosses the zero baseline at a higher binding energy (~13.5 eV), which we take to correlate with underlying changes in electronic structure at this point that are not, however, resolved as a distinct feature in the normal PES because of vibrational congestion. In this sense PECD contributes an additional probe of electronic structure changes.

For a given ionized orbital, the dynamical variation of its PECD with kinetic energy can be examined by variation of the photon energy. This provides a significant test for theoretical understanding and modeling of the dichroism on ionization. Therefore, we have made measurements for a range of photon energies extending from 8.85 eV (just above the first ionization threshold located at 8.7 eV) up to 26 eV, for both the R and S enantiomers. These results are summarized in Figure 1.4 as mean PECD values taken over the HOMO band (panel I) and the HOMO-1 band (panel II) as well as deeper orbitals (panels III and IV).

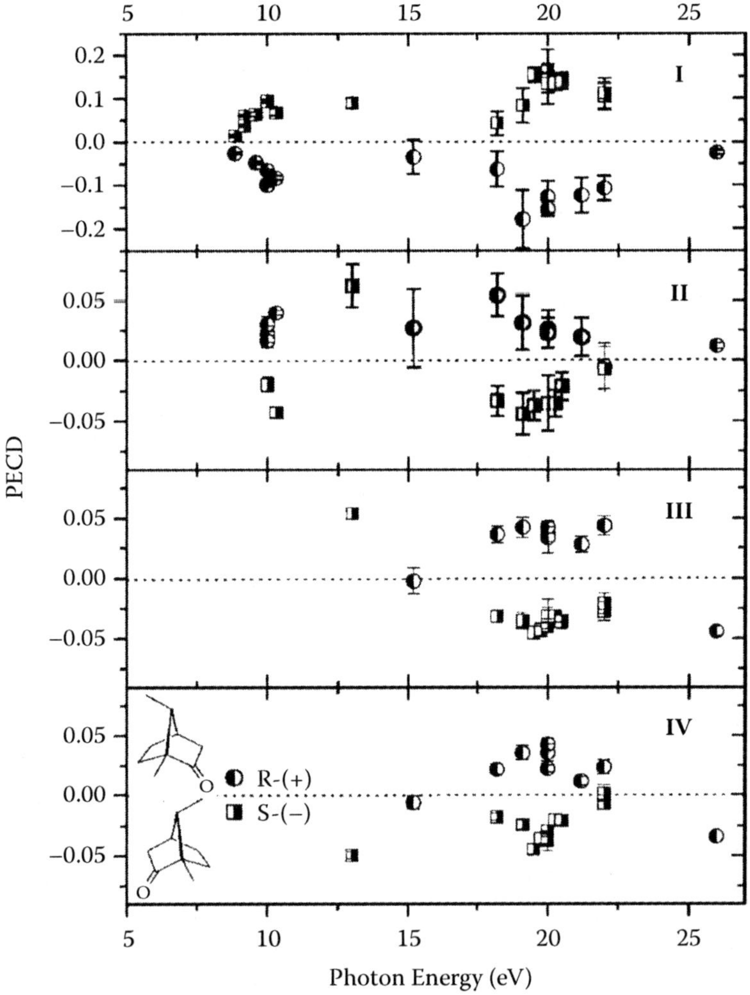


**FIGURE 1.4** Mean camphor PECD values for regions **I** to **IV** of the photoelectron spectrum (identified in Figure 1.1). Data for R and S enantiomers are displayed as a function of photon energy. Note the different scale used to plot region **I** (HOMO ionization) in the top panel. (Reproduced from L. Nahon, G. A. Garcia, C. J. Harding, E. A. Mikajlo, and I. Powis, *J. Chem. Phys.* 125 (2006): 114309. With permission.)

In the case of the HOMO, starting from a very low value at threshold, the PECD absolute magnitude reaches a first maximum around $h\nu$ = 10 eV, falls down to a near-zero value around 15 eV, and then passes through a broad second maximum. The mean dichroism in this region reaches its greatest magnitude, ~16%, around 20 eV. The HOMO-1 behaves similarly, although the PECD absolute magnitudes are generally reduced in this channel, and the sign of the PECD is opposite to the one of the HOMO. Note that this large resonant feature around 20 eV, mainly observed in the HOMO-1 channel PECD, and also observed as a marked dip in the β parameter variation,[7] is attributed to a shape resonance.[22]

For any given orbital the variation of its PECD with the electron kinetic energy (or correspondingly, with the photon energy) can itself be dramatic and strongly implies that PECD is also a final state (continuum) effect. This was also inferred from core-shell PECD (see, for instance, Hergenhahn et al.[27]), a situation for which the initial state is a localized, spherical (hence achiral) C 1*s* orbital and for which, therefore, the quite intense measured and calculated chiral PECD phenomenon must originate as a purely final state effect. PECD therefore appears to provide a powerful probe of the detailed shape of the asymmetric effective molecular potential experienced by the outgoing photoelectron.

## 1.5 PECD AS A PROBE OF STRUCTURE

### 1.5.1 Chiral Scattering of the Photoelectron

As more PECD data have been accumulated it has become increasingly apparent that the phenomenon offers a probe for detail extending beyond the simple identification of molecular chirality per se, or even the specific handedness (absolute configuration) of the enantiomer used in an experiment. In particular, calculated chiral parameters $b_1^{\{\pm 1\}}$ appear, quite generally, to respond strongly to even comparatively slight changes in the assumed molecular structure.

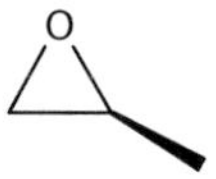

(S)-2-Methyl-oxirane

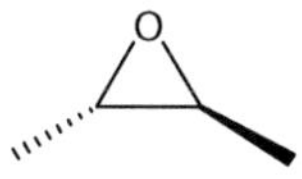

(2S,3S)-2,3-Dimethyl-oxirane

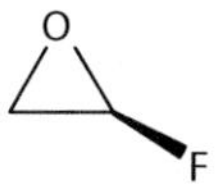

(S)-2-Fluoro-oxirane

(2S,3S)-2,3-Difluoro-oxirane

The first real suggestion of this sensitivity came in a set of calculations made using the B spline method that provided a comparative study[24] of four chiral oxiranes—the substituted methyl-, fluoro-, *trans*(1,2) dimethyl-, and *trans*(1,2) difluoro-oxiranes. Figure 1.5 reproduces the results obtained in this study for ionization of the oxygen lone pair orbitals. Situated on the C-O-C epoxy ring, this localized orbital is virtually identical in the four molecules. As can be seen in the figure, both the predicted

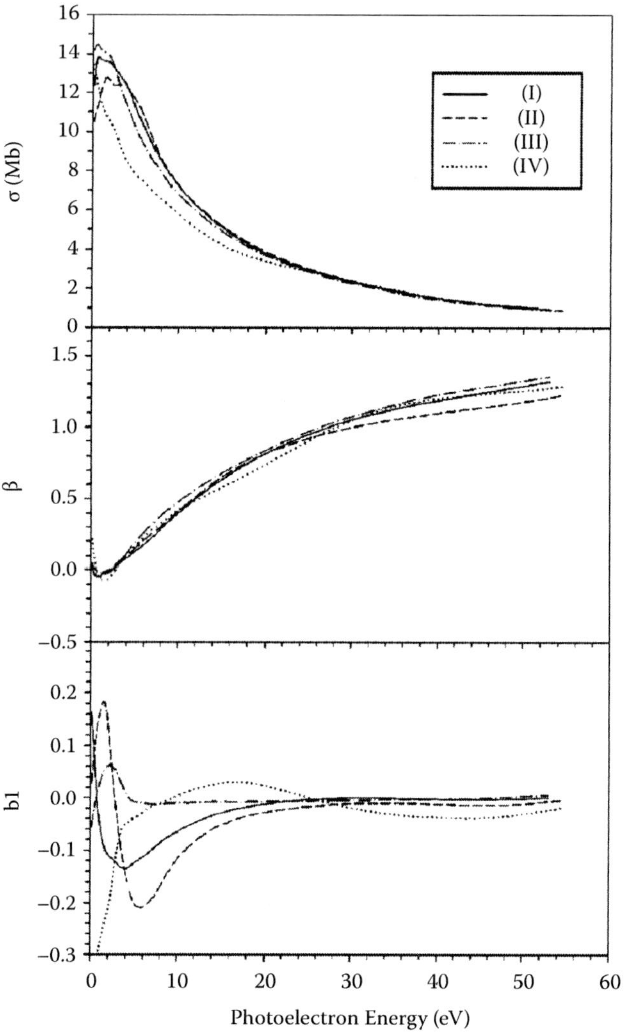


**FIGURE 1.5** Cross section σ, asymmetry parameter β, and chiral $b_1$ parameter for the oxygen valence lone pair orbital in S-methyl- (**I**), *trans*(1S, 2S)dimethyl- (**II**), S-fluoro- (**III**), and *trans*(1S, 2S) difluoro- (**IV**) oxirane. (Reproduced from M. Stener, G. Fronzoni, D. Di Tommaso, and P. Decleva, *J. Chem. Phys.* 120 (2004): 3284. With permission.)

ionization cross sections and beta anisotropy parameters for the four species are extremely similar. Yet the chiral $b_1$ dichroism parameters show much more divergent behavior, and reflect the differing substituents attached to the oxirane ring. Very similar behavior has been predicted for PECD in a series of substituted carvone molecules,[28] when either an O valence lone pair or C 1*s* core orbital (both localized at the carbonyl group in carvone) was ionized. As a bigger molecule, the substitutions considered in carvone were even further removed (by three to four atoms) from the ionizing orbital site than were those in the oxirane series. While in both studies the localized initial orbital appears little perturbed by the substitutions that were made, in the PECD phenomenon an influence is clearly exerted by substitutions occurring at near-neighbor sites and beyond.

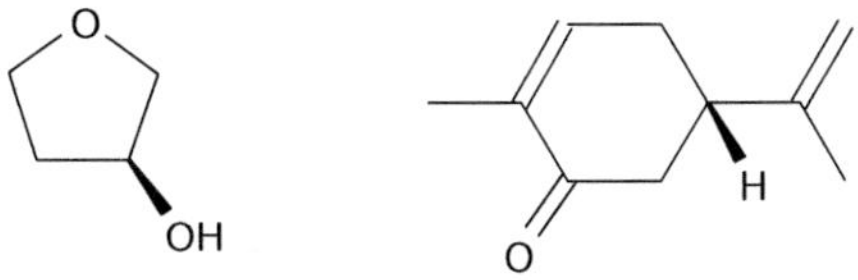

(S)-3-hydroxy-tetrahydrofuran (S)-Carvone

Analogous to this apparent sensitivity to electronic structure, a marked PECD dependence upon molecular conformation has also become evident. Attempts to model early experimental results on the valence PECD of 3-hydroxy-tetrahydrofuran[16] noted that only one of the two proposed conformers of this molecule allowed a reasonable degree of agreement between experiment and theory to be achieved. In a similar way, a combined experimental/theoretical study of the carbonyl C 1*s* core level PECD of carvone enantiomers found that very marked differences resulted from choosing either axial or equatorial locations for the isopropenyl tail group, and that lesser (though still significant) differences were displayed between the three rotational conformers available in each case.[28,29] This assumes a Boltzmann conformer population distribution led to improved agreement of the calculated behavior and experiment.

A powerful illustration of this conformational sensitivity was provided in a series of B spline calculations by Di Tommaso et al. that were performed for differing rotational conformations of the methyl group in methyl-oxirane.[23] These are shown in Figure 1.6. While the cross section is impervious to the methyl group orientation, the β parameter shows some small variation, but the chiral $b_1^{\{+1\}}$ parameter shows massive variations. For the latter, it is perhaps fortuitous that a Boltzmann average proves to return results close to the single fixed equilibrium geometry calculation that is more usually considered in modeling experimental data.

These clear differences in the response of the integrated cross sections, β parameter, and chiral $b_1^{\{\pm1\}}$ parameters to electronic and geometric changes of the *whole* molecule can be attributed to a different sampling of characteristics of the continuum photoelectron by these three dynamical measures. It is usual to consider the continuum function as an expansion in partial waves of specific angular momentum $\ell$, since $\ell$ itself is not a good quantum number in the molecular case. Associated with each $\ell$ wave will be a phase shift resulting from quite subtle interaction with the molecular ion potential. It is well known that the integrated cross section, σ, carries no information about these phase shifts, whereas *some* contributory terms to the β parameter (equivalently in our notation $b_2^p$) do, making the angular distribution potentially the more informative measure. On detailed examination of the theory underpinning Equation 1.2 it is found that while *every* contribution to $b_1^{\{\pm1\}}$ parameters now depends on relative phase shifts, there is yet another reason the chiral term is *highly* sensitive to molecular effects: the $b_1^{\{\pm1\}}$ chiral asymmetry parameters uniquely depend on interference terms between adjacent $\ell$ wave components of

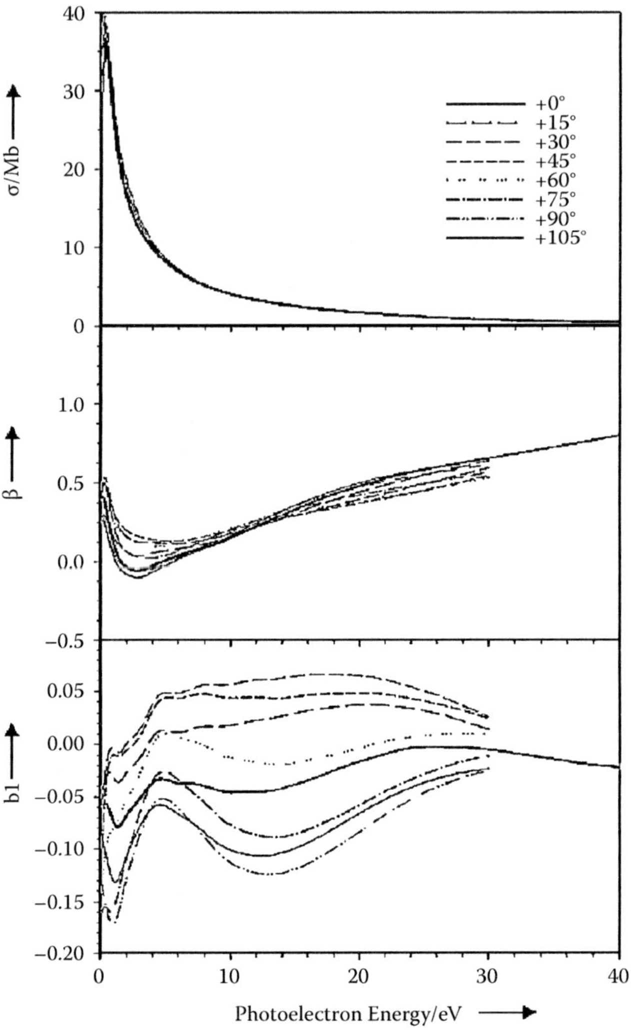


**FIGURE 1.6** Cross section σ, asymmetry parameter β, and chiral $b_1$ parameter for the 14a delocalized valence orbital in S-methyl-oxirane. Results are presented for a sequence of methyl group rotational conformers between 0° and 105°. (Reproduced from D. Di Tommaso, M. Stener, G. Fronzoni, and P. Decleva, *ChemPhysChem* 7 (2006): 924. With permission.)

the electron continuum function.[5,6] As a secondary consequence, the relative phase shifts between the adjacent $\ell$ waves enter $b_1^{\{\pm1\}}$ as a sine function—but for those contributions to $b_2^p$ where phases appear at all, they do so as cosine functions of the phase shifts between next-adjacent $\ell$ waves.[1,28]

These differences, especially the sine dependence, render $b_1^{\{\pm1\}}$ much more sensitive to even small (close to 0°) relative phase shifts that may be consequent on possibly subtle changes to the molecular shape and structure. One can thus also rationalize the apparent long-range, final state contributions found in PECD phenomena, and the slower rate of convergence with increasing $\ell$ that has been noted in computational studies. Overall, PECD may prove as important for the investigation of structure and fundamental electron molecule scattering dynamics in photoionization as for the examination of chirality per se.

### 1.5.2 Chemical Structure: Fenchone vs. Camphor

Having already considered the results of PECD studies of camphor we turn to a closely related molecule, fenchone. Fenchone differs only in the attachment site of two methyl groups, is perceived to have a similar camphoraceous odor, and is otherwise very similar to camphor.

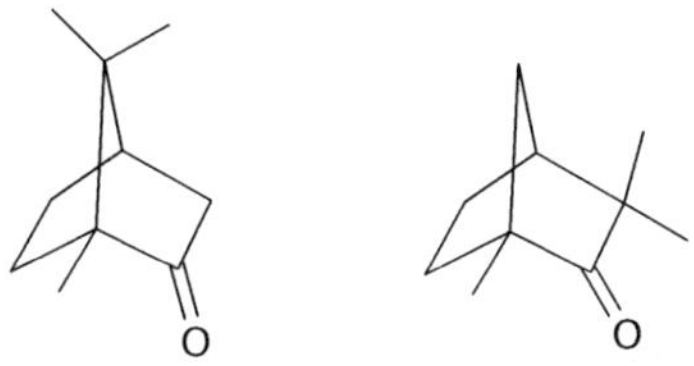


(1R,4R)-Camphor (1R,4s)-Fenchone

Figure 1.7 shows the $h\nu$ = 19.3 eV PECD and photoelectron spectrum extracted from a difference (L-CPL – R-CPL) image recorded with the DELICIOUS VMI

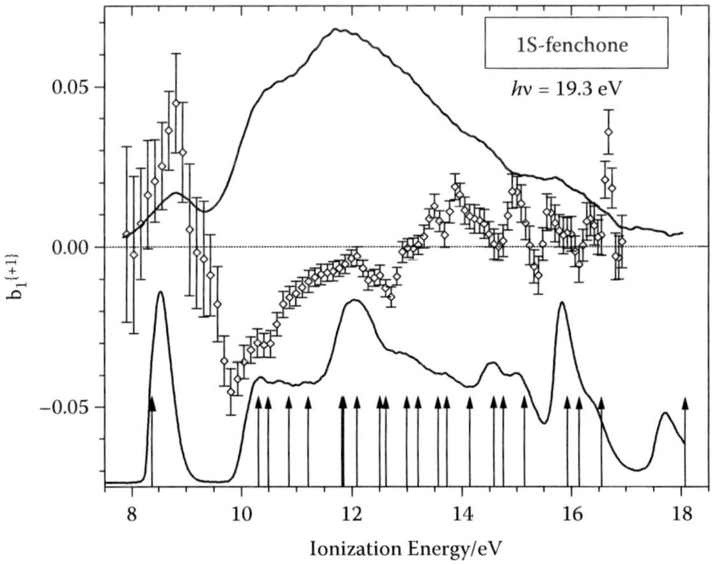


**FIGURE 1.7** PECD measurement for 1S-fenchone at $h\nu$ = 19.3 eV. The upper panel (background) shows the photoelectron spectrum extracted from the same 2D imaging data set, while the lower panel shows a higher-resolution valence PES recorded at $h\nu$ =95 eV. Also included are vertical markers indicating calculated vertical IPs. (Data taken from I. Powis, C. J. Harding, G. A. Garcia, and L. Nahon, *ChemPhysChem* 9 (2008): 475.)

instrument.[8] Also, for comparison, an independently recorded PES obtained at higher resolution has been included. There are a number of similarities with camphor. While, as before, the HOMO (carbonyl oxygen lone pair) orbital is well separated given the size of the molecule, the remainder of the orbital bands are partially overlapped, even in the higher-resolution spectrum. At the more modest energy resolution obtained in the VMI the congestion is more apparent. Nevertheless, although obtained simultaneously, and thus with identical energy resolution, the PECD spectrum is more obviously structured with features that can be correlated with the expected positions of unresolved bands. This "improvement" in resolution in the PECD spectrum results from the greater variation between $b_1^{\{\pm 1\}}$ parameters for adjacent orbitals as compared to cross sections, σ.

At this particular photon energy, it is also clear that the sign of the PECD switches between the HOMO and adjacent orbitals in fenchone, as was expressly noted at selected photon energies in our previous summary of camphor results (Section 1.4). A more detailed comparison between the PECD obtained from the similar HOMOs of fenchone and camphor is provided in Figure 1.8. Unfortunately, data below $h\nu$ = 13 eV were not obtained for fenchone owing to synchrotron beamline limitations at the time. While there is a weak, superficial resemblance in the PECD spectra in the region above $h\nu$ = 13 eV, there is also a very striking difference in magnitude seen in the experimental data, a difference that is also perfectly captured in the predictions of the CMS-Xα theoretical modeling. These calculations suggest that a much more profound difference could be expected below $h\nu$ = 13 eV. Note that such a difference between the PECD of fenchone vs. that of camphor is much less marked in the case of core-PECD[3] than here in the case of valence PECD, for an intrinsic reason that is not clear yet.

### 1.5.3 Glycidol Conformers

Glycidol ($C_3H_6O_2$) is an alcohol derivative of oxirane. It has nine possible conformations corresponding to rotations around the C-$CH_2$OH and COH bonds. The six lowest energies of these are shown in Figure 1.9. The **C1** and, to a lesser extent, the **C2** conformers are stabilized by the formation of intramolecular H-bonds from the hydroxyl to the ring oxygen. Calculated energy differences allow Boltzmann populations of nearly 100% **C1** at 50 K, rising toward 2:1 **C1**:**C2** at room temperature to be estimated.[4] In experiments conducted using the DELICIOUS VMI spectrometer, glycidol was seeded into a supersonic molecular beam intersecting the beam of circularly polarized VUV radiation in the spectrometer's ionization region.[9] Although the precise temperature of this source is not well characterized, it is clear that significant cooling may result; under conditions with high backing pressure cluster formation can be observed, although for these experiments milder expansion conditions were employed so that only monomers are being investigated.

The PECD spectra obtained were richly structured, allowing investigation of the photon energy dependence of the chiral $b_1^{\{+1\}}$ parameters on an orbital-by-orbital basis. Figure 1.10 presents the results obtained for the HOMO ionization. This orbital in the **C1** conformer has significant O 2*p* density but is not a simple lone pair, being of mixed, delocalized character. The experimental data in this figure can

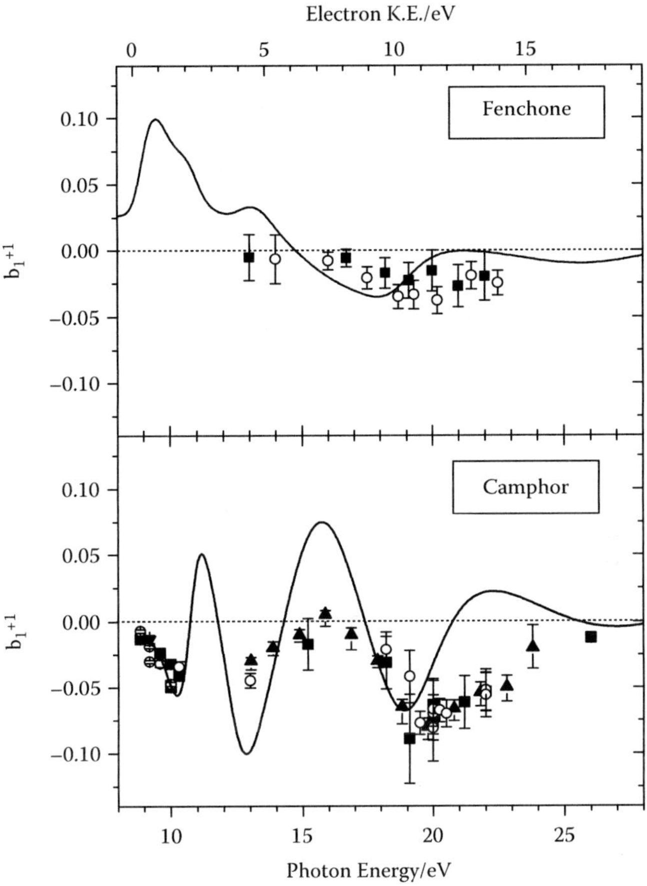


**FIGURE 1.8** Mean values deduced for the HOMOs of fenchone and camphor as a function of photon energy. Filled plotting symbols are experimental values for the 1R enantiomers. Open symbols were obtained for the 1S enantiomers, but have been negated for plotting and so are expected to follow the same trend lines. The curves show theoretically predicted energy dependence of the PECD. (Data for fenchone from I. Powis, C. J. Harding, G. A. Garcia, and L. Nahon, *ChemPhysChem* 9 (2008): 475; data for camphor from L. Nahon, G. A. Garcia, C. J. Harding, E. A. Mikajlo, and I. Powis, *J. Chem. Phys.* 125 (2006): 114309, and G. A. Garcia, L. Nahon, M. Lebech, J. C. Houver, D. Dowek, and I. Powis, *J. Chem. Phys.* 119 (2003): 8781; with additional data (▲) from T. Lischke, N. Böwering, B. Schmidtke, N. Müller, T. Khalil, and U. Heinzmann, *Phys. Rev. A* 70 (2004): 022507.)

be compared against the $b_1^{\{+1\}}$ parameters calculated for each of the six conformers shown in Figure 1.9. As anticipated, there is a massive divergence between the predicted PECD for the different conformations. Calculations for the conformers **C3**, **C4** provide a particularly poor agreement with experiment, the dichroism having the opposite sign to that observed. It is not unreasonable to discount a significant contribution from these structures under the conditions of our experiment.

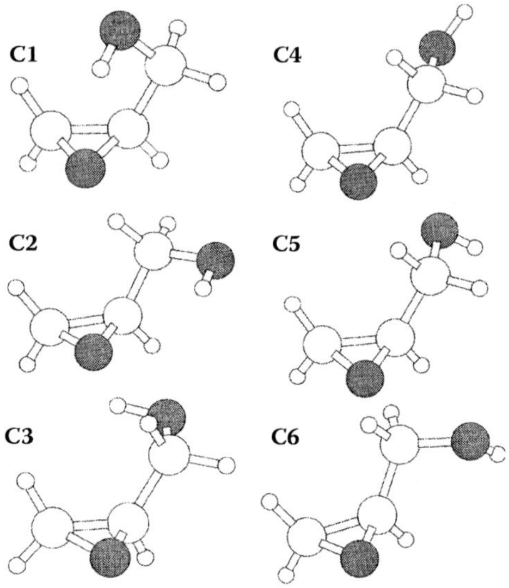


**FIGURE 1.9** The six lowest-lying conformers of glycidol. (Reproduced from I. Powis, C. J. Harding, S. Barth, S. Joshi, V. Ulrich, and U. Hergenhahn, *Phys. Rev. A* 78 (2008): 052501. With permission.)

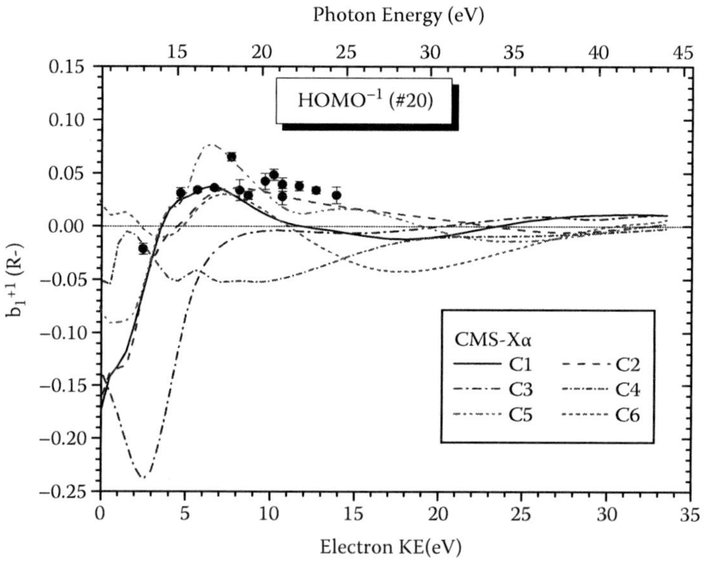


**FIGURE 1.10** Experimental and calculated PECD of the HOMO ionization of glycidol. Experimental data for R and S enantiomers are combined in a single figure by negating the S enantiomer values prior to plotting. Theoretical curves for each of the R enantiomers are plotted for each of the six conformers considered here. (Redrawn from data in A. Garcia, L. Nahon, C. J. Harding, and I. Powis, *Phys. Chem. Chem. Phys.* 10 (2008): 1628, with the inclusion of additional conformer data.)

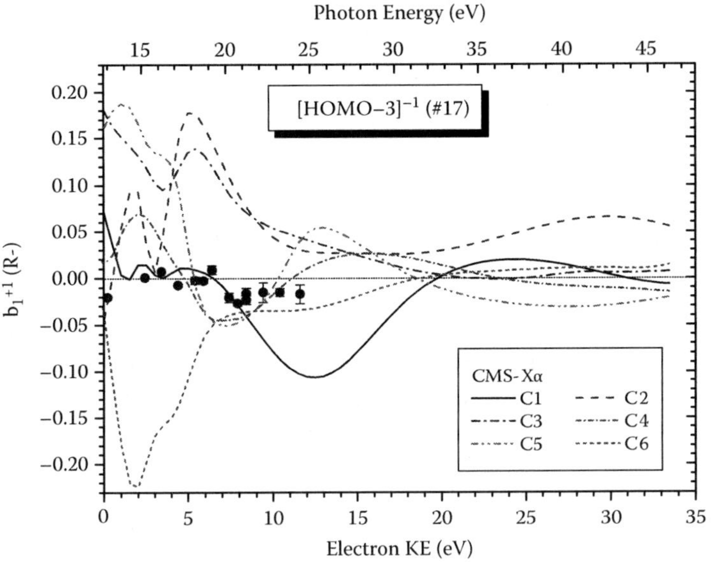


**FIGURE 1.11** As Figure 1.10 for the HOMO-3 (orbital 17) ionization of glycidol. (Redrawn from data in A. Garcia, L. Nahon, C. J. Harding, and I. Powis, *Phys. Chem. Chem. Phys.* 10 (2008): 1628, with the inclusion of additional conformer data.)

A similar set of results for ionization of the HOMO-3 (#17) orbital is presented in Figure 1.11. Now contributions in our molecular beam source from conformers **C5**, **C6** can be effectively dismissed; for both these geometries a large dichroism, though of opposite sign, is predicted within a few eV of threshold. Likewise **C2**, **C3** are predicted to have large positive $b_1^{\{+1\}}$ parameters at electron energies around 5 eV that are not in accord with experiment. This leaves **C1** or **C4** as the most plausible conformers for our experimental conditions, but as already argued, **C4** can probably be discounted for its poor performance in the HOMO comparison.

Similar comparisons between experimental PECD spectra and various postulated structures were made for other orbital ionizations,[9] and corroborate the strong inference that the dominant conformer in the pertaining molecular beam conditions is **C1**. This investigation has subsequently been extended to C 1*s* and O 1*s* core level PECD of this molecule,[4] albeit with different experimental sample conditions, but a similar conclusion concerning the dominance of the **C1** conformer was reached.

However, a general limitation of core level PECD in such molecules stems from the very similar core binding energy shifts displayed at many of the individual atomic sites. As a result, many of the individual orbital sites are fully overlapped in the X-ray photoelectron spectrum (XPS), meaning that only an averaged value for the PECD can be observed. The case of glycidol perfectly exemplifies this—the three C $1s^{-1}$ ionizations cannot be individually resolved in the XPS, nor can the two O $1s^{-1}$ ionizations, so that there are no opportunities to obtain truly site-specific PECD data for this molecule.[4] In contrast, the VUV-PES provides a range of fully and partially resolvable valence orbital ionizations, with varying character and degrees of (de-)

localization. While we have emphasized the final state scattering contribution to PECD, there is also a significant influence of the initial orbital characteristics (see Section 1.4). Correspondingly, the complete valence PECD data set spanning several orbital ionizations is, in general, anticipated to be much richer and varied in detail, providing an enhanced opportunity for a full, relatively unambiguous interpretation and assignment of structure, conformation, etc.

Although discounting contributions from **C3**, **C4**, **C5** under typical molecular beam or room temperature equilibrium conditions, it is interesting to note the great differences in predicted PECD of these three conformers as seen, for example, in Figures 1.10 and 1.11. This particular subgroup shares the same *anti*-$O_H$-C-C-$O_R$ conformation (Figure 1.9), and differs only in the torsional angle of the hydroxyl H-atom around the C-OH bond. Effectively then, the rotational conformation of a single H-atom in glycidol is seen to be capable of inducing dramatically large changes in the predicted chiral $b_1^{\{\pm1\}}$ parameters.

The glycidol calculations that have been discussed to this point utilize for the neutral molecule description an angular basis on the H-atoms consisting only of *s*-type functions—effectively a minimal (angular) basis set on the H. While the ionized state description uses a greatly expanded angular basis on all atoms (to account for scattering of the departing photoelectron into much higher $\ell$ partial waves), the reliability of the neutral state description may be reviewed. The dominant **C1** conformer clearly owes much of its relative stability to the formation of an intramolecular O-H–O hydrogen bond completing a five-membered ring structure. This suggests that a basis set incorporating polarization functions to capture the electrostatic H-bond interactions ought to be considered.

A series of comparative calculations for PECD of the **C1** conformer was therefore undertaken in which the neutral molecule basis was increased. In all cases the first-row atomic basis ran to *f* functions ( $\ell_{\max} = 3$ as in the original calculation), but the H-atom basis was now extended to include up to *d* ( $\ell = 2$ ) functions. Not unexpectedly, the most significant impact of allowing for H-bond-induced polarization in the neutral molecule was found for ionization of those orbitals having electron density in the vicinity of the five-membered ring structure. Most benefit derived from additional polarization functions added on the hydroxyl H-atom itself. In Figure 1.12 we compare the original **C1** PECD calculations for the HOMO and HOMO-1 ionizations, utilizing a minimal $\ell_{\max} = 0$ H-atom basis in the neutral molecule, with calculations where the hydroxyl H-bonded atom has additional $\ell = 1,2$ functions. In all cases the continuum state description utilizes $\ell_{\max} = 18,10,8$ for, respectively, the asymptotic, first-row atom, and H-atom regions.

Modest, but nonetheless useful improvements to the calculated $b_1^{\{+1\}}$ parameters result from the inclusion of polarization functions on the relevant H-bonded atom for these outer orbital ionizations, though for some of the other orbitals with different density distributions through the molecule the differences may be much less significant.[9] Much of the interesting detail of PECD that has been discussed previously can be ascribed to final state scattering effects that allow the whole molecular interior to be probed by the departing photoelectron. But this sensitivity to H-bond-induced polarization is a clear indication of an *initial* state effect that is responding to the characteristics and distribution displayed by the ionizing orbital.

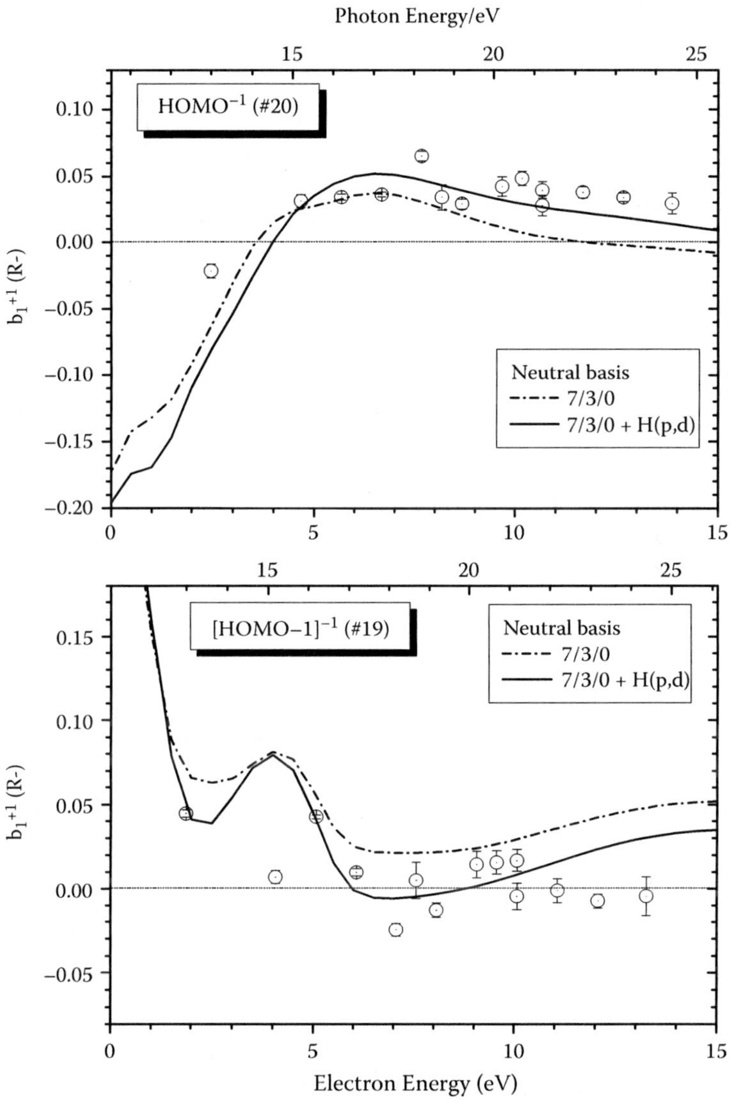


**FIGURE 1.12** Comparison of experimental PECD for outer orbitals of R-glycidol with CMS-Xα calculations made by either using a minimal H-atom basis or augmenting with (p,d) polarization functions on the hydroxyl H-atom. (Data taken from G. A. Garcia, L. Nahon, C. J. Harding, and I. Powis, *Phys. Chem. Chem. Phys.* 10 (2008): 1628.)

## 1.6 REFLECTIONS ON PECD

### 1.6.1 Applications and Developments

PECD measurements, as outlined here, currently have the capability to assist in the orbital characterization and assignment for species of interest and, not least, identification of the absolute chiral configuration can be expected should this be *a priori* unknown. It is likely that these capabilities will prove to be of particular relevance in the study of smaller prebiotic and biomolecules, which are frequently chiral, but which also likely have sufficiently congested vibronic structure to impede electronic assignment in their conventional photoelectron spectra. PECD spectra offer some improved prospects for improving the effective resolution and revealing underlying orbital structure. At the same time, clear information may be provided on the actual population of plausible conformers of an isolated molecule sample. Commonly for such molecules, conformer populations are determined by small energetic differences that may lie close to the achievable accuracy limit in many feasible calculations; experimental verification is clearly important in such circumstances. These facets are quite well exemplified in a recent study of the amino acid derivative alaninol by Turchini et al.[30]

The important role of vibrational motion in influencing photoionization dynamics generally has long been recognized in work with small, symmetric molecules, particularly in connection with shape resonance phenomena,[31] although it is still far from usual for vibronic effects to be explicitly included in theoretical treatments of such systems.[32] From current evidence discussed here, it is not, however, unreasonable to infer that vibrations, especially large-amplitude motions such as the multiple torsional and wagging modes that can be more commonly expected in larger molecules, may play an enhanced role in achieving a complete understanding of PECD effects. This then poses a unique challenge for computational PECD modeling for typical cases, but one that can be addressed thanks to the ongoing increases in computational power available.

Perhaps a more immediate interest lies with the further adaptation of the experimental technique that can be anticipated in the light of technical developments, especially in sources and detectors, that enhance the overall method. At root this allows one to build upon the potential sensitivity afforded by the unprecedented magnitude of the PECD asymmetry, and upon its remarkable sensitivity to aspects of the molecular structure. But while isolated molecule, gas phase studies provide important benchmarking for computational tools, allowing fundamental understanding to be placed on a secure footing, more insights must surely follow from a more interventionist experimental manipulation of the molecular environment. Recent step improvements in VUV SR photon flux are permitting us greater freedom to exploit the supersonic molecular beam sample inlet to this end. At a minimum, this and alternative source developments provide opportunities to control effective sample temperature and so manipulate conformer populations. Moreover, sudden expansion cooling may allow less stable conformer populations to be frozen in, permitting their characterization under idealized, isolated conditions.

Perhaps of even greater long-term potential, the use of well-established molecular beam clustering techniques in conjunction with PECD opens the door to studying chiral molecular interactions and recognition. We recall here the key point, that the PECD phenomenon as a photoionization-based technique is effectively universal to chiral systems and has no restrictions, such as requirement for the presence of specific chromophore groupings.

Actually, there is another advantage to be reaped as a consequence of this being an ionization-based measurement: the corresponding molecular ion can be detected simultaneously with the energy/angle-resolved electron, and this means that there is scope in a mixed cluster source to identify the precursor species for each electron if the molecular ion is mass analyzed. For instance, the "molecular handshake" between two different masses' monomers could be investigated enantioselectively by PECD with the contributions from homo- and heterodimers of various sizes being disentangled by mass analysis of the corresponding ion.

The authors are currently employing just such an angle-resolved photoelectron-photoion coincidence (ARPEPICO) technique in a modified VMI spectrometer (DELICIOUS2) that is equipped with an ion time-of-flight (TOF) spectrometer opposing the VMI photoelectron spectrometer.[33] The field that extracts the electrons is used to propel the ions into a linear TOF analyzer. The temporal correlation established by the delayed detector coincidence between the electron and the longer flight time ion serves to identify the electron-ion pair originating from a single ionization event, while the precise delay interval permits the ion mass to be established. We are thus able to extract PECD data from mass-selected samples. Implementation of this coincidence detection provides an ability to filter out spurious/background signal, hence increasing the reliability of the data, but the principal benefit is an ability to look at size-selected cluster species, at least in the near-threshold region below any fragmentation onsets. Currently the ARPEPICO technique has allowed PECD from homogeneous dimers through to heptamers to be recorded. For the future, more complex and larger molecular systems can be tackled. Microsolvation effects on electronic and geometric structure might thus be addressed to better understand the influence of solvation shells in solution phase CD. Bigger polymeric systems may also prove to be accessible to measurement. For such systems, possible caging effects or even microcrystallization (local ordering) may dramatically modify or conceivably further enhance the PECD observed on the monomer, just as has been shown for the instance of a giant CD in absorption seen with chiral molecule nanoparticles.[34]

### 1.6.2 A Photophysical Approach Contributing to Life's Homochirality?

The origin of life's homochirality—the fact that in the living world all amino acids are L-type and that the ribose sugars of nucleic acids are all D-type, while both are generally synthesized in the lab as a racemic mixture—is one of the most puzzling issues scientist have been dealing with since Louis Pasteur, who intuited that such a biomolecular asymmetry was a signature of life.[35]

Some amino acids, the chiral building blocks of proteins, which probably played a key role in the appearance of life on Earth, were discovered in carbonaceous meteorites,[36] with enantiomeric excesses (e.e.) and isotopic composition[37] indicating an extraterrestrial origin of biomolecular asymmetry. They were also synthesized in the laboratory by simulating interstellar/circumstellar ice photochemistry.[38] Therefore, assuming in the following an interstellar or circumstellar origin of elementary building blocks of life such as amino acids, one is looking for an asymmetric bias to which these species would have been exposed during their journey toward Earth that would have induced a significant e.e. Note that such an e.e. does not have to reach 100% (pure homochirality) since even a small but significant e.e. can be amplified by autocatalytic reactions leading to near homochirality.[39]

The most likely scenario for the origin of life's homochirality probably belongs to the abiotic category, in which one considers homochirality (or at least a noticeable enantiomeric excess) as a starting point, a necessary condition for the development of life via chemical evolution. The numerous possible scenarios cannot all be described here, but the reader will find reviews published elsewhere.[40] We will focus on theories based upon asymmetric photon-induced processes induced by CPL, and whose validity in the context of the search of the origin of life's homochirality lies in the fact that several CPL astronomical sources have been reported.[41] Among these are asymmetric photochemical processes in which, because of CD in absorption, one enantiomer absorbs more UV light, and can therefore be more photolyzed, leading to a measurable e.e., up to 2.6% in solid-state leucine.[42]

PECD could be an alternative photophysical process. Indeed, amino acids are expected to exhibit the same qualitative behavior as camphor, as calculated already in the case of alanine.[18] Therefore, the case of camphor is interesting in terms of the qualitative understanding of the relevance of PECD as a possible origin of life homochirality. Indeed, if one refers to the images of Figure 1.3, and integrates the relative flux of photoelectrons in the two half spaces "forward" and "backward" (as defined by the photon propagation axis), then, remembering that exchanging enantiomers produces the same asymmetry effect as swapping light polarization, for a given light helicity (L-CPL or R-CPL) there is a 7% excess of electrons coming from the S enantiomer in the forward direction and from the R enantiomer in the backward direction. Because of momentum conservation, this asymmetry in the photoelectron angular emission will be accompanied by an opposite asymmetry of the associated recoiling ion. Therefore, a circularly or partially circular polarized VUV radiation of a given helicity photoionizing a racemic mixture of gas phase randomly oriented enantiomers of amino acids in an interstellar molecular cloud could induce, via the PECD effect, an asymmetric recoiling distribution of the molecular ions, leading to an enantiomeric excess of the ion in a given direction. Depending on the full integration over the whole VUV spectrum, taking into account the sign of the PECD for each orbital, the intensity of each PES band, the recoil velocity transferred from the corresponding photoelectron, and the sign of the photon helicity, as well as the respective positions of the light source, the molecular cloud, and Earth, this would lead to a nonzero e.e. in favor of one or the other enantiomer traveling in a given direction of space. Note that considering the quite low excess energy deposited into the system, only a few eV above the ionization thresholds for the Lyman $\alpha$ energy, there would be very little fragmentation of the ion,[43] so that the produced ions would be mainly parent ions. These ions with a given e.e. would then get neutralized by adsorption on micrometeorites, for instance, and could therefore "seed" Earth with an asymmetric distribution of amino acid enantiomers.

## ACKNOWLEDGMENTS

We are very grateful to Gustavo Garcia, who, as a graduate student with IP and then as a postdoctoral fellow and a permanent scientist with LN, played a very important role in the experimental investigations presented in this chapter, in particular for the spectrometer design, data acquisition, and data treatment. We also thank the Royal Society and the British Council/Alliance (EGIDE) program for collaborative awards.

## REFERENCES

1. I. Powis. 2008. *Adv. Chem. Phys.* 138:267.
2. J. Fidler, P. M. Rodger, and A. Rodger. 1993. *J. Chem. Soc. Perkin Trans.* 2:235.
3. V. Ulrich, S. Barth, S. Joshi, U. Hergenhahn, E. A. Mikajlo, C. J. Harding, and I. Powis. 2008. *J. Phys. Chem. A* 112:3544.
4. I. Powis, C. J. Harding, S. Barth, S. Joshi, V. Ulrich, and U. Hergenhahn. 2008. *Phys. Rev. A* 78:052501.
5. B. Ritchie. 1976. *Phys. Rev. A* 13:1411.
6. I. Powis. 2000. *J. Chem. Phys.* 112:301.
7. L. Nahon, G. A. Garcia, C. J. Harding, E. A. Mikajlo, and I. Powis. 2006. *J. Chem. Phys.* 125:114309.
8. I. Powis, C. J. Harding, G. A. Garcia, and L. Nahon. 2008. *ChemPhysChem* 9:475.
9. G. A. Garcia, L. Nahon, C. J. Harding, and I. Powis. 2008. *Phys. Chem. Chem. Phys.* 10:1628.
10. G. A. Garcia, L. Nahon, C. J. Harding, E. A. Mikajlo, and I. Powis. 2005. *Rev. Sci. Inst.* 76:053302.
11. A. T. J. B. Eppink and D. H. Parker. 1997. *Rev. Sci. Inst.* 68:3477.
12. G. A. Garcia, L. Nahon, and I. Powis. 2004. *Rev. Sci. Inst.* 75:4989.
13. N. Böwering, T. Lischke, B. Schmidtke, N. Müller, T. Khalil, and U. Heinzmann. 2001. *Phys. Rev. Lett.* 86:1187.
14. T. Lischke, N. Böwering, B. Schmidtke, N. Müller, T. Khalil, and U. Heinzmann. 2004. *Phys. Rev. A* 70:022507.
15. S. Turchini, N. Zema, G. Contini, G. Alberti, M. Alagia, S. Stranges, G. Fronzoni, M. Stener, P. Decleva, and T. Prosperi. 2004. *Phys. Rev. A* 70:014502. S. Stranges, S. Turchini, M. Alagia, G. Alberti, G. Contini, P. Decleva, G. Fronzoni, M. Stener, N. Zema, and T. Prosperi. 2005. *J. Chem. Phys.* 122:244303. G. Contini, N. Zema, S. Turchini, D. Catone, T. Prosperi, V. Carravetta, P. Bolognesi, L. Avaldi, and V. Feyer. 2007. *J. Chem. Phys.* 127:124310.
16. A. Giardini, D. Catone, S. Stranges, M. Satta, M. Tacconi, S. Piccirillo, S. Turchini, N. Zema, G. Contini, T. Prosperi, P. Decleva, D. Di Tommaso, G. Fronzoni, M. Stener, A. Filippi, and M. Speranza. 2005. *ChemPhysChem* 6:1164.
17. V. McKoy, T. A. Carlson, and R. R. Lucchese. 1984. *J. Phys. Chem.* 88:3188.
18. I. Powis. 2000. *J. Phys. Chem. A* 104:878.
19. D. Dill and J. L. Dehmer. 1974. *J. Chem. Phys.* 61:692. J. W. Davenport. 1976. Dissertation, University of Pennsylvania.
20. K. H. Johnson. 1973. *Adv. Quant. Chem.* 7:143.
21. Y. Hikosaka, J. H. D. Eland, T. M. Watson, and I. Powis. 2001. *J. Chem. Phys.* 115:4593.

22. M. Stener, D. D. Tommaso, G. Fronzoni, P. Decleva, and I. Powis. 2006. *J. Chem. Phys.* 124:024326.
23. D. Di Tommaso, M. Stener, G. Fronzoni, and P. Decleva. 2006. *ChemPhysChem* 7:924.
24. M. Stener, G. Fronzoni, D. Di Tommaso, and P. Decleva. 2004. *J. Chem. Phys.* 120:3284.
25. D. Toffoli, M. Stener, G. Fronzoni, and P. Decleva. 2002. *Chem. Phys.* 276:25.
26. E. E. Rennie, I. Powis, U. Hergenhahn, O. Kugeler, G. Garcia, T. Lischke, and S. Marburger. 2002. *J. Elec. Spec. Rel. Phen.* 125:197.
27. U. Hergenhahn, E. E. Rennie, O. Kugeler, S. Marburger, T. Lischke, I. Powis, and G. Garcia. 2004. *J. Chem. Phys.* 120:4553.
28. C. J. Harding and I. Powis. 2006. *J. Chem. Phys.* 125:234306.
29. C. J. Harding, E. A. Mikajlo, I. Powis, S. Barth, S. Joshi, V. Ulrich, and U. Hergenhahn. 2005. *J. Chem. Phys.* 123:234310. I. Powis. 2008. *Chirality* 20:961.
30. S. Turchini, D. Catone, G. Contini, N. Zema, S. Irrera, M. Stener, D. D. Tommaso, P. Decleva, and T. Prosperi. 2009. *ChemPhysChem*, 10:1839.
31. J. L. Dehmer, D. Dill, and A. C. Parr. 1985. In *Photophysics and photochemistry in the vacuum ultraviolet*, ed. S. P. McGlynn, G. L. Findley, and R. H. Huebner, 341. Dordrecht: D. Reidel.
32. M. Hoshino, R. Montuoro, R. R. Lucchese, A. De Fanis, U. Hergenhahn, G. Prumper, T. Tanaka, H. Tanaka, and K. Ueda. 2008. *J. Phys. B At. Mol. Opt. Phys.* 41:085105. R. Montuoro, R. R. Lucchese, J. D. Bozek, A. Das, and E. D. Poliakoff. 2007. *J. Chem. Phys.* 126:244309.
33. G. A. Garcia, H. Soldi-Lose, and L. Nahon. 2009. *Rev. Sci. Inst.* 80:023102.
34. J. Paul, A. Dorzbach, and K. Siegmann. 1997. *Phys. Rev. Lett.* 79:2947.
35. L. Pasteur. 1853. *C.R. Acad. Sci.* (Paris) 37:162.
36. D. P. Glavin and J. P. Dworkin. 2009. *Proc. Natl. Acad. Sci. USA* 106:5487. J. R. Cronin and S. Pizzarello. 1997. *Science* 275:951. M. H. Engel and S. A. Macko. 1997. *Nature* 389:265.
37. S. Pizzarello, M. Zolensky, and K. A. Turk. 2003. *Geochim. Cosmochim. Acta* 67:1589.
38. G. M. Munoz Caro, U. J. Meierhenrich, W. A. Schutte, B. Barbier, A. Arcones Segovia, H. Rosenbauer, W. H. P. Thiemann, A. Brack, and J. M. Greenberg. 2002. *Nature* 416:403. M. P. Bernstein, J. P. Dworkin, S. A. Sandford, G. W. Cooper, and L. J. Allamandola. 2002. *Nature* 416:401.
39. T. Shibata, J. Yamamoto, N. Matsumoto, S. Yonekubo, S. Osanai, and K. Soai. 1998. *J. Am. Chem. Soc.* 120:12157. C. Girard and H. B. Kagan. 1998. *Angew. Chem. Int. Ed.* 37:2923.
40. U. Meierhenrich, B. Barbier, R. Jacquet, A. Chabin, C. Alcaraz, L. Nahon, and A. Brack. 2001. In *Exo-/Astro-Biology* 496:167. A. G. Griesbeck and U. L. Meierhenrich. 2002. *Angew. Chem. Int. Ed.* 41:3147. A. T. Borchers, P. A. Davis, and M. E. Gershwin. 2004. Exp. Biol. Med. 229:21. U. Meierhenrich. 2008. *Amino acids and the asymmetry of life*. Berlin: Springer.
41. J. Bailey, A. Chrysostomou, J. H. Hough, T. M. Gledhill, A. McCall, S. Clark, F. Menard, and M. Tamura. 1998. *Science* 281:672.
42. U. J. Meierhenrich, L. Nahon, C. Alcaraz, J. H. Bredehoft, S. V. Hoffmann, B. Barbier, and A. Brack. 2005. *Angew. Chem. Int. Ed.* 44:5630.
43. G. A. Garcia, L. Nahon, and I. Powis. 2003. *Int. J. Mass. Spec.* 225:261.
44. G. A. Garcia, L. Nahon, M. Lebech, J. C. Houver, D. Dowek, and I. Powis. 2003. *J. Chem. Phys.* 119:8781.

# 2 High-Resolution Microwave Spectroscopy of Chiral Molecular Contact Pairs

*Xunchen Liu and Yunjie Xu*

**CONTENTS**

2.1 Introduction....27
2.2 Experimental Details....28
2.3 Theoretical Approaches....30
2.4 Chiral Molecular Contact Pairs....31
2.4.1 Chirality Recognition (Permanent-Permanent Chirality Interactions)....31
2.4.2 Chirality Induction (Permanent-Transient Chirality Interactions)....33
2.4.3 Chirality Synchronization (Transient-Transient Chirality Interactions)....36
2.5 Concluding Remarks....37
References....37

## 2.1 INTRODUCTION

High-resolution microwave (MW) spectroscopy, in combination with a supersonic jet expansion, has been used widely in the last two decades to characterize the intermolecular interactions in the complexes consisting of atoms and simple molecules.[1–3] These spectroscopic studies of hydrogen (H)-bonded complexes and van der Waals clusters have greatly enriched our understanding of intermolecular interactions and provided the essential steps to build a link between the properties of bulk matter and the properties of the constituent atoms and molecules. For example, the pure rotational studies of helium clusters doped with a polar linear molecule have provided a unique opportunity to probe superfluidity, a formally bulk property, at the molecular level with atom-by-atom resolution.[4] A list of noncovalently bound molecular systems investigated using high-resolution spectroscopy has been compiled by Novick.[3]

A particularly interesting aspect of intermolecular interactions is the so-called chirality recognition effect. Chirality recognition refers to the omnipresent, fascinating ability of nature to discriminate between right- and left-handed forms (called enantiomers) of a chiral molecule. For example, our nose can distinguish the left-handed carvone, a naturally occurring compound found in caraway seeds, that smells like cumin, from the right-handed carvone, an extract from spearmint that smells like mint. The past ten years have witnessed the beginning and the significant development of using jet-cooled low-resolution spectroscopy to probe the intermolecular forces responsible for chirality recognition at the molecular level.[5] This area of research was pioneered by the groups of Zehnacker and Giardini-Guidoni in the ultraviolet region[6] and by Suhm and coworkers in the infrared (IR) region.[7,8]

High-resolution MW spectroscopy offers great promises for the studies of chiral molecular systems. The spectroscopic instrument utilized in the high-resolution MW spectroscopic measurements is a pulsed Fourier-transform (FT) MW spectrometer. Such a spectrometer provides a very high-resolution capability, typically less than a few kHz, and high sensitivity. The kHz spectral resolution offers a great advantage for the studies of relatively large and highly complex chiral clusters, which often have an amazingly large number of possible conformers. With such a resolution, one can easily tell apart conformers with the subtlest differences in their structures. It also allows the detections of very narrow hyperfine splitting, such as those due to nuclear quadrupole coupling or spin rotation interactions. These hyperfine patterns are often extremely useful in identifying the molecular species and assisting the spectral assignments. FTMW spectroscopy has also been used to determine relative energy differences as small as one-tenth of a kcal $mol^{-1}$.[9] This can lead to a clear experimental energetic ordering of the observed chiral conformers. Furthermore, FTMW spectroscopy is applicable for any molecular system with a permanent electrical dipole moment, even a system with an extremely small dipole moment. For example, rotational spectra of the mixed rare gas clusters, whose dipole moments are solely induced by the weak van der Waals interactions, have been measured using FTMW spectroscopy.[9,10] In summary, the rotational spectroscopic studies of chiral molecular systems can provide accurate structural and dynamical information of individual conformers that may not be obtainable from the low-resolution electronic and FTIR measurements.

In this chapter, the advantages and challenges of using high-resolution MW spectroscopy to study chiral molecular systems are discussed. In particular, we describe some general approaches one can use to achieve unambiguous assignments of rotational spectra of chiral molecular systems with many potential conformers. The remainder of this chapter is organized as follows. In the next section we present the experimental technique used to measure rotational spectra of chiral clusters and the assignment procedures. In Section 2.3, the complementary computational calculations are described. The chiral molecular systems investigated so far using high-resolution MW spectroscopy are discussed in Section 2.4. Concluding remarks are given in Section 2.5.

## 2.2 EXPERIMENTAL DETAILS

The spectroscopic instrument used is a Balle–Flygare-type[11] pulsed molecular beam FTMW spectrometer. This technique utilizes the coherent excitation of the molecular clusters and the subsequent detection of the coherent molecular emission.[1] General descriptions of the current state-of-the-art designs can be found in a few reviews[1,2] and book chapters.[12] The chiral molecular complexes are generated by pulsing a sample mixture, containing a trace amount (<1%) of the target molecules in 1 ~ 10 bar neon or helium backing gas, through a pinhole nozzle (General Valve Series 9) with a typical nozzle orifice of 0.5 ~ 0.8 mm. For some studies, such as in the case of the 2-fluoroethanol (FE) dimer,[13] it was found that a modified nozzle cap with a 0.8-mm nozzle orifice and a 3.0 mm long exit channel provided a better sensitivity. The full width at half height of a well-resolved line is about 10 ~ 20 kHz, and the uncertainty in the frequency measurements is 1 ~ 2 kHz. The estimated effective rotational temperature in a neon or helium expansion is about 1 ~ 5 K. The effective conformational temperature for the H-bonded binary systems, such as the series of complexes between oxirane (Ox), propylene oxide (PO), trans-dimethyl oxirane (DO), and ethanol (EtOH), has a typical value of ~60 K, considerably higher than the corresponding rotational temperature.[14]

While it is possible to optimize the sample mixture for the production of a certain chiral molecular complex, a large number of other complexes are still simultaneously produced in a jet expansion. This, coupled with the fact that chiral molecular clusters have in general many possible conformers, tends to generate a very dense rotational spectrum with a forest of transitions. To pick out a set of rotational transitions belonging to a particular conformer is therefore a highly challenging task. The assignment process can be aided by a number of systematic approaches. To reduce the number of possible candidates for an initial assignment, transitions of any monomers and the $\text{monomer}_x$–$\text{rare gas}_y$ (x, y = 0, 1, 2, ...) clusters have to be eliminated. Often, the rotational spectra of the chiral monomers have already been investigated and the clusters containing rare gas atoms can be identified by switching to a different backing rare gas. For the chiral molecular contact pairs whose binding partners are with permanent chirality, such as the PO dimer, one can use an enantiomeric pure sample instead of a racemic sample to identify lines belonging to the hetero- or homochiral binary complexes by observing the disappearance or increase in their intensities, respectively. If a chiral molecular system contains any quadrupole nuclei, such as $^{14}$N, one can expect to observe nuclear quadrupole hyperfine structures with distinct patterns, which can be further used for the quantum number assignments. The magnitude of the transition dipole moment can be estimated experimentally by comparing the so-called optimized $\pi/2$ microwave pulse length to that of a molecule with known dipole moment. A preliminary set of the assigned transitions is then used in a spectroscopic fit. The standard deviation of the fit is expected to be comparable to the experimental uncertainty, and any further transitions predicted are expected to be observed within a few kHz. Because of the extremely high accuracy of the measured frequencies, there is hardly any room for a misassignment.

To correlate a particular set of transitions to a certain conformer, one can in principle carry out extensive isotopic studies and derive the actual structures from the isotopic structural analyses. This approach is rarely used in reality, because very few chiral molecules are available in enriched isotopes and the amount of work involved is tremendous. Instead, one can take advantage of the theoretical modeling (see Section 2.3). Often, one can correlate a set of transitions observed to a particular conformer by comparing the experimental and calculated rotational constants, if the rotational constants of the anticipated conformers are predicted to be quite different. In addition, one can utilize the observed and calculated *a*-, *b*-, and *c*-dipole moments and the nuclear quadrupole coupling constants if the molecular system under consideration has any quadrupole nuclei, for further discrimination.[15] The latter strategy was successfully demonstrated in the rotational study of a natural amino acid L-threonine.[15]

## 2.3 THEORETICAL APPROACHES

To aid the spectroscopic searches and assignments, high-level *ab initio* calculations are often carried out prior to the experimental studies. The starting geometries of the potential conformers are often based on chemical intuition from other related systems or snapshots[16] generated by molecular dynamics simulation packages such as AMBER.[17] The initial screening calculations are typically carried out at a low level of theory such as HF/3-21G or B3LYP/6-31G.[14,18] The final geometry optimizations of chiral molecular systems are usually performed using the second-order Møller–Plesset (MP2) perturbation theory[19] with the 6-311++G(d,p) or similar basis sets.[20] The calculated rotational constants are, in turn, used to predict the rotational transition frequencies. Empirically, the MP2/6-311++G(d,p) or the related Pople-type basis sets have been found to provide astonishingly good predictions for the conformational geometries in a number of studies of chiral complexes.[13,14,21–24] The typical differences between the (best) calculated and the experimental rotational constants are less than 10% for the systems investigated so far. Although 10% may seem like a manageable error, one should keep in mind that the differences among the rotational constants of different conformers may also be of similar magnitude. Furthermore, the differences in the predicted rotational constants using two similar size basis sets, for example, 6-311++G(d,p) and aug-cc-pVDZ, can also be of similar order of magnitude.[25] So the experimental data are essential in getting the exact geometry of a particular conformer.

The calculations can also provide additional information about the electric dipole moment components, nuclear quadrupole coupling constants, and preliminary relative stabilities of the conformers. It is necessary to include basis set superposition error (BSSE) and zero-point energy (ZPE) corrections in evaluating the relative stabilities. Such corrections are regularly of similar magnitude as the energy differences among different conformers, i.e., the energies responsible for chirality recognition.[13,14] For the medium-size basis sets, such as 6-311++G(d,p), it was found that the BSSE correction tended to overcorrect, and the empirical half BSSE correction may be used instead. In general, a difference of a few kJ/mol may be expected at the MP2/6-311++G(d,p) level when compared to the experimental data. Single-point energy calculations using larger basis sets, such as MP2/aug-cc-pVTZ,[26] at the previously optimized geometry, can provide a better prediction of the electronic structure of the complex. As a result, the energy ordering and the dipole moment components predicted in this manner are in general more accurate.[13] In addition, theoretical calculations also provide information about the relative magnitude of contributions to the chirality recognition energy from several different sources, such as the conformational energies of the monomers, the deformation energies of the binding subunits, and the intermolecular interaction energies between the binding partners.

## 2.4 CHIRAL MOLECULAR CONTACT PAIRS

The chiral molecular contact pairs investigated using FTMW spectroscopy so far can be categorized into three subgroups based on whether the molecular subunits involved carry a permanent steorogenic center. Here we use the notation chirality recognition, chirality induction, and chirality synchronization, introduced by Zehnacker and Suhm.[5] These subgroups include chirality recognition in the 2-butanol dimer,[21] the glycidol dimer,[24] and the PO dimer,[22] where both molecular subunits involved carry a permanent steorogenic center; chirality induction in PO-EtOH,[22] DO-EtOH,[14] and PO-FE,[13] where only one binding partner carries a permanent steorogenic center; and chirality synchronization in the EtOH[27] and FE[13] dimers, where both binding partners possess only transient chirality. Chirality recognition in a single chiral molecule, such as butan-2,3,-diol,[28] is a special case of chirality recognition, where the recognition is established through an intramolecular H-bond and two distinct forms, i.e., homochiral (*RR* or *SS*) and heterochiral (*RS* or *SR*), exist. This phenomenon is better known as diastereoisomerism, and this category of molecular systems is not included in the discussions below.

### 2.4.1 Chirality Recognition (Permanent-Permanent Chirality Interactions)

The observation of the rotationally resolved spectrum of the 2-butanol dimer is the first high-resolution spectroscopic study of chirality recognition reported.[21] The 2-butanol monomer is one of the simplest chiral molecules that can be regarded as truly organic. It has a decent vapor pressure at room temperature and a relatively large electric dipole moment, making it an ideal candidate for a pure rotational study. Nine local minima, rising from the three conformations of the central C-C bond and the three conformations of the C-O bond, were identified for the 2-butanol monomer in the theoretical investigations.[29] Only three of them were detected experimentally in a jet expansion using FTMW spectroscopy.[30] Based on this fact and the assumption that the dimer would be held together by strong H-bonds involving the OH functional groups, one may expect at least nine dimer conformations. Four lowest-energy conformers, corresponding to two hetero- and two homochiral complexes, were located theoretically. The two heterochiral conformers were predicted to be near prolate tops with a large *a*-dipole moment. One can therefore expect an eye-catching spectral pattern with a roughly constant (B + C) spacing for the sequential *J* transitions. While the most stable heterochiral dimer was detected and assigned, the other heterochiral dimer, about 4 kJ/mol less stable, was expected to be sparely populated in the jet expansion and was not observed. The two homochiral species, about 1.5 to 2 kJ/mol less stable, on the other hand, were expected in the jet expansion when the enantiomeric pure 2-butanol sample was used. They were, however, not assigned, partially because of their less distinctive spectral patterns. This first study demonstrates the promises and challenges in probing chirality recognition using high-resolution MW spectroscopy.

The dimer and larger oligomers of glycidol are among the first chirality recognition systems investigated using FTIR spectroscopy.[7] Glycidol contains a rigid oxirane ring and an OH functional group. The two dominant monomeric conformations, i.e.,

M1 and M2, contain an intramolecular O-H···O H-bond each. The FTMW rotational study of the glycidol dimers was recently published by Caminati and coworkers.[24] The two most stable homochiral binary conformers, stabilized by two, relatively strong O-H···O H-bonds, were unambiguously identified. One of them consists of two M1 subunits, while the other one has a M1 and a M2 subunit. The third most stable homochiral binary conformer is made of two M2 subunits. It was predicted to be of similar stability and have large *a*- and *b*-dipole moments, but was not observed in their spectral search. This was attributed to a kinetic-statistic factor. Since the ratio of the monomeric species M1 and M2 is about 4:1 at the preexpansion temperature, the probability for the M2 monomers to encounter each other is much lower than that for the M1 monomers. This study is the first high-resolution spectroscopic detection of a permanent chiral contact pair consisting of two different monomeric conformers.

The PO dimer is a unique case of chirality recognition. While at least one H-bond is present in all the other chiral molecular systems studied using spectroscopy, the PO dimer relies exclusively on the secondary C-H···O H-bonds and the feeble van der Waals interactions. This allows a rare opportunity to focus the attention on the weak secondary H-bonds and their roles in chirality self-recognition. There are two important starting questions:

1. Will the PO dimer lock itself in several specific positions, or will it be a very floppy system with a relatively flat minimum and large-amplitude motions?
2. If it does exist in a number of different conformations, can FTMW spectroscopy distinguish the subtle differences among them?

Even though PO is a simple rigid chiral molecule, if one ignores the methyl group internal rotation, which is expected to be quenched upon complexation, the number of possible binary conformers is still unexpectedly large. This is because the directional requirement for the secondary H-bonds and van der Waals interactions is much less stringent than that for the primary H-bonds. Twelve most stable conformers were predicted for the PO dimer, where each PO binding partner acts simultaneously as a proton acceptor and donor. Surprisingly, the binding energy of the PO dimer, connected through four secondary H-bonds, is ~15 kJ/mol, similar to that of PO-$H_2O$ (~21 kJ/mol),[25(a)] with a classic H-bond. Even more remarkable is that six of the twelve PO binary conformers, which were predicted to have strong *a*-type transitions, were detected and unambiguously assigned. The geometries of the six conformers observed are given in Figure 2.1, where the secondary H-bonds are indicated. The direct detection of the six distinct conformers implies that although each secondary H-bond is relatively weak compared to a primary H-bond, the combined strength is enough to lock the binding subunits in a specific position. The observation suggests that the combined effect of the secondary H-bonds and van der Waals interactions can be significant, and as such, they may be expected to play an important role in many biological processes since secondary H-bonds are very common and are in abundance in large biological molecules. The ability to model the combined effect of these feeble interactions accurately will be important in correctly predicting the outcome of a bioprocess.

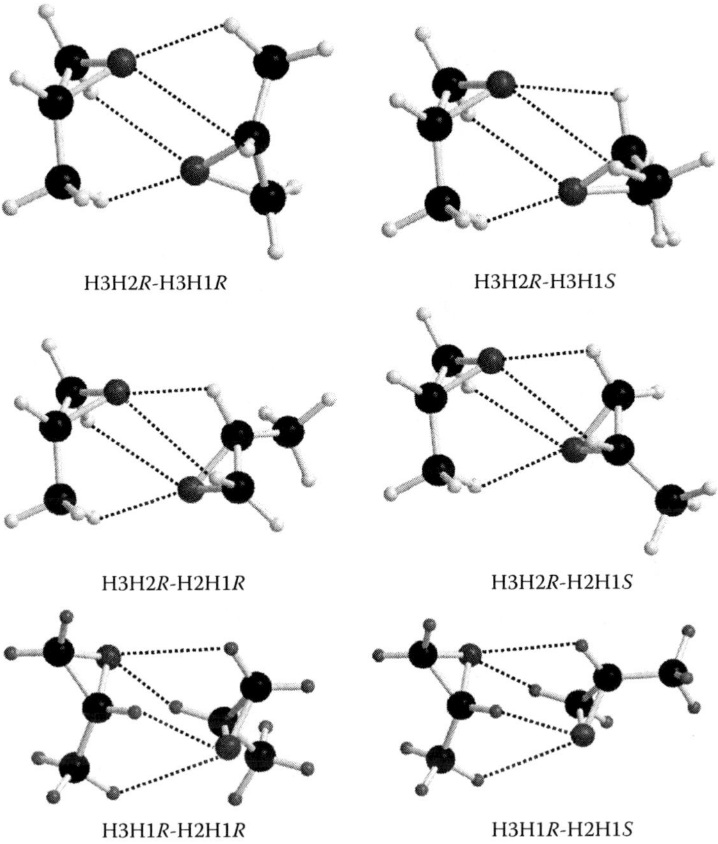


**FIGURE 2.1** The geometries of the six observed homo- and heterochiral PO dimers. The naming notation is of the form HxHy*R*-Hx'Hy'*R* or *S*, where *R* or *S* specifies the chirality of the monomer subunit and the letters x, y, and x', y' take the integer values of 1, 2, or 3, indicating that the corresponding hydrogen atoms are from the CH, $CH_2$, or $CH_3$ sites, respectively. For example, H3H2*R*-H2H1*S* represents a heterochiral dimer with four intermolecular H-bonds where $CH_3$ and $CH_2$ of *R*-PO each contributes one H-atom to the O-atom of *S*-PO, and the other two H-atoms from $CH_2$ and CH of *S*-PO point toward the O-atom of *R*-PO. The four secondary intermolecular hydrogen bonds are indicated.

### 2.4.2 Chirality Induction (Permanent-Transient Chirality Interactions)

Some molecules, such as EtOH and FE, can rapidly tunnel between their two mirror imaged chiral conformations. They are said to exhibit *transient* chirality. For EtOH, rotation about the O-C bond gives rise to three different conformations: *gauche*+ (*G*+), *gauche*- (*G*–), and *trans* (*T*), with the $\tau_{HOCC}$ at about +60, –60, and 180°. A schematic diagram, showing qualitatively a one-dimensional tunneling path that connects these three conformers, is given in Figure 2.2a. The *T* conformation is 0.47 kJ/mol[31] more stable than the mirror image pair: *G*+ and *G*–. Such transiently chiral enantiomers, i.e., *G*+ and *G*– EtOH, cannot be isolated or distinguished, unless the

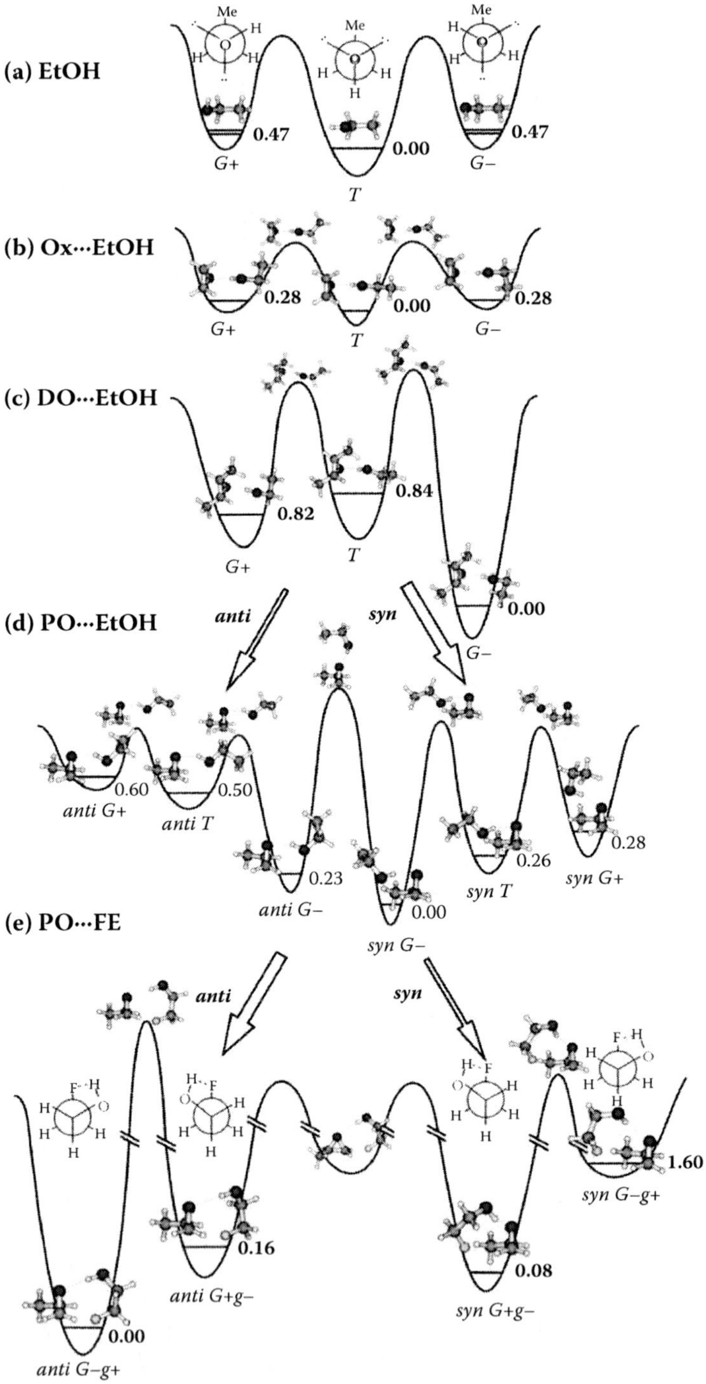
(a) EtOH
Me
H
H
H
Me
H
H
H
Me
H
H
H
0.47
0.00
0.47
G+
T
G–
(b) Ox···EtOH
0.28
0.00
0.28
G+
T
G–
(c) DO···EtOH
0.82
0.84
G+
T
0.00
G–
(d) PO···EtOH
anti
syn
0.60
0.50
anti G+
anti T
0.23
anti G–
0.00
syn G–
0.26
syn T
0.28
syn G+
(e) PO···FE
anti
syn
F
H
O
H
H
H
H
H
F
O
H
H
H
H
H
F
O
H
H
H
H
F
H
O
H
H
H
0.00
anti G–g+
0.16
anti G+g–
0.08
syn G+g–
1.60
syn G–g+

**FIGURE 2.2** (Color Figure 2.2 follows page 46.) The ground vibrational energy levels of the conformers of (a) EtOH, (b) Ox···EtOH, (c) DO···EtOH, (d) PO···EtOH, and (e) PO···FE as a function of $\tau_{HOCC}$ in EtOH (see text for details). FE is labeled with *G*+, *G*-, and *T* for $\tau_{OCCF}$, and *g*+, *g*-, and *t* for the $\tau_{HOCC}$ torsional angle, following the notation of refs. 13 and 18, and only *compact* G conformations, i.e., *G*+*g*- and *G*-*g*+, are indicated. The energy levels for EtOH are the experimental values taken from Ref. 31, while the rest are the calculated values obtained at the MP2/6-311++G(d,p) levels with ZPE corrections (in kJ/mol). Please note that the relative energy orderings of conformers predicted for (d) and (e) at this level of theory differ from the experimental ones. The G tunneling pair (not drawn to scale) is degenerated in EtOH and in Ox···EtOH, while the degeneracy is lifted in DO···EtOH and PO···EtOH. The fatter arrows indicate either *syn* or *anti*-conformers are preferred. *syn* or *anti* specify that the H-bonding partner is on the same or the opposite side as the methyl group of PO, respectively. The connecting one dimensional potential curves are only qualitative. The transition states were calculated at the B3LYP/6-311++G(d,p) level of theory. All conformers listed except *syn G*-*g*+ of PO···FE were detected experimentally.

degeneracy is lifted by interactions with another chiral identity. The FTMW studies of the binary complexes formed between EtOH and a series of oxirane-based molecules,[14,22] namely, Ox, PO, and DO, allow one to examine chirality induction in some detail.

The Ox molecule is achiral with $C_{2v}$ symmetry. The oxirane ring bisects the homotopic lone electron pairs at oxygen. The H-bonding from the hydroxyl group of EtOH to the lone electron pairs of Ox produces three conformers: Ox···*G*+ EtOH and Ox···*G*– EtOH, that are mirror images of each other, and Ox···*T* EtOH (Figure 2.2b). Conversely, DO is a chiral molecule with $C_2$ symmetry and homotopic lone pairs at oxygen. The enantiomeric pair (*G*– and *G*+) of EtOH can now be discriminated through diastereomeric interactions with the chiral DO molecule. For example, the *G*– conformation of ethanol was found to be favored by *R,R*-DO over the *G*+ and *T* conformations through the asymmetric spatial hindrance caused by the methyl groups of DO (see Figure 2.2c). PO is a chiral molecule with $C_1$ symmetry. Its oxygen lone pairs are diastereotopic because of the asymmetric environment provided by the methyl group. In PO···EtOH, the recognition process occurs through both diastereofacial and diastereomeric interactions. Altogether, six local minima were predicted for PO···EtOH. All six of them were detected and assigned in the FTMW study, with an experimental stability ordering of *syn G*– < *anti G*– < *syn T* = *syn G*+ < *anti G*+ < *anti T*.[22]

The addition of methyl functional group(s) to oxirane ring was shown to strengthen the O-H···O hydrogen bond because it enhances the electron density at the oxirane oxygen, and therefore its hydrogen bond acceptor capability. The FTMW studies show that the binding energy increases on average by about 1.3 $kJmol^{-1}$ for the addition of the first methyl group and by about 2.4 $kJmol^{-1}$ for the second.[14] In general, the addition of methyl groups raises the tunneling barriers between different conformers. It also changes the symmetry of the potential energy surfaces, as can be seen in Figure 2.2c and d. In PO···EtOH, the diastereomeric and diastereofacial chirodiastaltic energies were estimated to be ~0.2–0.3 kJ/mol. The preference of EtOH to form a H-bond with PO from the same side as the methyl group of PO, was attributed to the secondary interaction of the methyl group with the hydroxyl oxygen.

In DO···EtOH, the addition of the second methyl group eliminates the diastereofacial interactions, while it reinforces the diastereomeric interactions to 0.8 kJ/mol by enhancing the secondary interactions with the EtOH.

In the related FTMW study of PO···FE, the effects of fluorination in chirality induction were examined.[13] FE has nine conformations with a wider stability span than the three conformations of EtOH. FE predominantly takes on the *compact G* conformations (Figure 2.2e), whereas all three conformations of EtOH are populated in a jet expansion with a slight preference for the *T* conformation. Such preference of FE was attributed to the intramolecular H-bonding between F and H-O.[32] Out of the 14 local minima predicted for PO···FE, only three of them were detected in the jet expansion and assigned. All three of them consist solely of the *compact G* FE subunits, and the energy ordering was determined experimentally to be *anti G–g+* ≈ *anti G+g–* < *syn G+g–*.[13] Because of the strong electron withdrawing property of the fluorine group, the acidity of the neighboring hydroxyl group increases, leading to its higher binding ability in H-bonding. The stability was found to increase by ~5 kJ/mol going from PO···EtOH to PO···FE.[13] A major feature uncovered in the study is that the $CH_2F$-$CH_2$- chain of FE turns toward the PO molecule, allowing the C-H···F-C interaction, which is absent in the PO···EtOH complex. Therefore, the diastereofacial discrimination in the PO···FE complex favors the *anti* structures, in which the C-F bond mainly interacts with the α-H of PO, instead of the *syn* structures, in which the less acidic methyl β-H are involved. In PO···EtOH, on the other hand, the *syn* structures are favored due to the secondary H-bonds between the hydroxyl oxygen and the methyl group.

### 2.4.3 Chirality Synchronization (Transient-Transient Chirality Interactions)

For the chiral molecular contact pairs showing the transient-transient chirality interactions or chirality synchronization,[5] each transiently chiral subunit synchronizes its adaptation of a particular transiently chiral conformation with its partner in order to bind selectively to each other. The dimers of EtOH[27] and FE[13] probed using FTMW spectroscopy are two such examples. We will focus on the comparison of these two systems.

The EtOH dimer has attracted much theoretical attention,[33] aiming to predict its stable gas phase conformations. The IR absorption[34] and FTMW studies[27] established that the three most stable binary EtOH conformers are the *associated* conformers, consisting of the *G* and *T* conformations of EtOH. Fluorination of EtOH brings several major effects in the corresponding dimer. Just as in the case of PO···FE, only the *compact G* FE subunits (Figure 2.2)[13,18] are utilized for the most stable FE dimers observed, whereas both the *G* and *T* conformations of EtOH are used in the constructions of the three most stable binary EtOH conformers observed. This observation and the related *ab initial* calculations indicate that the relative stability of the FE conformers plays a major role in the relative stability of the binary FE conformers. For the EtOH dimers, however, this is not the case because of the much smaller stability differences among the monomeric conformers of EtOH. The EtOH dimer can only form the *associated* conformers since EtOH itself does not have an intramolecular H-bond. The opposite is the case for the FE dimer since the formation of the intermolecular C-F···H-O H-bonds in the *inserted* conformers contributes significantly to their extra stability. The calculations of the distortion energies of the H-bond donors and acceptors show that the *inserted* conformers with the C-F···H-O H-bonds experience less distortion than the *associated* ones without such bonds. These observations are of interest in light of the controversy on the existence of H-bonds between the C-F and –OH/–NH groups.[35] Finally, in the FE dimers, the directional binding of the hydroxyl group of the H-bond donor to the diastereotopic lone pairs at oxygen of the H-bond acceptor leads to the *compact* (c) or *open* (o) distinction, where the two subunits either tilt toward or point away from each other.[18] The fact that all the FE dimers observed are *compact* indicates a strong diastereofacial discrimination capability of the oxygen lone pairs in the FE dimers.

## 2.5 CONCLUDING REMARKS

The high-resolution MW spectroscopic studies of chiral molecular contact pairs reviewed here demonstrate that this technique is well suited for probing structural, energetic, and dynamical properties of such systems. These studies have provided detailed and quantitative information about molecular recognition involving chiral and transiently chiral molecules. Currently, there are efforts to push the boundary of the molecular systems investigated to even larger contact pairs with several functional groups and to larger oligomers beyond binary contact pairs. All these efforts aim to methodically increase the complexity of the chiral molecular systems investigated to better mimic chirality recognition in biological systems.

## REFERENCES


1. Y. Xu, J. Van Wijngaarden, and W. Jäger. 2005. *Int. Rev. Phys. Chem.* 24:301–38.
2. E. Arunan, S. Dev, and P. K. Mandal. 2004. *Appl. Spectrosc. Rev.* 39:131–81.
3. S. Novick. Bibliography of rotational spectra of weakly bound complexes. http://www.wesleyan.edu/chem/faculty/novick/vdw.html.
4. (a) J. Tang, Y. Xu, A. R. W. McKellar, and W. Jäger. 2002. *Science* 297:2030–33. (b) Y. Xu, W. Jäger, J. Tang, and A. R. W. McKellar. 2003. *Phys. Rev. Lett.* 91:163401/1–4. (c) A. R. W. McKellar, Y. Xu, and W. Jäger. 2006. *Phys. Rev. Lett.* 97:183401/1–4.
5. A. Zehnacker and M. A. Suhm. 2008. *Angew. Chem. Int. Ed.* 47:6970–92.
6. K. Le Barbu, V. Brenner, Ph. Millié, F. Lahmani, and A. Zehnacker-Rentien. 1998. *J. Phys. Chem. A* 102:128–37. A. Latini, D. Toja, A. Giardini-Guidoni, S. Piccirillo, and M. Speranza. 1999. *Angew. Chem. Int. Ed.* 38:815–17.
7. N. Borho, T. Häber, and M. A. Suhm. 2001. *Phys. Chem. Chem. Phys.* 3:1945–48. N. Borho and M. A. Suhm. 2002. *Phys. Chem. Chem. Phys.* 4:2721–32. N. Borho and M. A. Suhm. 2003. *Org. Biomol. Chem.* 1:4351–58.
8. T. B. Adler, N. Borho, M. Reiher, and M. A. Suhm. 2006. *Angew. Chem. Int. Ed.* 45:3440–45.
9. Y. Xu and W. Jäger. 1997. *J. Chem. Phys.* 106:7968–80.
10. Y. Xu and W. Jäger. 1997. *J. Chem. Phys.* 107:4788–96.
11. (a) T. J. Balle and W. H. Flygare. 1981. *Rev. Sci. Instrum.* 52:33–45. (b) U. Andresen, H. Dreizler, J.-U. Grabow, and W. Stahl. 1990. *Rev. Sci. Instrum.* 61:3694–99.

12. Y. Xu and W. Jäger. 2008. Microwave rotational spectroscopy. *Enclyclopedia of Inorganic Chemistry.* ed. R.H. Crabtree. Chichester: John Wiley. 227–42.
13. X. Liu, N. Borho, and Y. Xu. 2009. *Chem. Eur. J.* 15:270–77.
14. N. Borho and Y. Xu. 2007. *Phys. Chem. Chem. Phys.* 9:4514–20.
15. J. L. Alonso, C. Pérez, M. E. Sanz, J. C. López, and S. Blanco. 2009. *Phys. Chem. Chem. Phys.* 11:617–27.
16. G. Yang and Y. Xu. 2008. *Phys. Chem. Chem. Phys.* 10:6787–95. M. Losada, P. Nguyen, and Y. Xu. 2008. *J. Phys. Chem. A* 112:5621–27.
17. D. A. Case, et al. 2006. *AMBER 9*. San Francisco, CA: University of California.
18. T. Scharge, C. Emmeluth, T. Häber, and M. A. Suhm. 2006. *J. Mol. Struct.* 786:86–95.
19. J. S. Binkley and J. A. Pople. 1975. *Int. J. Quantum. Chem.* 9:229–36.
20. R. Krishnan, J. S. Brinkley, R. Seeger, and J. A. Pople. 1980. *J. Chem. Phys.* 72:650–54.
21. A. K. King and B. J. Howard. 2001. *Chem. Phys. Lett.* 348:343–49.
22. N. Borho and Y. Xu. 2007. *Angew. Chem.* 119:2326–29. N. Borho and Y. Xu. 2007. *Angew. Chem. Int. Ed.* 46:2276–79.
23. Z. Su, N. Borho, and Y. Xu. 2006. *J. Am. Chem. Soc.* 128:17126–31.
24. A. Maris, B. M. Giuliano, D. Bonazzi, and W. Caminati. 2008. *J. Am. Chem. Soc.* 130:13860–61.
25. (a) Z. Su, Q. Wen, and Y. Xu. 2006. *J. Am. Chem. Soc.* 128:6755–60. (b) Z. Su and Y. Xu. 2007. *Angew. Chem.* 119:6275–78. Z. Su and Y. Xu. 2007. *Angew. Chem. Int. Ed.* 46:6163–66.
26. T. H. Dunning, Jr. 1989. *J. Chem. Phys.* 90:1007–23.
27. J. P. I. Hearn, R. V. Cobley, and B. J. Howard. 2005. *J. Chem. Phys.* 123:134324.
28. J. P. I. Hearn and B. J. Howard. 2007. *Mol. Phys.* 105:825–39.
29. H. Hagemann, J, Mareda, C. Chiancone, and H. Bill. 1997. *J. Mol. Struct.* 410–411:357–60.
30. A. K. King and B. J. Howard. 2001. *J. Mol. Spectrosc.* 205:38–42.
31. J. C. Pearson, K. V. L. N. Sastry, E. Herbst, and F. C. D. Lucia. 1997. *J. Astrophys.* 480:420–31.
32. D. A. Dixon and B. E. Smart. 1991. *J. Phys. Chem.* 95:1609–12.
33. (a) L. Gonzalez, O. Mo, and M. Yanez. 1999. *J. Chem. Phys.* 111:3855–61. (b) V. Dyczmons. 2004. *J. Phys. Chem. A* 108:2080–86. (c) M. Ehbrecht and F. Huisken. 1997. *J. Phys. Chem. A* 101:7768–77.
34. C. Emmeluth, V. Dyczmons, T. Kinzel, P. Botschwina, M. A. Suhm, and M. Yáñez. 2005. *Phys. Chem. Chem. Phys.* 7:991–7.
35. (a) J. A. K. Howard, V. J. Hoy, D. O'Hagan, and G. T. Smith. 1996. *Tetrahedron* 52:12613–22. (b) J. D. Dunitz and R. Taylor. 1997. *Chem. J. Euro.* 3:89–98. (c) T. J. Barbarich, C. D. Rithner, S. M. Miller, O. P. Anderson, and S. H. Strauss. 1998. *J. Am. Chem. Soc.* 121:4280–81.

# 3 Infrared and Raman Detection of Transient Chirality Recognition in the Gas Phase *The Case of Ethanol*

*Martin A. Suhm*

**CONTENTS**

3.1 Introduction ........ 39
3.2 Parity Violating Forces ........ 40
3.3 Torsional Chirality in Ethanol ........ 40
3.4 Fluorination Effects ........ 41
3.5 Argon Collisions ........ 42
3.6 Argon Attachment ........ 42
3.7 Water Attachment ........ 43
3.8 Dimerization ........ 43
3.9 Chiral Molecule Attachment ........ 44
3.10 Experimental Techniques ........ 44
3.11 Conclusions ........ 45
Acknowledgments ........ 46
References ........ 46

## 3.1 INTRODUCTION

Imagine a simple chiral molecule in its R configuration close to $T = 0$ K in free space. Within the Born–Oppenheimer approximation, its nuclei will be somewhat delocalized from the lowest-energy structure due to zero-point vibrations. A small fraction of the nuclear frame probability density will extend into transitional configurations, which are equally far from the R and its mirror image S configuration. By continuity and symmetry, there must then also be probability density in the fully developed S configuration region. In other words, the ground state of the molecule will be a superposition of R- and S-like configurations. It will be symmetric with respect to stereomutation and thus achiral in this approximation. There will also be a nearby stationary state with a wavefunction that changes sign at configurations that are equally far from R and S. Like the symmetric ground state, this antisymmetric ground state is achiral. If we prepare the molecule in a low-energy R configuration, it is not in a stationary state. Within this approximation, it will oscillate between R and S states with a full period $\tau$, which is the inverse of the energy splitting $\Delta E$ between the antisymmetric and the symmetric state, divided by Planck's constant, $h$.[1] If the probability for transitional configurations is low enough, this oscillation will be slow—ideally, one can store the R molecules in bottles. This is the case for a high barrier toward stereomutation, as we know it from molecules with "asymmetric carbon" atoms. If the barrier is low, such as in many pyramidal amines, one cannot store the R molecules in bottles. The timescale on which such a flexible, transiently chiral molecule may still be considered to be chiral is governed by quantum mechanical tunneling and depends strongly on the height and width of the barrier for stereomutation. The width also involves the mass, which has to be dislocated during the inversion process. This simple model of a chiral molecule with a more or less high stereomutation barrier provides the starting point for the present chapter, in which perturbations of increasing complexity will be introduced to the chiral molecule and the consequences of these interactions with other systems will be discussed. If the binding partner is also chiral, the term *chirality recognition* is used. Depending on the stereomutation speed of the binding partner, chirality discrimination, chirality induction, and chirality synchronization phenomena will be discussed.

## 3.2 PARITY VIOLATING FORCES

The subtle asymmetry of nature with respect to the two mirror image structures of a molecule has been discussed before in detail.[1] For very high stereomutation barriers, such parity-violating forces can completely destroy the equivalence of R and S states of the molecule.

## 3.3 TORSIONAL CHIRALITY IN ETHANOL

In the following, we will address molecules with much lower barriers, and it is instructive to think of ethanol as an example. Ethanol can exist in two enantiomeric *syn* or *gauche* (g±, torsional CCOH angle of the OH group relative to the carbon backbone around ±60°) forms as well as in a 0.5-kJ/mol more stable *anti* or *trans* (t, torsional CCOH angle around 180°) state,[2] which is achiral (see the scheme in the top part of Figure 3.1). Thus, the vibrational ground state will be mostly concentrated in the t well, whereas excited stationary states may have significant amplitude in the two g wells, either with symmetric ($g^+$) or with antisymmetric ($g^-$) amplitudes.[2] Note the difference between $g^+$, which represents the symmetric energy eigenstate with positive parity, and g+, which denotes the localized state where the OH group points to the right of the ethanol backbone with a torsional angle of +60°. In a supersonic jet expansion of ethanol entrained in an inert carrier gas such as helium, one can stabilize and detect the t and $g^{\pm}$ species at low temperature, e.g., via their OH stretching fundamentals. This

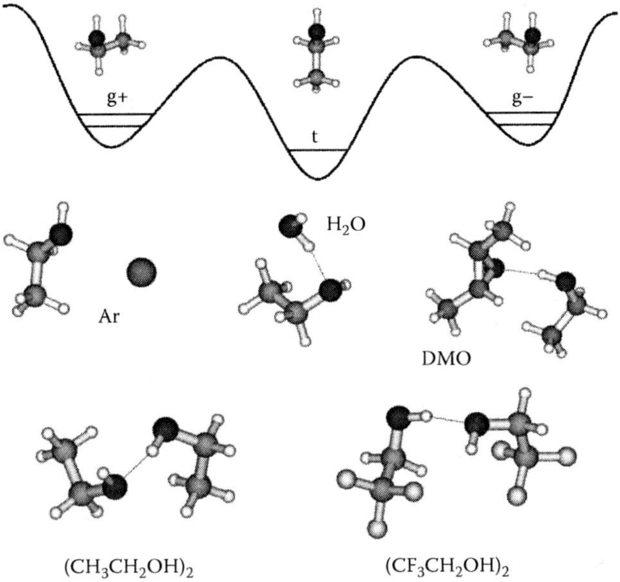


**FIGURE 3.1** (Color Figure 3.1 follows page 46.) Gauche-trans isomerism in ethanol monomers (top) and complexes of ethanol with Ar, water, trans-2,3-dimethyloxirane (DMO), and itself, as well as the dimer of 2,2,2-trifluoroethanol.

works particularly well in the spontaneous Raman scattering spectrum, because it is dominated by closely spaced $\Delta J = 0$ rovibrational transitions and allows for a clear spectral separation.[3] From the relative intensities, one can determine an effective conformational temperature of the ethanol gas expansion, taking into account the somewhat different scattering cross sections of the two conformations (Figure 3.2). Depending on the expansion conditions, conformational temperatures in the 10 to 100 K range may be easily prepared. Note that there is no spectral tunneling doublet in the g signal, although the barrier between the g+ and g− forms is low. Indeed, the conversion between the g forms is quite fast (10 ps for the full period), but it does not change much with OH stretching excitation, and therefore the Raman transitions for the symmetric and antisymmetric $g^{\pm}$ states overlap almost perfectly (Figure 3.2). On a timescale of ≈1 ps, ethanol can thus be caught in a transiently chiral state. On a longer timescale, the chiral structures will stereomutate.

## 3.4 FLUORINATION EFFECTS

By fully fluorinating the methyl group in ethanol, the gauche conformations are stabilized by a kind of intramolecular hydrogen bond, whereas the trans conformation turns into a high-lying minimum or transition state structure.[4] Stereomutation is now slowed down to 170 ps (for the full period). If only one or two terminal hydrogen atoms in ethanol are replaced by fluorine, the racemization process between the

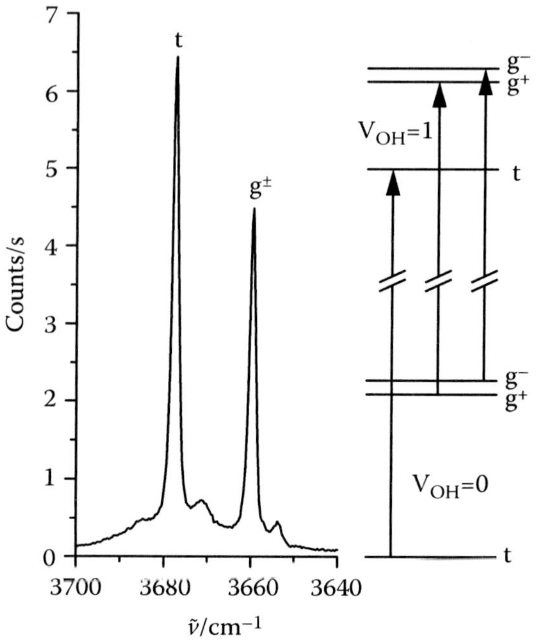


**FIGURE 3.2** Raman spectrum of a supersonic jet expansion of ethanol in helium in the OH stretching range,[3] showing the trans and superimposed gauche$^{\pm}$ transitions and an energy level scheme.

mirror image gauche forms is probably even shifted into the μs regime, because it now involves significant heavy atom motion.[5]

## 3.5 ARGON COLLISIONS

The elastic collision of an Ar atom with an ethanol molecule may be viewed as an event that catches ethanol molecules in a chiral state, because the Ar atom covers several atomic diameters during 1 ps at ambient temperature. Thus, for a spectator moving with the Ar atom, gauche ethanol molecules will typically have chiral character during the encounter. This is even more the case for trifluoroethanol and its partially fluorinated variants. Inelastic collisions with Ar can enhance the conversion between different conformations, the mechanism by which conformational cooling in supersonic jets happens.

## 3.6 ARGON ATTACHMENT

At low temperatures, Ar atoms may become more persistently bound to ethanol and its derivatives, thus forming van der Waals complexes with a vibrational period of ≈1 ps.[6] In such complexes, the stereomutation is usually quenched to a large degree, thus increasing the lifetime of the transiently chiral species by many orders of magnitude.

Unfortunately, the complex of Ar with gauche ethanol has so far not been detected by microwave spectroscopy. Instead, the more stable complex involving the achiral trans ethanol has been observed.[6] This complex is also chiral, because the Ar atom preferentially binds side-on to minimize the distance to all three heavy atoms of the ethanol (Figure 3.1). The two faces of the trans ethanol molecule are enantiotopic, and the Ar thus creates an R or an S complex, depending on the attachment site. By switching molecular faces on a timescale of 0.6 μs, the van der Waals complex undergoes a slow stereomutation process, but spectator atoms will definitely see a chiral species during their passage.

## 3.7 WATER ATTACHMENT

A more interesting binding partner for ethanol is water. In principle, it could act as a hydrogen bond acceptor toward ethanol, but it was recently shown by Raman jet spectroscopy that the more stable complex, in which water acts as a hydrogen bond donor, is observed instead.[7] Furthermore, it was experimentally verified that the hydrogen-bond-donating water molecule forces the ethanol molecule into a preferred gauche conformation (Figure 3.1). Among the two lone electron pair choices that gauche ethanol has available at its oxygen atom, water preferentially binds to the one that points in the direction of the molecular backbone. This allows for a compact water-ethanol dimer structure in which the water oxygen can also be coordinated through a weak C-H hydrogen bond. Evidently, this structure is chiral, but its stereomutation period remains unknown, as the complex has not been observed by high-resolution microwave spectroscopy, so far. The inversion process is likely to be slow. Thus, the water-ethanol complex is a simple intermolecular example for the quenching of stereomutation in a transiently chiral molecule. It is also a simple example for adaptive aggregation or induced fit, because ethanol changes its preferred conformation in order to best accommodate a water molecule.

## 3.8 DIMERIZATION

What happens if two ethanol molecules form a complex? Even if both remain in their preferred achiral trans conformation, a chiral complex will still form, because the hydrogen bond donor molecule can choose between two lone electron pairs of the acceptor. This is also the case if one of the two molecules opts for a gauche conformation. If both molecules are in a gauche conformation, one must distinguish between homoconfigured and heteroconfigured ethanol pairs. In the former case, both gauche forms have the same chirality or helicity, whereas in the latter case, they have opposite handedness. It was recently shown that a compact homoconfigured gauche ethanol dimer (Figure 3.1) is the most stable of all combinations.[8] Like in a handshake (but unlike the case of a pair of masticating cattle[9]), there is thus a preference for combining objects of the same chirality in ethanol dimer. Unlike handshakes, which by convention involve right hands, the ethanol combinations g+g+ and g–g– have the same probability. They may even interconvert into each other, but the pathway is quite involved and the ground state stereomutation splitting is probably exceedingly small. In any case, ethanol dimer provides an example for chirality synchronization,[10] because the two molecular building blocks synchronize their transiently chiral conformations. One should note that the energetic preference for this homochiral structure is very subtle. Even in a cold supersonic jet expansion, at least three other dimer conformations coexist, two of which have also been structurally characterized.[11] By improving the conformational relaxation through the addition of efficient argon collision partners, the homoconfigured gg dimer population is sufficiently enhanced. Apart from the chirality synchronization issue, ethanol dimer also provides a case of adaptive aggregation. Like in the ethanol-water complex, the ethanol units change their preferred conformation from trans to gauche in order to form the most stable complex. When the three terminal hydrogen atoms in ethanol are replaced by fluorine atoms, the g conformations are intrinsically stabilized and no need for adaptive aggregation arises. Remarkably, the resulting trifluoroethanol dimer is found exclusively in the homoconfigured gg form[4] (see Figure 3.1). No other dimer conformation could be detected, despite at least one nearly isoenergetic structure, which is heteroconfigured but involves a different hydrogen bond pattern. Thus, the case of chirality synchronization is much more pronounced in trifluoroethanol dimer. In contrast, partially fluorinated ethanols can be stabilized in a number of dimer conformations,[5] and one may speak of regular chirality discrimination because of the slow stereomutation of the participating monomers.

## 3.9 CHIRAL MOLECULE ATTACHMENT

So far, a slowdown of the stereomutation process has been achieved, but left- and right-handed ethanol molecules still have the same abundance in the generated complexes. This changes when the binding partner for ethanol is "permanently" chiral, such as in the case of 2,3-dimethyl oxirane (DMO), if the two methyl groups point in different directions of the ring plane (see Figure 3.1). Now, one of the two gauche forms of ethanol achieves a snug fit to the oxirane when it binds to one of the equivalent oxygen lone pairs, whereas the other one does not.[12] The energy difference leads to a different abundance of one of the gauche ethanol forms in the mixed dimers, which has been detected by microwave spectroscopy.[12] This is a simple example for chirality induction,[10] which plays a prominent role in asymmetric synthesis using chiral catalysts to introduce new stereocenters in organic molecules.

## 3.10 EXPERIMENTAL TECHNIQUES

The phenomena discussed in the preceding sections may be qualitatively predicted by quantum chemical calculations that include adequate treatments of electron correlation. Their quantitative description by electronic structure theory and internuclear quantum dynamics still represents a major challenge.[10] Therefore, experimental cluster spectroscopy is of paramount importance in this field. Ethanol and its simple homologues and binding partners do not lend themselves easily to electronic double-resonance techniques.[10] Direct infrared absorption,[8,13] microwave excitation,[6,11,12] and

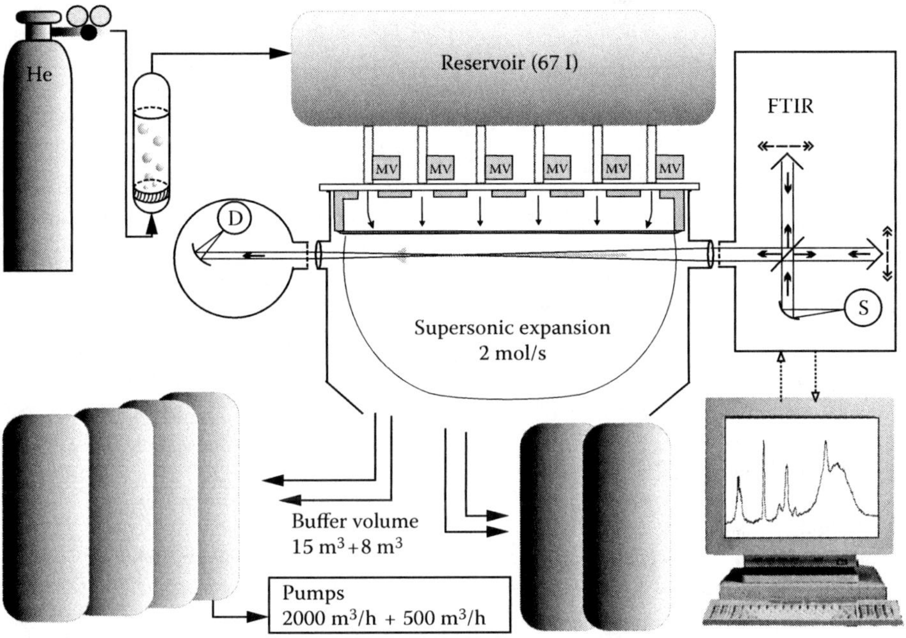


**FIGURE 3.3** High-throughput pulsed slit jet cluster spectrometer based on synchronized Fourier-transform infrared (FTIR) detection.[13] It is optimized for the detection of monomer and dimer spectra of volatile compounds from 200 to 8,000 $cm^{-1}$.

Raman scattering[4,7] are therefore methods of choice. Direct absorption profits from high-throughput pulsed supersonic nozzles (Figure 3.3). Microwave spectroscopy is indispensable for slow stereomutation processes and rigorous structural constraints (see Chapter 2). Spontaneous Raman scattering is one of the most recent advances in the field of hydrogen-bonded clusters and offers improved band resolution and spatial resolution, even if symmetry does not render it complementary to infrared spectroscopy.

## 3.11 CONCLUSIONS

This chapter has discussed increasingly strong perturbations to ethanol, a preferentially achiral but conformationally flexible molecule.[14] Such perturbations may lead to a stabilization of chiral conformations, to a slowdown of their stereomutation rate, and finally to a preference for one of the mirror images. These transformations are achieved by subtle intermolecular interactions with molecules of increasing complexity, and their effects are probed by infrared, microwave, and Raman spectroscopy in supersonic jets. In addition to more spectacular effects between permanently chiral molecules such as $\alpha$-hydroxy esters,[10] these subtle manifestations of chirality recognition underscore the power of gas phase spectroscopy in unraveling the forces that control the shape of flexible molecules in condensed phases.

## ACKNOWLEDGMENTS

Contributions from group members and colleagues are reflected in the list of references and are gratefully acknowledged, as is financial support from the FCI and the DFG.

## REFERENCES

1. Quack, M., Stohner, J., Willeke, M. 2008. High-resolution spectroscopic studies and theory of parity violation in chiral molecules. *Annu. Rev. Phys. Chem.* 59:741.
2. Pearson, J. C., Brauer, C. S., Drouin, B. J. 2008. The asymmetric top—asymmetric frame internal rotation spectrum of ethyl alcohol. *J. Mol. Spectrosc.* 251:394.
3. Wassermann, T. N., Zielke, P., Lee, J. J., Cézard, C., Suhm, M. A. 2007. Structural preferences, argon nanocoating, and dimerization of n-alkanols as revealed by OH stretching spectroscopy in supersonic jets. *J. Phys. Chem. A* 111:7437.
4. Scharge, T., Cézard, C., Zielke, P., Schütz, A., Emmeluth, C., Suhm, M. A. 2007. A peptide co-solvent under scrutiny: self-aggregation of 2,2,2-trifluoroethanol. *Phys. Chem. Chem. Phys.* 9:4472. Scharge, T., Luckhaus, D., Suhm, M. A. 2008. Observation and quantification of the hydrogen bond effect on O-H overtone intensities in an alcohol dimmer. *Chem. Phys.* 346:167.
5. Scharge, T., Wassermann, T. N., and Suhm, M. A. 2008. Weak hydrogen bonds make a difference: dimers of jet-cooled halogenated ethanols. *Z. Phys. Chem.* 222:1407.
6. Maris, A., Caminati, W., Velino, B., Andrews, C. M., Howard, B. J. 2004. Free and pulsed jet rotational spectra and van der Waals motions of ethanol-argon. *Chem. Phys. Lett.* 399:39.
7. Nedić, M., Wassermann, T. N., Xue, Z., Zielke, P., Suhm, M. A. 2008. Raman spectroscopic evidence for the most stable water/ethanol dimer and for the negative mixing energy in cold water/ethanol trimers. *Phys. Chem. Chem. Phys.* 10:5953.
8. Emmeluth, C., Dyczmons, V., Kinzel, T., Botschwina, P., Suhm, M. A., Yáñez, M. 2005. Combined jet relaxation and quantum chemical study of the pairing preferences of ethanol. *Phys. Chem. Chem. Phys.* 7:991.
9. Jordan, P., Kronig, R. de L. 1927. Movements of the lower jaw of cattle during mastication. *Nature* 120:807.
10. Zehnacker, A., Suhm, M. A. 2008. Chirality recognition between neutral molecules in the gas phase. *Angew. Chem. Int. Ed.* 47:6970.
11. Hearn, J. P. I., Cobley, R. V., Howard, B. J. 2005. High-resolution spectroscopy of induced chiral dimers: a study of the dimers of ethanol by Fourier transform microwave spectroscopy. *J. Chem. Phys.* 123:134324.
12. Borho, N., Xu, Y. 2007. Molecular recognition in 1:1 hydrogen-bonded complexes of oxirane and *trans*-2,3-dimethyloxirane with ethanol: a rotational spectroscopic and *ab initio* study. *Phys. Chem. Chem. Phys.* 9:4514.
13. Borho, N., Suhm, M. A., Le Barbu-Debus, K., Zehnacker, A. 2006. Intra- vs. intermolecular hydrogen bonding: dimers of alpha-hydroxyesters with methanol. *Phys. Chem. Chem. Phys.* 8:4449.
14. Emmeluth, C., Dyczmons, V., Suhm, M. A. 2006. Tuning the hydrogen bond donor/acceptor isomerism in jet-cooled mixed dimers of aliphatic alcohols. *J. Phys. Chem. A* 110:2906.

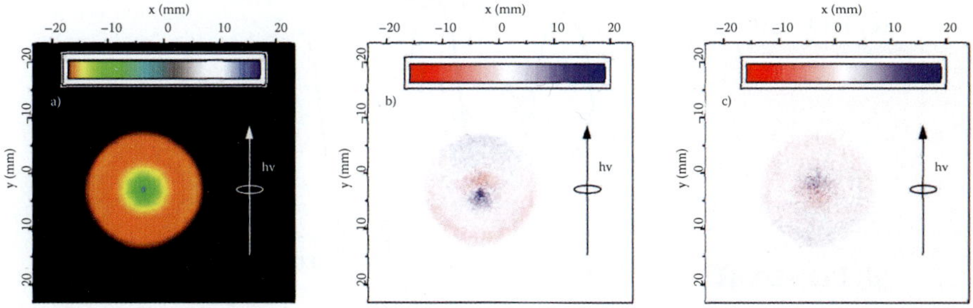


**FIGURE 1.2** (a) Raw photoelectron image of R camphor taken at $h\nu$ = 10.3 eV. The two rings correspond to the two outermost orbitals that can be ionized. (b) Untreated subtraction between R camphor images recorded with different light helicity (left minus right circularly polarized light). The color scale reveals the forward/backward asymmetry along the propagation direction of the light, with an opposite sign for the two electronic states. (c) Idem with the S enantiomer. The intensity distribution patterns are antisymmetric with respect to the switch of enantiomers, as predicted by theory. (Reproduced from L. Nahon, G. A. Garcia, C. J. Harding, E. A. Mikajlo, and I. Powis, *J. Chem. Phys.* 125 (2006): 114309. With permission.)

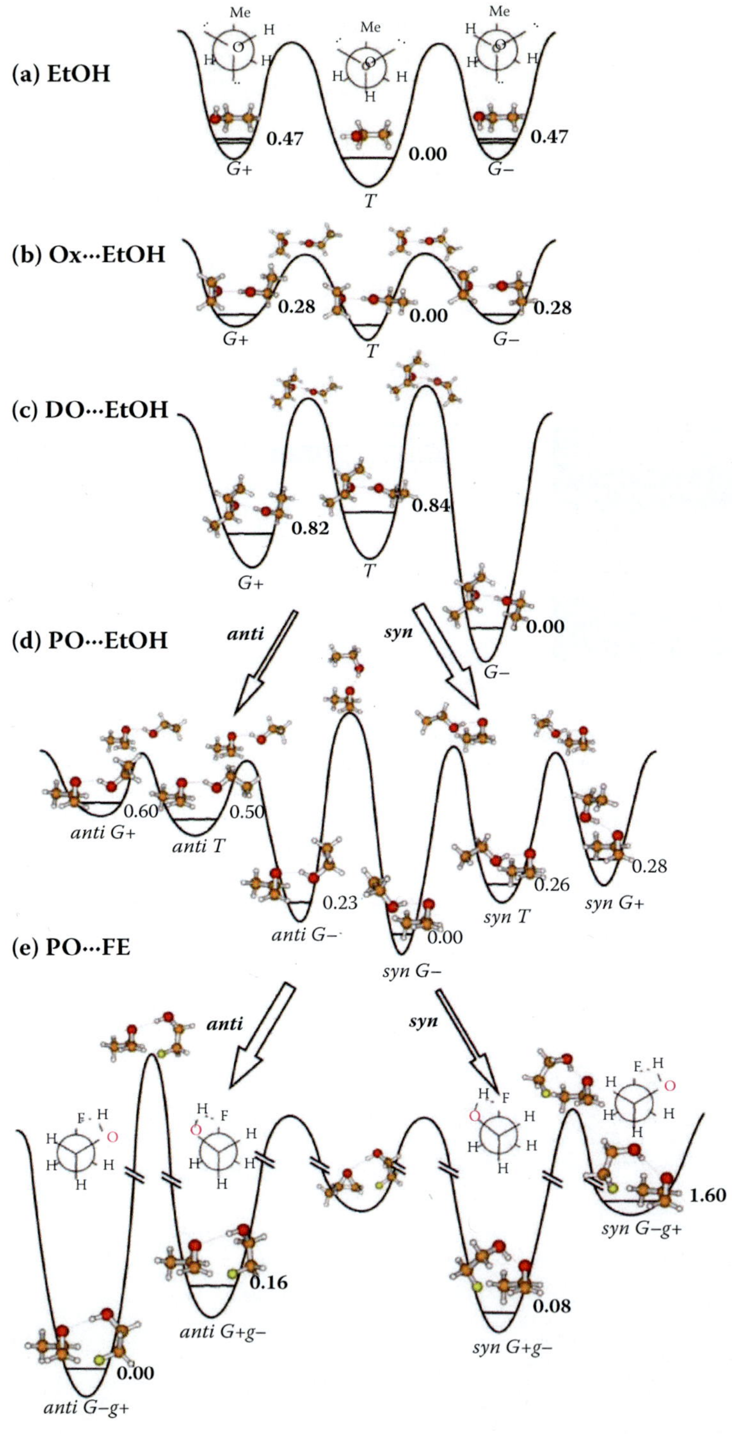

(a) EtOH
Me
H
H
H
O
G+
0.47
T
0.00
G−
0.47
(b) Ox···EtOH
G+
0.28
T
0.00
G−
0.28
(c) DO···EtOH
G+
0.82
T
0.84
G−
0.00
(d) PO···EtOH
anti
syn
anti G+
0.60
anti T
0.50
anti G−
0.23
syn G−
0.00
syn T
0.26
syn G+
0.28
(e) PO···FE
anti
syn
anti G−g+
0.00
anti G+g−
0.16
syn G+g−
0.08
syn G−g+
1.60

**FIGURE 2.2** The ground vibrational energy levels of the conformers of (a) EtOH, (b) Ox···EtOH, (c) DO···EtOH, (d) PO···EtOH, and (e) PO···FE as a function of $\tau_{HOCC}$ in EtOH (see text for details). FE is labeled with *G+*, *G-*, and *T* for $\tau_{OCCF}$, and *g+*, *g-*, and *t* for the $\tau_{HOCC}$ torsional angle, following the notation of refs. 13 and 18, and only *compact* G conformations, i.e., *G+g-* and *G-g+*, are indicated. The energy levels for EtOH are the experimental values taken from Ref. 31, while the rest are the calculated values obtained at the MP2/6-311++G(d,p) levels with ZPE corrections (in kJ/mol). Please note that the relative energy orderings of conformers predicted for (d) and (e) at this level of theory differ from the experimental ones. The G tunneling pair (not drawn to scale) is degenerated in EtOH and in Ox···EtOH, while the degeneracy is lifted in DO···EtOH and PO···EtOH. The fatter arrows indicate either *syn* or *anti*-conformers are preferred. *syn* or *anti* specify that the H-bonding partner is on the same or the opposite side as the methyl group of PO, respectively. The connecting one dimensional potential curves are only qualitative. The transition states were calculated at the B3LYP/6-311++G(d,p) level of theory. All conformers listed except *syn G-g+* of PO···FE were detected experimentally.

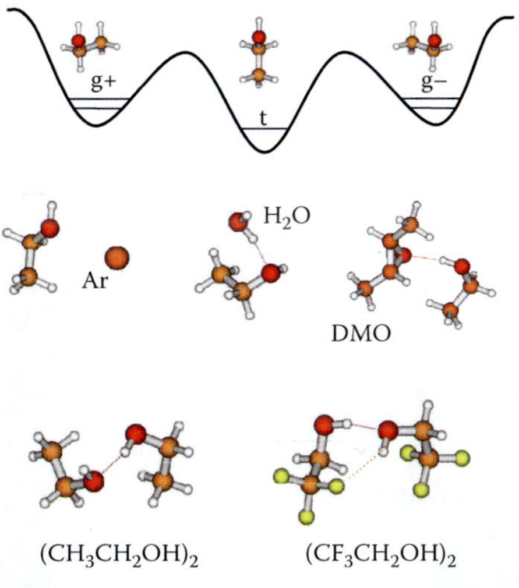


**FIGURE 3.1** Gauche-trans isomerism in ethanol monomers (top) and complexes of ethanol with Ar, water, trans-2,3-dimethyloxirane (DMO), and itself, as well as the dimer of 2,2,2-trifluoroethanol.

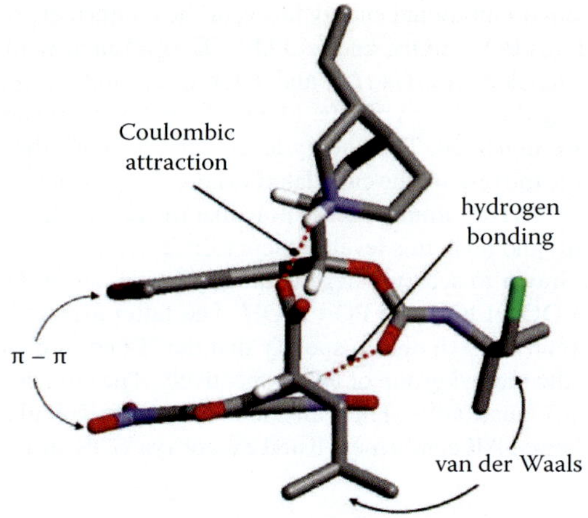


**FIGURE 10.2** Annotated x-ray crystal structure of *L*-DNB-Leu (bottom) in complex with chloro-tBuCQN (top). (Reprinted from Schug, K.A., Frycak, P., Maier, N.M., Lindner, W., *Anal. Chem.* 77 (2005): 3660–70. Copyright 2005, American Chemical Society. With permission.)

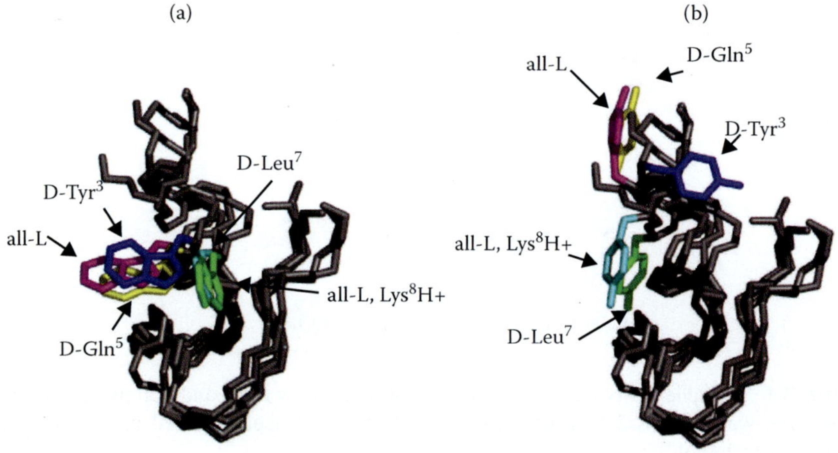


**FIGURE 11.1** Gas-phase structures of Trp-cage (NLYIQWLKDGGPSSGRPPPS) dications calculated as an average trajectory at 305 K shown in comparison (best fit) with each other. The positions of the critical Tyr$^3$ and Trp$^6$ side chains are shown in panels (a) and (b), respectively. Gln$^5$ residue was protonated if not indicated otherwise. The second proton is located on Arg$^{16}$ in all cases. (Reproduced from Patriksson, A., Adams, C. M., Kjeldsen, F., Raber, J., van der Spoel, D., Zubarev, R. A., *Int. J. Mass Spectrom.* 248 (2005): 124–35. With permission.)

# 4 The Role of Deformation Energy of Bifunctional Entities on the Formation of Diastereoisomers

*Katia Le Barbu-Debus*

## CONTENTS

4.1 Introduction ..... 47
4.2 Experimental Section ..... 48
4.3 Theoretical Section ..... 49
4.4 Main Results ..... 50
4.4.1 Characteristic Structures ..... 51
4.4.1.1 Addition (Figure 4.2a) ..... 51
4.4.1.2 Insertion (Figure 4.2b) ..... 51
4.4.1.3 Head-to-Head (Figure 4.2c) ..... 51
4.4.1.4 Cycle (Figure 4.2d) ..... 51
4.4.2 α-Methyl-2-Naphthalenemethanol/Amino Alcohols ..... 52
4.4.3 Methyl Mandelate/Methyl Lactate (MeMan/MeLac)[11] ..... 54
4.4.4 Cis-1-Amino-2-Indanol/Methyl Lactate (AI/MeLac)[12] ..... 56
4.5 Conclusion ..... 57
References ..... 58

## 4.1 INTRODUCTION

In 1995 we pioneered the study of chiral recognition in supersonic expansion experiments. The experiments rest on the formation of neutral jet-cooled complexes between chiral molecules and aim at understanding the nature of the forces responsible for chiral recognition. In what follows, we will focus on chiral recognition in neutral complexes in their electronic ground state, namely, on structural aspects from both an experimental and a theoretical point of view. The experiments depend on the formation of weakly bound diastereoisomer complexes involving a chiral chromophore (the selectand) in one of its pure enantiomers, and each enantiomer of a chiral complexing agent. Interactions with the selectand result in a shift of the energy levels, both in the ground state and in the excited state, which in turn results in a shift of the $S_0$-$S_1$ absorption spectrum. We profit from the enantioselectivity of this shift to evidence the formation of heterochiral (RS) and homochiral (RR) diastereoisomers.[1–3] The enantioselectivity of the $S_0$-$S_1$ spectrum also allows us to resort to double-resonance IR-UV spectroscopy,[4,5] to obtain vibrational spectra that can be directly related to the structures. We have focused on the region of 3 μm to obtain information on the ν(CH), ν(NH), and ν(OH) stretch modes, the last ones being very sensitive to complexation[6] and solvation.[7] In parallel to these experimental approaches, quantum chemical calculations have been used to model the vibrational spectroscopy of the ground state. The comparison between theoretical and experimental results allows us, most of the time, to assign an experimentally observed spectrum to a calculated structure. However, the assignment of calculated structures to experimentally observed complexes encounters some difficulties. Some of them are specific to the theoretical methods used (exploration of the potential energy surface, harmonic frequencies calculations); others pertain to the characteristics of experimental conditions (kinetic vs. thermodynamic cooling).

Our main contribution to the domain has been to determine criteria that can be used to confidently relate the calculated complexes to the experimentally observed ones. Indeed, pinhole jet experiments clearly differ from solution or other gas-phase experiments. The molecules or complexes are formed at low temperature, which makes entropic contribution negligible. On the other hand, the cooling and complex formation processes are not thermodynamically controlled. The main goal in the present chapter is to evidence the importance of kinetic factors on the diastereoisomers formation, hence on chiral recognition. The guiding line is to show that the diastereoisomers, formed in classical pinhole supersonic expansion experiments, are not necessarily related to the global minimum of the potential energy surface due to the impossibility of overcoming a barrier during the cooling process. In what follows, we will show that this barrier can be directly related to what we will call deformation energy. In order to evidence this reality, we will revisit in this chapter most of our results obtained during the last four years.[5,8–12]

## 4.2 EXPERIMENTAL SECTION

The experimental setup combines the use of a supersonic expansion with laser spectroscopy. The supersonic jet rests on an adiabatic expansion of the carrier gas seeded with the molecules of interest, which results in a strong cooling of the internal degrees of freedom down to a few K. These conditions also allow formation of complexes that would not be stable at room temperature. The first step of our experimental approach is to record the $S_0$-$S_1$ electronic spectrum by laser-induced fluorescence. The use of supersonic expansion makes these spectra much more simple and easier to handle than those obtained in gas phase or in solution because of the cooling of internal degrees of freedom. By using pure enantiomers of the complexing agent, we are able to record separately the excitation spectrum of both diastereoisomers.

The second step consists in recording vibrational spectra resorting to IR/UV double resonance. This technique rests on the use of two tunable lasers, namely, IR and UV sources. The UV laser is fixed on a transition observed in the excitation spectrum and is a probe of the ground-state population of the entity (molecule, complex, diastereoisomer, isomer, etc.) under study. The IR laser is scanned in the energy range of interest (most of the time between 3,000 and 3,600 $cm^{-1}$). When its frequency is resonant with a vibrational level of the probed species, a depletion of its ground state happens, which manifests itself by a decrease of the fluorescence signal. If the IR laser is not resonant with a vibrational level of the entity under study, and if it is resonant with that of an other species present in the jet, nothing occurs—the fluorescence signal stays at its maximum (100%). By scanning the IR laser frequency, the spectrum of the probed species is recorded as a series of dips in the fluorescence signal. By changing the UV laser frequency and fixing it to another transition observed on the excitation spectrum, the vibrational spectrum of a second species can be recorded. Eventually we can record separately the vibrational spectrum of all isomers present in the jet.

## 4.3 THEORETICAL SECTION

One of the major difficulties encountered when studying molecular complexes with numerous degrees of freedom is to fully explore the potential energy surface (PES). Moreover, the complexes studied here contain an aromatic chromophore, which makes them highly polarizable. For this reason, it is important to take dispersion into account. We have therefore followed a two-step procedure that meets these two requirements. First, the global exploration of the potential energy surface is performed by a semiempirical approach in which only the intermolecular interactions are optimized. Second, a local optimization of all the degrees of freedom of the most stable local minima is performed using *ab initio* methods. The harmonic frequencies are also calculated, which gives access to both zero-point energy of the system (ZPE) and the calculated vibrational spectra.

The methodology used in the semiempirical step has been developed by Claverie[13] and extended by Brenner et al.[14] It is based on the perturbation exchange theory and rests on a description of the interaction energy as a sum of four terms (electrostatic, repulsion, polarization, and dispersion), which can be written analytically as a function of the intermolecular distances, combined with a whole exploration of the potential energy surface. This method has the advantage of evaluating well the dispersion component. When necessary, this exploration step has been performed for the different conformers of both the selectand and the solvent.

The *ab initio* step aims at a full optimization of all degrees of freedom. The methods used (DFT or correlated methods) have evolved in parallel with the workstation power. For the first systems described hereafter, we have optimized the structures coming out from the semiempirical part at the B3LYP/6-31G(d,p) level of theory. At this level of theory, we have also calculated the frequencies associated with the optimized structures and their BSSE by means of the Boys and Bernardi[15] counterpoise method. The dissociation energies (DEs) relative to the closest isomer will be given in what follows as the sum of the binding energy, half of the BSSE, and the ZPE. For more recent studies, optimization of the complexes has been done at the MP2/6-31G(d,p) level of theory. As the size of the systems still precludes the frequencies' calculation at this level of theory, the frequencies are calculated at the B3LYP/6-31(d,p) level on the optimized structure at this level.

In addition, we have also calculated the deformation energy (Edef), which can be defined as the difference in energy of the subunit in its optimized geometry and in the geometry it adopts in the complex. It is directly related to the deformation that a given subunit has to undergo to allow the formation of the complex. In a given diastereoisomer, each subunit has its own deformation energy. In what follows, we will give the higher of the two values.

For the final attribution, three factors are considered. We obviously take into consideration the correspondence between observed and calculated spectra, and the thermodynamics, i.e., the stability of the calculated complexes. But kinetic factors, related to the use of supersonic expansion, are also very important. In this chapter, we will demonstrate that we are able to quantify them in terms of deformation energy and give some examples as an illustration.

## 4.4 MAIN RESULTS

Our recent work has focused on chiral discrimination between enantiomers of bifunctional molecules, amino alcohol and α-hydroxyester, which share the characteristic of presenting an intramolecular hydrogen bond. These enantiomers will be discriminated by either a monofunctional selectand (alcohol) or bifunctional ones. In what follows we will discuss results obtained for systems of increasing complexity. All the molecules discussed here are given in Figure 4.1. Before going into details, we will give an overview of the complexes that are likely to be formed. Four kinds of structures have been calculated; they are depicted in Figure 4.2 and described hereafter.

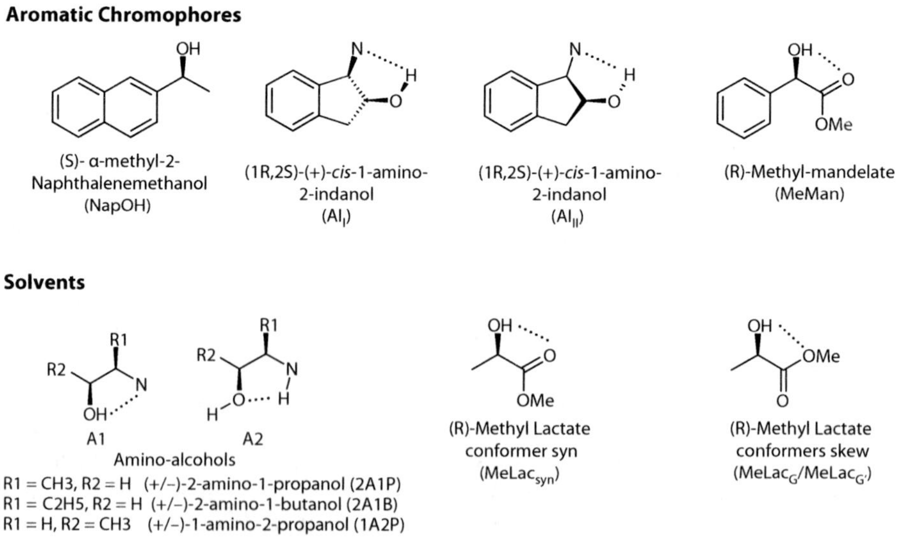


**FIGURE 4.1** Schemes of the molecules discussed in the text.

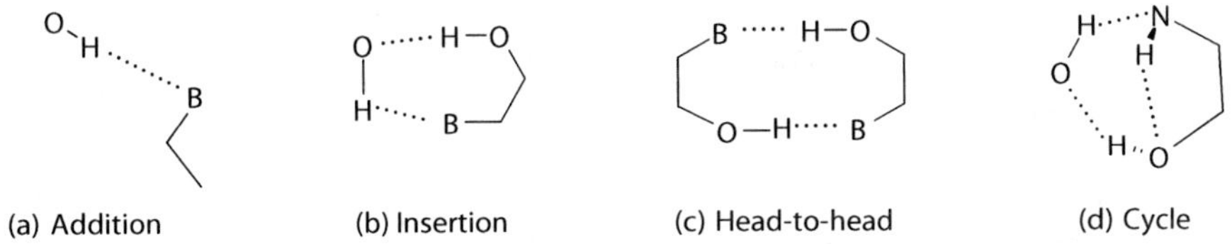


**FIGURE 4.2** Schemes of the characteristic structures of the calculated complexes: (a) addition, (b) insertion, (c) head to head, and (d) cycle.

## 4.4.1 Characteristic Structures

### 4.4.1.1 Addition (Figure 4.2a)

In the addition structure, one of the OH donor groups of the system forms a hydrogen bond toward an acceptor group (B for Lewis base) of the second subunit. Examples of addition structures are given in Figure 4.3b and c. They will be denoted "addition OH_B," where B stands for OH, $NH_2$, $OCH_3$, or O=C. When the molecule, which contains the donor part, is monofunctional, none of the entities undergo large modification of their starting geometry, as no intramolecular hydrogen bond has to be disrupted. Consequently, the Edef associated with the diastereoisomer formation is low. Conversely, when the donating molecule is bifunctional, an intramolecular hydrogen bond has to be broken to allow the formation of the intermolecular hydrogen bond, which leads to higher Edef.

### 4.4.1.2 Insertion (Figure 4.2b)

In the case of insertion complexes, the intramolecular hydrogen bond of one of the subunits is disrupted to allow inserting an OH group of the second molecule. Examples of insertion structures are given in Figures 4.3a and 4.5a and d. In other words, the intramolecular hydrogen bond is replaced by two intermolecular hydrogen bonds. The Edef is often high for the molecule that undergoes opening of its intramolecular hydrogen bond. When two bifunctional molecules are involved in this kind of complex, each molecule can be inserted into the internal hydrogen bond of its partner. This leads to two types of insertion complexes. The nomenclature used will be "insertion_A: conformers involved" if the chromophore is accepting the insertion and "insertion_I: conformers involved" if the chromophore is inserting into the intramolecular hydrogen bond of the second subunit.

### 4.4.1.3 Head-to-Head (Figure 4.2c)

When the diastereoisomer is composed of two bifunctional subunits, the intramolecular hydrogen bonds of the two subunits can be disrupted to allow the formation of two intermolecular hydrogen bonds. This results in a so-called head-to-head structure, an example of which is given in Figure 4.5b. The Edef of this structure is high, even higher than that of the insertion complexes, for both subunits.

### 4.4.1.4 Cycle (Figure 4.2d)

The cycle structure is only observed for systems containing at least one bifunctional molecule with functions acting simultaneously as hydrogen bond donors and acceptors, like an amino alcohol. It is characterized by the formation of two intermolecular hydrogen bonds. The OH group of the amino alcohol acts as a donor in a hydrogen bond toward the OH group of its partner, which in turns acts as a donor toward the nitrogen lone pair of the amino alcohol. The intramolecular hydrogen bond of the amino alcohol remains intact, while the intramolecular hydrogen bond of its partner is either slightly or totally opened.

### 4.4.2 α-Methyl-2-Naphthalenemethanol/Amino Alcohols

We will first describe the structures of the complexes involving a monofunctional chromophore, α-methyl-2-naphthalenemethanol (NapOH). It has one donor function, the hydroxyl group, and two acceptor sites, the O atom and the π ring. The interaction of the R enantiomer of NapOH has been studied with various chiral amino alcohol molecules: 2-amino-1-propanol (2A1P),[5] 1-amino-2-propanol (1A2P), and 2-amino-1-butanol (2A1B).[8] These amino alcohols exist in two conformers: a "closed" form A1 and an "open" form A2 (see Figure 4.1). The A1 conformer displays an OH...N intramolecular hydrogen bond, while A2 displays a NH...O intramolecular hydrogen bond and a free hydroxyl group. At the B3LYP/6-31G(d,p) level of theory, the differences in energy between A1 and A2 are 2.5, 1.9, and 2.1 kcal/mol for the 2-amino-1-propanol (2A1P),[5] 1-amino-2-propanol (1A2P), and 2-amino-1-butanol (2A1B) molecules, respectively.[8]

The complexes between NapOH and the two conformers A1 and A2 of the above-mentioned amino alcohols have been taken into account in the calculations. Three main families of complexes have been identified: the insertion_I: A1, the addition OH_OH: A1, and the addition OH_$NH_2$: A2 structures. These families are depicted in Figure 4.3 for the NapOH/1A2P system. The insertion_I: A1 family is obtained by inserting the OH group of NapOH into the intramolecular hydrogen bond of the A1 conformer of the amino alcohol to allow the formation of two intermolecular hydrogen bonds. These complexes are the most stable calculated forms but also have high Edef (around 3 kcal/mol). The addition OH_OH: A1 structure consists in forming an intermolecular hydrogen bond between the hydroxyl group of NapOH and the O atom of the amino alcohol taken in its A1 conformer. Finally, the addition OH_$NH_2$: A2 family is obtained by forming an intermolecular hydrogen bond between the OH

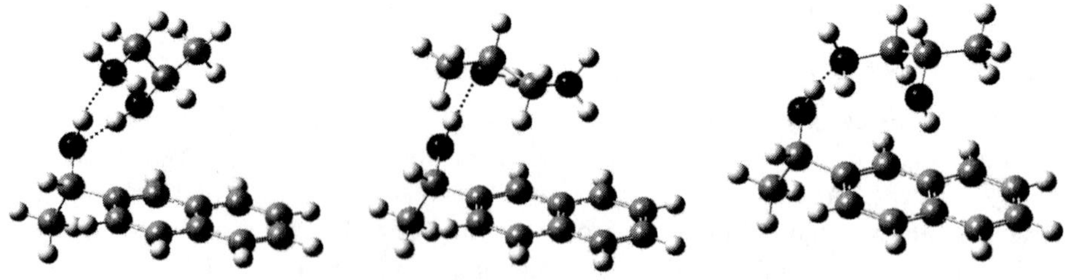

(a) Insertion I: A1 (b) Addition OH_OH: A1 (c) Addition OH_$NH_2$: A2

**FIGURE 4.3** Complexes structures calculated for the NapOH/1-aminopropan-2-ol: (a) insertion I: A1, (b) addition OH_OH: A1, and (c) addition OH_$NH_2$: A2.

group of NapOH and the N atom of amino alcohol taken in its A2 conformer. These two last complexes have low Edef (less than 0.5 kcal/mol) as the intramolecular bond of the amino alcohol is not disrupted.

The main result on these systems is that the insertion_I: A1 family has never been observed experimentally despite its high binding energy. Figure 4.4 illustrates this result: the observed spectra labeled $SS_I$ and $SS_{II}$ are typical of what was observed for these systems. We can clearly see that none of them fit the calculated spectrum associated with the insertion_I: A1 complex. Conversely, the addition OH_OH: A1 complex is observed in all the NapOH/amino alcohol complexes we have studied. The addition $OH\_NH_2$: A2 appears less systematically. Indeed, it is never observed for heterochiral complexes and only appears in homochiral ones. There is therefore a strong chiral discrimination in the nature of the hydrogen bond formed. Binding on the amino group can occur in homochiral complexes only, and not in heterochiral pairs.

The fact that we observe the less stable complex (addition OH_OH: A1) to the detriment of the most stable one (insertion_I: A1) is at first sight surprising. This observation has been directly related to the cooling process taking place in our experimental conditions. It seems to evidence that the cooling of molecules and complexes in a supersonic expansion, and especially pinhole experiments, is mainly kinetically controlled. The idea is that the opening of an intramolecular hydrogen bond that allows the formation of an insertion complex is the limiting factor and is directly related to the barrier for complex formation; this can be quantified by the deformation energy.[11] For the complexes between NapOH and amino alcohols presented here, the smallest value of the calculated deformation energy corresponding to a nonobserved complex is 2.77 kcal/mol. It is obtained for the complex between NapOH and the nonchiral amino-ethanol molecule. So the question raised here is

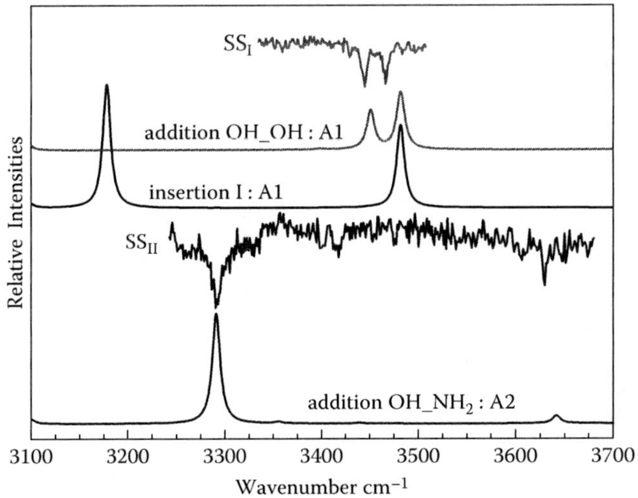


**FIGURE 4.4** Experimental vibrational spectra obtained for the homochiral complexes between NapOH and 1-aminopropan-2-ol ($SS_I$ and $SS_{II}$) compared with calculated vibrational spectra of addition OH_OH: A1, insertion I and addition $OH\_NH_2$: A2.

to estimate the upper limit in Edef, which makes the complex formation possible. Is it of the order of the 2.77 kcal/mol value or is it smaller? To answer this question, we have undertaken a systematic study of hydrogen-bonded complexes involving bifunctional chiral molecules.

### 4.4.3 Methyl Mandelate/Methyl Lactate (MeMan/MeLac)[11]

This example and the following one are a little more complicated because the two molecules in interaction are bifunctional. The isolated methyl lactate (MeLac) molecule has been widely studied previously.[16–23] It exists mainly in three conformers denoted hereafter as Syn, G, and G', with Syn being the most stable. The relative energy of G and G' with respect to Syn is 1.74 and 1.90 kcal/mol at MP2/6-31G(d,p) level of theory, respectively. By the same way, the methyl mandelate (MeMan) molecule exists in two conformers denoted hereafter as Syn and G, the latter one being 1.60 kcal/mol higher in energy.[24] For both molecules, the Syn conformer displays an intramolecular hydrogen bond between the hydroxyl and the carbonyl group of the molecule, while the G (or G') conformer displays an intramolecular hydrogen bond between the hydroxyl and the methoxy group of the molecule. In our work, we have calculated all the complexes formed from the Syn conformer of MeMan (the only one observed in our experimental conditions) and the three conformers of MeLac.[11] Optimization of the structure of the diastereoisomers has been performed at the MP2/6-31G(d,p) level of theory, and only their binding energy will be given, as no ZPE values have been calculated at this level.

Five structures have been obtained when complexing R-MeMan with the S-Syn conformer of MeLac. First, two insertion structures have been calculated. In terms of dissociation energy, they appear as the first (9.09 kcal/mol; Figure 4.5a) and the fourth (7.86 kcal/mol; Figure 4.5d) minima. Their Edef values are 2.16 and 2.01 kcal/mol, respectively. Comparison between experimental and theoretical vibrational spectra rules out these structures as candidates for the observed experimental spectrum. In other words, a Edef higher than 2 kcal/mol precludes the formation of the complexes. The head-to-head structure is the second most stable structure (9.01 kcal/mol; Figure 4.5b) and displays very large Edef (3.51 kcal/mol), which in view of the results described above makes the formation of this complex impossible. The third structure is an addition complex (Figure 4.5c). Its dissociation and deformation energies amount to 8.73 and 1.83 kcal/mol, respectively. Once again, the spectrum associated with this configuration of the RS diastereoisomer does not match the experimental spectrum. We have therefore calculated the complexes formed with the G and G' conformers of MeLac. In that case, we have obtained four insertion structures (Figure 4.5f–i). It appears that only the fourth most stable structure, for which the MeMan hydroxyl group inserts into the intramolecular hydrogen bond of the G' conformer of MeLac (insertion_I: G'; Figure 4.5i), displays a relatively low Edef (1.05 kcal/mol), and that its vibrational spectrum fits satisfyingly the experimental results.

The results obtained for the RR diastereoisomer parallel those described for RS. The topology of the hydrogen bond network is very similar for the two diastereoisomers. They only differ by the relative position of the phenyl ring of MeMan and

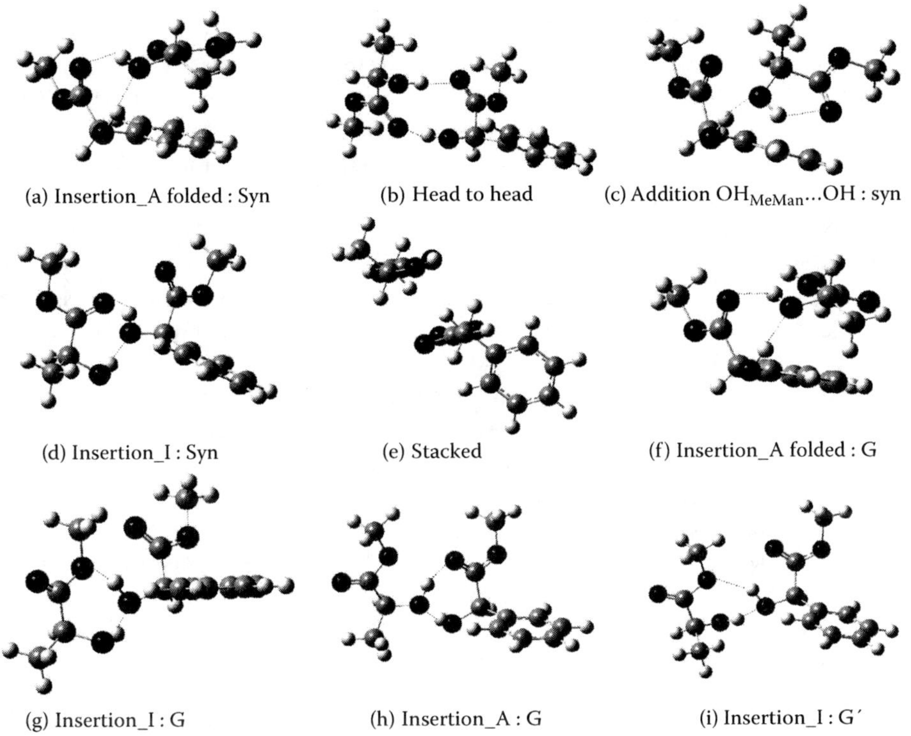


**FIGURE 4.5** Calculated structures for heterochiral complexes between methylmandelate and methyl lactate: (a) insertion_A folded: Syn, (b) head to head, (c) addition $OH_{MeMan}\cdots$ OHsyn, (d) insertion_I: Syn, (e) stacked, (f) insertion_A folded: G, (g) insertion_I: G, (h) insertion_A: G, and (i) insertion_I : G'.

the MeLac molecule relative to the lactate plane of MeMan. In the RR complex, the MeMan phenyl ring and the MeLac molecule are on the same side of the MeMan lactate plane. Therefore, the subtle balance between dispersion and electrostatic forces is slightly in favor of dispersion. In contrast, when they are on opposite sides, as is the case in RS, the balance tends to be slightly in favor of electrostatics. Consequently, the hydrogen bonds, even if the same in nature for both diastereoisomers, are slightly tighter in the RS complex, which results in a slightly larger red shift of the observed vibrational frequencies.

The important conclusion is that the complexes associated with the observed spectra for both RS and RR diastereoisomers are those obtained by inserting MeMan into the intramolecular hydrogen bond of the G' conformer of MeLac. Their Edef values are 1.05 and 1.33 kcal/mol, respectively. The fact that we can observe them experimentally shows that the upper limit of Edef is higher than 1.33 kcal/mol. Moreover, we have shown that the addition $OH_{MeMan}$_OHsyn (Edef = 1.83 kcal/mol) and the insertion_I:G (Edef = 1.89 kcal/mol) complexes are not observed. So with the study of this system we can consider that the limit in Edef that allows the formation of the complexes is somewhere between 1.3 and 1.9 kcal/mol.

### 4.4.4 Cis-1-Amino-2-Indanol/Methyl Lactate (AI/MeLac)[12]

Here again both molecules are bifunctional. The chromophore of the system is the 1-amino-2-indanol molecule. It exists in two conformations ($AI_I$ and $AI_{II}$) in the supersonic expansion[24]; they are depicted in Figure 4.1. Both of them display an intramolecular hydrogen bond directed from the hydroxyl group toward the amino group. They are related through the puckering motion of the indane plane: in $AI_I$ (the most stable) the OH group is in equatorial position and the $NH_2$ group is in axial position, while in $AI_{II}$, the OH and $NH_2$ groups are in the axial and equatorial position, respectively. The difference in energy between these two conformers is 0.48 kcal/mol. All complexes between these two conformers and the three conformers of MeLac have been calculated. The dissociation energies together with the deformation energies are given in Table 4.1 for the complexes with Edef lower than 1.9 kcal/mol. We can notice that no complex with the $AI_I$ conformer appears in this table, which is due to the fact that it always has huge Edef.

We have observed experimentally two complexes for each diastereoisomer; one of them is almost identical for the two diastereoisomers, and the other one is chirality dependent. The chirality-dependent complex for the heterochiral diastereoisomer has been shown to be due to the formation of an addition $OH_{MeLac}...OH_{AI}$: $AI_{II}$-G structure, while it is due to formation of a cycle: $AI_{II}$-G complex for the homochiral pair. They have very different spectroscopic fingerprints but share in common low Edef: 1.21 and 0.48 kcal/mol, respectively. The chirality dependence of these structures arises from the favorable interaction between the $C^*H$ group of the asymmetric center of MeLac and the aromatic ring, which can be obtained for one diastereoisomer namely, RS for addition and SS for the cycle (see Figure 4.6)

**TABLE 4.1**
**Binding and Deformation Energies in kcal/mol of the AI/MeLac System**

| Diastereoisomer | Complex | BE | Edef |
|---|---|---|---|
| RS | Insertion_A: $AI_{II}$-G | 11.38 | 1.81 |
| | Addition $OH_{MeLac}...OH_{AI}$: $AI_{II}$-G | 11.14 | 1.21 |
| | Insertion_A: $AI_{II}$-G*bis* | 10.87 | 1.57 |
| | Cycle bifurcated: $AI_{II}$-G' | 10.63 | 1.63 |
| | Head to head: $AI_{II}$-G | 10.23 | 1.81 |
| | Cycle: $AI_{II}$-G | 10.12 | 0.77 |
| RR | Cycle: $AI_{II}$-G | 11.61 | 0.48 |
| | Insertion_A: $AI_{II}$-G' | 11.48 | 1.81 |
| | Insertion_A: $AI_{II}$-G'*bis* | 10.93 | 1.62 |
| | Cycle: $AI_{II}$-Syn | 10.66 | 0.58 |
| | Cycle bifurcated: $AI_{II}$-G | 10.26 | 1.66 |

*Note:* Results are obtained at MP2/6-31G(d,p) level of theory. Binding energies are given relative to the closest fragment.

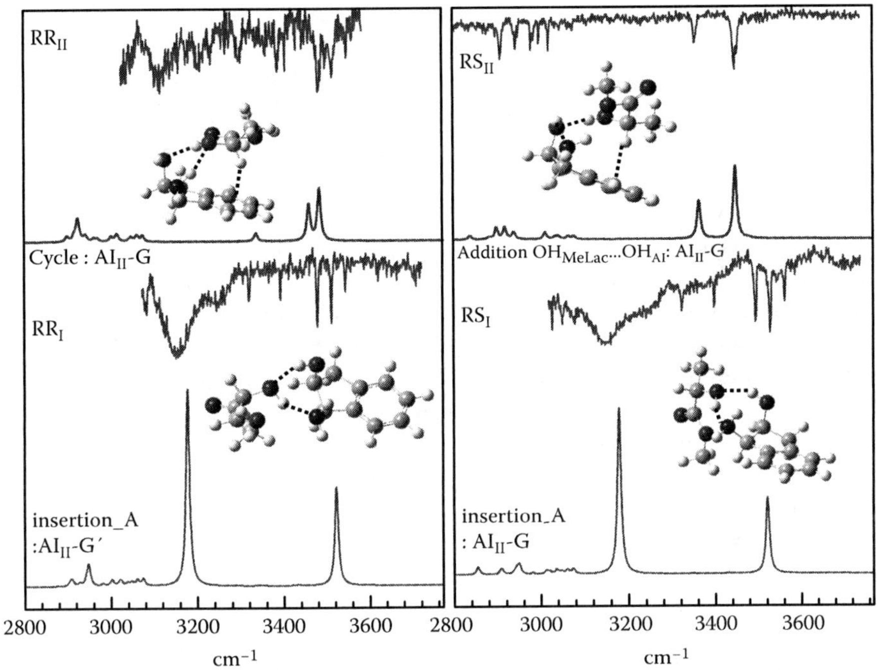


**FIGURE 4.6** Experimental and calculated vibrational spectra together with the geometry associated with the observed complexes between 2-amino-1-indanol and methyl lactate.

and not for the other. The chirality-independent structure corresponds to the insertion_A: AI$_{II}$-G for the heterochiral complex and the insertion_A: AI$_{II}$-G' for the homochiral diastereoisomer. Their Edef values are comprised between 1.8 and 1.6 kcal/mol. Because of the rigidity of this structure, the chiral centers of AI and MeLac are kept far apart, which explains the lack of sensitivity of this structure to chirality.

As we have definitely observed complexes with Edef of 1.6 kcal/mol, we can safely assume that the Edef value that precludes formation of complexes in our experimental conditions is between 1.6 and 1.9 kcal/mol.

## 4.5 CONCLUSION

We have tried to demonstrate in this chapter that supersonic expansion cooling of complexes is governed by what we have called deformation energy. This means that the formation of a complex is feasible if none of the molecules at play has to undergo a too large deformation of its geometry. We have also succeeded in demonstrating that this value is between 1.6 and 1.9 kcal/mol. We must mention, however, that this value is true when the deformation corresponds to the opening of an intramolecular hydrogen bond and when the complexes are bound by weak intermolecular interactions. The deformation of the subunits has been shown to play an important role also in ionic clusters. For example, Miller and Lisy[25] mentioned that the large barrier to subsequent cluster rearrangement precluded formation of the minimum energy isomer for $Li^+(H2O)_4Ar$ ionic systems. However, the limit of 1.6 to 1.9 kcal/mol can certainly not be transposed to systems for which the binding (dissociation) energy is a hundred times greater than the deformation energy. Maybe a ratio between the binding energy and the deformation energy would be more physically meaningful and applicable to a larger set of systems.

Last, we should underline that the kinetic control of complex formation is one of the very important factors for chiral recognition in bifunctional molecules. Indeed, the most strongly bound complexes, like head-to-head or insertion complexes, often involve large deformation energies. On the other hand, their large stability arises from multiple intermolecular hydrogen bonds, which make the structure rigid, and hence not very enantioselective. This is nicely illustrated in the example of AI-MeLac: the observed insertion structure has larger deformation energy than the observed cycle or addition structure and is much less sensitive to stereochemical factors.

This notion of deformation energy could also be of interest in the concept of thermodynamical enantioselectivity vs. kinetic enantioselectivity as discussed in gas-phase ion-molecule reactions.

## REFERENCES


1. Alrabaa, A. R., Breheret, E., Lahmani, F., Zehnacker, A. 1995. *Chem. Phys. Lett.* 237:480.
2. Al Rabaa, A., Le Barbu, K., Lahmani, F., Zehnacker-Rentien, A. 1997. *J. Phys. Chem. A* 101:3273.
3. Al Rabaa, A., Le Barbu, K., Lahmani, F., Zehnacker-Rentien, A. 1997. *J. Photochem. Photobiol. A Chem.* 105:277.
4. Le Barbu, K., Lahmani, F., Zehnacker-Rentien, A. 2001. A spectroscopic study discrimination in jet-cooled van der Waals complexes: The role of conformational isomerism as studied by double resonance spectroscopy. In *Physical chemistry of chirality*, ed. J. Hicks. Amsterdam: Elsevier, 225.
5. Le Barbu, K., Lahmani, F., Zehnacker-Rentien, A. 2002. *J. Phys. Chem. A* 106:6271.
6. Broquier, M., Lahmani, F., Zehnacker-Rentien, A., Brenner, V., Millie, P., Peremans, A. 2001. *J. Phys. Chem. A* 105:6841.
7. Seurre, N., Sepiol, J., Lahmani, F., Zehnacker-Rentien, A., Le Barbu-Debus, K. 2004. *Phys. Chem. Chem. Phys.* 6:4658.
8. Seurre, N., Sepiol, J., Le Barbu-Debus, K., Lahmani, F., Zehnacker-Rentien, A. 2004. *Phys. Chem. Chem. Phys.* 6:2867.
9. Seurre, N., Le Barbu-Debus, K., Lahmani, F., Zehnacker, A., Borho, N., Suhm, M. A. 2006. *Phys. Chem. Chem. Phys.* 8:1007.
10. Le Barbu-Debus, K., Guchhait, N., Zehnacker-Rentien, A. 2007. *Phys. Chem. Chem. Phys.* 9:4465.
11. Le Barbu-Debus, K., Broquier, M., Mahjoub, A., Zehnacker-Rentien, A. 2008. *J. Phys. Chem. A* 112:9731.
12. Le Barbu-Debus, K., Broquier, M., Mahjoub, A., Zehnacker-Rentien, A. 2009. *Phys. Chem. Chem. Phys.* DOI: 10.1039/b906834a.
13. Claverie, P. 1978. *Intermolecular interactions from diatomics to biopolymers*, ed. B. Pullman. New York: Wiley.
14. Brenner, V., Zehnacker, A., Lahmani, F., Millie, P. 1993. *J. Phys. Chem.* 97:10570.
15. Boys, S., Bernardi, F. 1970. *Mol. Phys.* 19:553.

16. Gigante, D. M. P., Long, F. J., Bodack, L. A., Evans, J. M., Kallmerten, J., Nafie, L. A., Freedman, T. B. 1999. *J. Phys. Chem. A* 103:1523.
17. Freedman, T. B., Lee, E., Nafie, L. A. 2000. *J. Phys. Chem. A* 104:3944.
18. Borba, A., Gomez-Zavaglia, A., Lapinski, L., Fausto, R. 2004. *Vibrational Spectros.* 36:79.
19. Borho, N., Suhm, M. A. 2004. *Phys. Chem. Chem. Phys.* 6:2885.
20. Ottaviani, P., Velino, B., Caminati, W. 2006. *Chem. Phys. Lett.* 428:236.
21. Zielke, P., Suhm, M. A. 2006. *Phys. Chem. Chem. Phys.* 8:2826.
22. Borho, N., Xu, Y. J. 2007. *Phys. Chem. Chem. Phys.* 9:1324.
23. Losada, M., Xu, Y. J. 2007. *Phys. Chem. Chem. Phys.* 9:3127.
24. Le Barbu-Debus, K., Lahmani, F., Zehnacker-Rentien, A., Guchhait, N. 2006. *Chem. Phys. Lett.* 422:218.
25. Miller, D. J., Lisy, J. M. 2008. *J. Am. Chem. Soc.* 130:15381.

# 5 Chiral Recognition in Mass Spectrometry, Focusing on FAB Mass Spectrometry

*Motohiro Shizuma*

## CONTENTS

5.1 Introduction ........................................................ 61
5.2 FABMS/Relative Peak Intensity (RPI) Method ........................................................ 63
5.3 FABMS/Enantiomer-Labeled (EL) Guest Method ........................................................ 63
5.3.1 Basic Concept of FABMS/EL Guest Method ........................................................ 63
5.3.2 Isotope Effects on $I_R/I_{S\text{-}dn}$ Values ........................................................ 66
5.3.3 Dependency of $I_R/I_{S\text{-}dn}$ Values on Initial Concentrations of Chiral Host and Chiral Guest Pair ........................................................ 67
5.3.4 Dependency of $I_R/I_{S\text{-}dn}$ Values on Temperature ........................................................ 69
5.3.5 Dependency of $I_R/I_{S\text{-}dn}$ Values on Matrix ........................................................ 72
5.3.6 Application of FABMS/EL Guest Method: Screening of New Chiral Hosts ........................................................ 72
5.4 FABMS/Enantiomer-Labeled (EL) Host Method ........................................................ 75
5.4.1 Basic Concept of FABMS/EL Host Method ........................................................ 75
5.4.2 Relationship between FABMS/EL Host Method and the Guest Method ........................................................ 76
5.5 ESI and MALDI MS/Enantiomer-Labeled (EL) Method ........................................................ 81
5.6 Conclusion ........................................................ 82
References ........................................................ 82

## 5.1 INTRODUCTION

As is well known, mass spectrometry (MS) is insensitive to chirality difference. For example, the mass spectra of an enantiomeric pair of *R* and *S* amino acid methyl esters are identical. Therefore, in order to provide solutions to such chiral recognition problems, a contrivance should be devised for conventional mass spectrometry.

Generally, the differences in diastereomeric intermolecular interactions between each enantiomer of a guest and a chiral host (enantiopure) are detected by spectroscopic or chromatographic methods for chiral recognition. Here, chiral hosts are defined as molecules containing convergent binding sites, and chiral guests as ions or molecules that are captured by the chiral hosts, according to Cram's definition in host-guest chemistry.[1] In the case of nuclear magnetic resonance (NMR) spectroscopy, chemical shift differences induced by adding a chiral-solvating agent such as Parker reagent are used to elucidate the information.[2] Optical resolution using gas or liquid chromatography can be regarded as a practical application of the differences in the diastereomeric interactions of each enantiomer with a chiral stationary phase.[3,4] In the case of MS, since the mass of each diastereomer ion is identical, the peak intensity is the only source of diastereomeric differentiation. To evaluate diastereomeric differentiation by MS, there are mainly two methods. One is to evaluate the relative peak intensity of the diastereomer ions against an internal standard, having different mass from the diastereomer ion in MS. Another is to evaluate the difference in the diastereomer ion stability on the basis of the relative intensity of the product ion by collision-induced dissociation (CID) of the diastereomer ion as the precursor ion in MS/MS. The kinetic method by Tao et al. is one of the typical techniques for evaluating chirality difference using CID in MS/MS.[5]

Noncovalent bonding diastereomeric complex ions of the chiral host and guest are detected easily in soft-ionization MS, such as fast atom bombardment (FAB),[6] electrospray ionization (ESI),[7] matrix-assisted laser desorption ionization (MALDI),[8] etc. As an early example, Baldwin et al. detected the chirality effects on the diastereomeric complexation of alkyl tartrates using FABMS.[9] The FAB mass spectra of ethyl tartrate (ET) and the 1:1 mixture of isopropyl D-tartrate ($IPT_D$) and deuterated isopropyl L-tartrate ($IPT_L$-$d_{14}$) were measured, and the protonated dimers $(ET + IPT_D + H)^+$ and $(ET + IPT_L\text{-}d_{14} + H)^+$ were detected. The relative peak intensity of the dimer ions changed linearly, depending on the optical purity of ET. The results suggested the future possibility of quantitative chiral detection by FABMS.

In order to quantitatively treat the peak intensity data of the diastereomeric complex ion, Sawada et al. proposed new procedures using FABMS, the relative peak intensity (RPI) method in 1992[10] and the enantiomer/labeled guest (or host) method in 1994.[11] In this chapter, chiral recognition in FABMS is described using the host-guest complexation systems of chiral crown ether (host) with the ammonium ion of an amino acid ester (guest) via charge-dipole electrostatic intermolecular interactions as typical examples (Chart 5.1). The following abbreviations for the ammonium

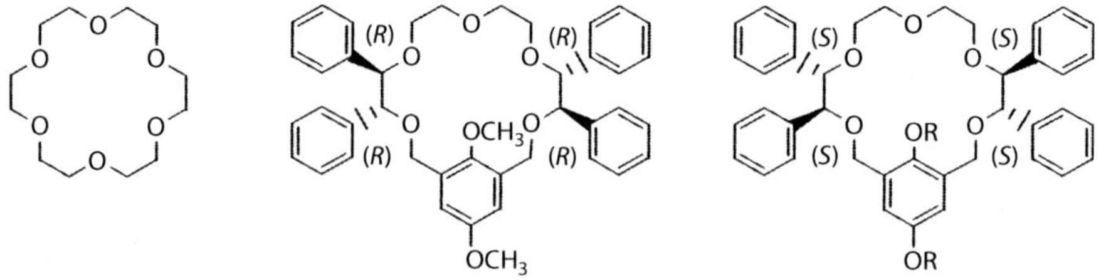

**1** (Achiral host)

**2a**

**2b** ($R = CH_3$)

**2b′** ($R = CH_3$)

**CHART 5.1** Achiral and chiral hosts (**1**, **2a**, and **2b**), ion of amino acid esters are used: A-O-$R^+$ ($X^-$)—A, amino acid moiety; R, alkyl group of the ester moiety; $X^-$, counter anion.

## 5.2 FABMS/RELATIVE PEAK INTENSITY (RPI) METHOD

In FABMS, the samples are dissolved in a liquid matrix such as glycerol, diethanolamine, 3-nitrobenzyl alcohol (NBA), etc. An accelerated neutral atom such as xenon runs against the matrix solution to ionize the sample. As the ionization energy is low, noncovalent host-guest complex ions can be produced in the gas phase without decomposition.

An equimolar amount of a reference achiral host ($H_{Ref}$) is added to a target chiral host (H), and the 1:1 mixture of the hosts is complexed with a target chiral guest ($G_R^+$ or $G_S^+$) under competitive conditions (Figure 5.1). The relative peak intensity value of the two resulting complex ions $(H + G_R)^+$ and $(H_{Ref} + G_R)^+$ [or $(H + G_S)^+$ and $(H_{Ref} + G_S)^+$] observed in one FAB mass spectrum is called the relative intensity, $I(H + G_R)^+/I(H_{Ref} + G_R)^+ = RPI_R$ [ or $I(H + G_S)^+/I(H_{Ref} + G_S)^+ = RPI_S$] value. The ratio of the $RPI_R$ and $RPI_S$ values is regarded as a new measure that indicates the degree and *R*/*S* selectivity of chiral recognition (RPI method).[10,12,13]

The FAB mass spectra are shown in Figure 5.1 as a typical example of the RPI method. Hosts **1**, **2a**, and each enantiomer of the guest Phe-O-$^i$Pr$^+$ ($Cl^-$) were used as the achiral host, the enantiopure chiral host, and chiral guest, respectively (Chart 5.1). In this case, the $RPI_R$, $RPI_S$, and the $RPI_R/RPI_S$ values were 0.073, 0.045, and 1.6, respectively.

One of the advantages of this method is that available reference hosts can be chosen. However, it has several problems, as follows:[14] (1) Two consecutive measurements with great caution are essentially required under the same machine and same substrate-concentration conditions. (2) In addition, adequate selection of the reference achiral compounds is essential to obtain the proper RPI values. When the association abilities between chiral and achiral hosts toward a chiral guest are much different, proper RPI values are not obtained. Also, when the efficiency of ionization and transferability to the gas phase of $(H + G_R)^+$ and those of $(H_{Ref} + G_R)^+$ [or $(H + G_S)^+$ and $(H_{Ref} + G_S)^+$] shows much difference, the proper RPI values are not obtained. By using an adequate reference, the FABMS/RPI method becomes a powerful tool for evaluating chirality difference.[14,15]

Vékey and coworkers improved the RPI method and evaluated the chiral recognition ability of pyridino-crown ether derivatives with high accuracy.[16,17] Davey et al. examined enantioselectivity in the complexation of permethylated cyclodextrins by the RPI method.[18] Also, Liang et al. (FAB/Fourier-transform ion cyclotron resonance (FTICR)/MS) and Krishna et al. (liquid secondary ionization MS) examined chiral discrimination on the basis of the RPI values.[19,20]

## 5.3 FABMS/ENANTIOMER-LABELED (EL) GUEST METHOD

### 5.3.1 Basic Concept of FABMS/EL Guest Method

Sawada et al.[21] proposed a new way to observe a 1:1 diastereomeric host-guest complexation, the FABMS/EL guest method, as an improvement on the above RPI

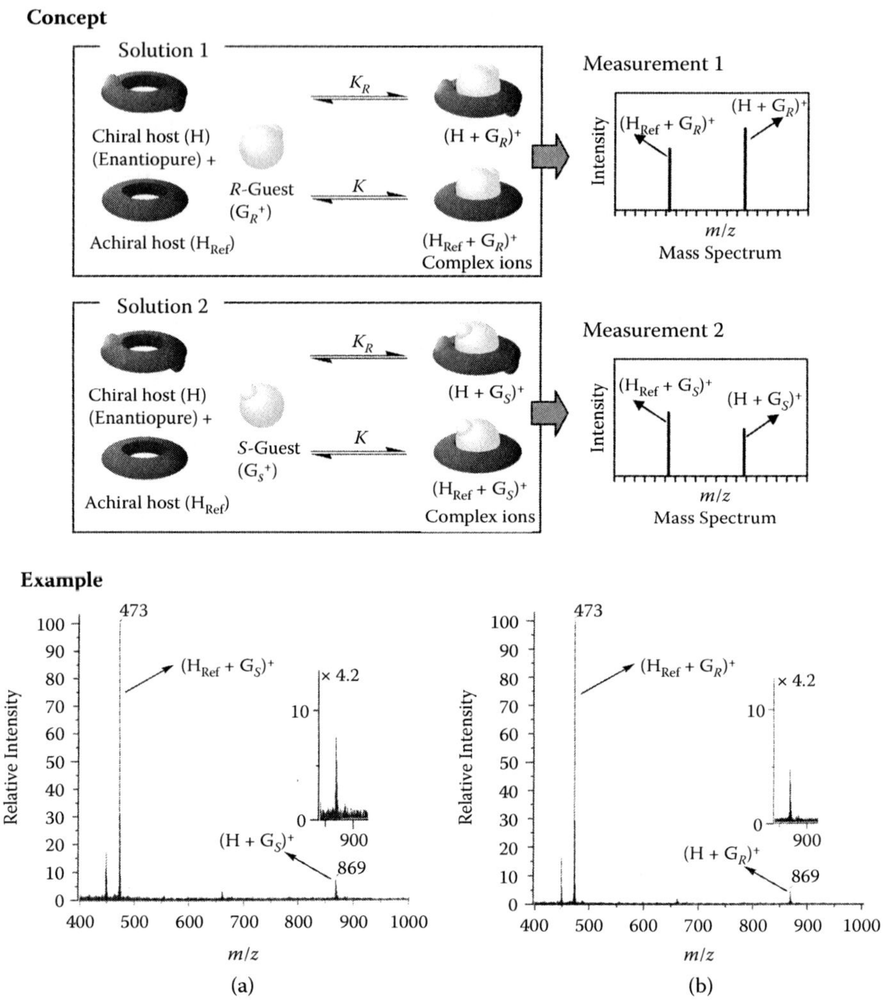


**FIGURE 5.1** Illustration of concept of the FABMS/RPI method and the typical mass spectra. Top: A chiral host, an $R$ enantiomer of guest ion ($G_R^+$), and an achiral reference ($H_{Ref}$). The RPI value is shown as $I(H + G_R)^+/I(H_{Ref} + G_R)^+$. Bottom: A chiral host (H), an $S$ enantiomer of guest ion ($G_S^+$), and an achiral reference ($H_{Ref}$). The RPI value is shown as $I(H + G_S)^+/I(H_{Ref} + G_S)^+$. The ratio of the RPI values represents the chiral recognition ability of a host. (a) Chiral host (H), **2a**; achiral host ($H_{Ref}$), **1**; guest ($G^+(X^-)$): (*R*)-Phe-O-$^i$Pr$^+$ ($Cl^-$). (b) H, **2a**; $H_{Ref}$, **1**; $G^+(X^-)$, (*R*)-Phe-O-$^i$Pr$^+$ ($Cl^-$). [H]:[$H_{Ref}$]:[$G^+$] = 0.0833:0.0833:0.0833 (M) in NBA. (Dr. M. Shizuma measured the mass spectra for this book.)

technique. In the improved method, one enantiomer of a chiral guest is isotopically labeled, and an equimolar amount of a mixture of a labeled and an unlabeled enantiomeric guest was complexed with a chiral host. Therefore, the new method is called the enantiomer-labeled (EL) guest method. The basic concept is illustrated in Figure 5.2, in which the $S$ enantiomer of the chiral guest is labeled with deuterium

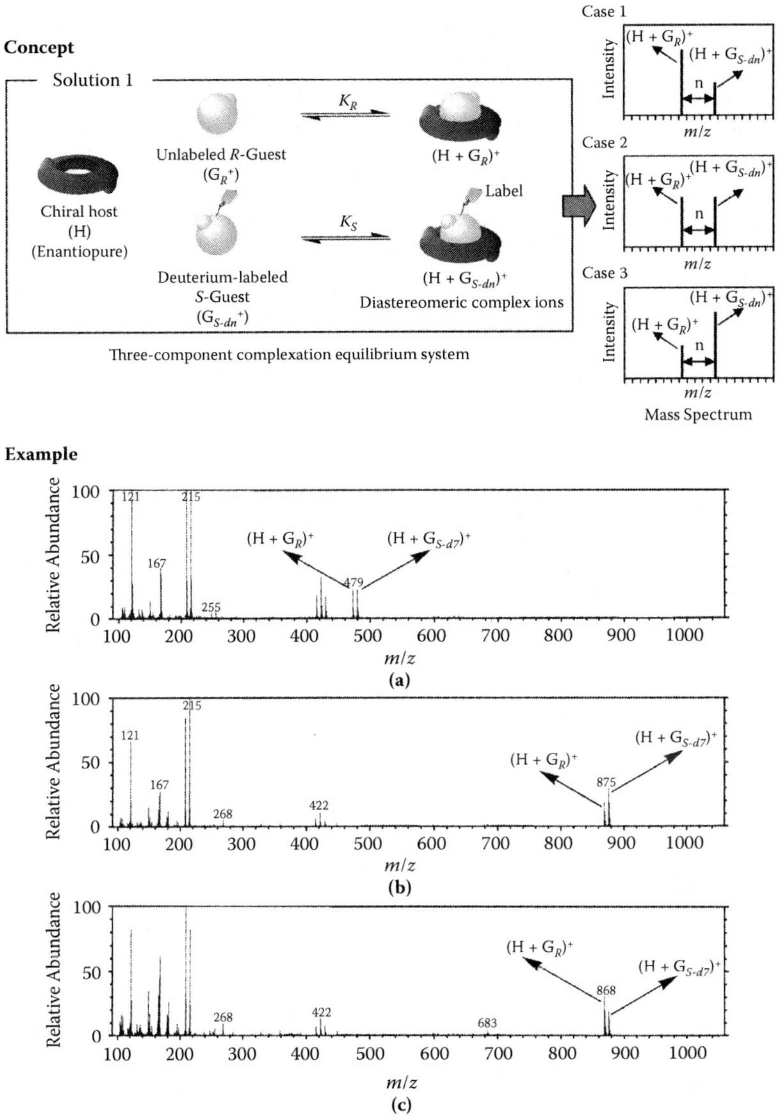


**FIGURE 5.2** Illustration of concept of the FABMS/EL guest method and the typical mass spectra. In the case of a host with an enantiomeric pair of deuterium-labeled/unlabeled guests, the resulting diastereomeric host-guest complex ions are distinguished in the mass spectrum, which exhibits peaks with different mass numbers (*m/z*) by n (n = number of deuterium atoms). The mass spectrum is classified into three cases according to the relative peak intensity of the complex ions: case 1, *R* preference; case 2, *S* preference; case 3, no preference. (a) Host (H), **1**; guest ($G^+ X^-$), Phe-O-$^i$Pr$^+$ ($Cl^-$); (b) H, **2a**; ($G^+ X^-$), Phe-O-$^i$Pr$^+$ ($Cl^-$); (c) H, **2b**; ($G^+ X^-$), Phe-O-$^i$Pr$^+$ ($Cl^-$). The *S* enantiomer of the guest was labeled with deuterium atoms (n = 7). (Mr. H. Adachi (Osaka University) measured the mass spectra for this book.)

atoms as an example. The primary advantage gained in the use of a deuterium isotope tag is that it allows us to determine straightforwardly the *R*/*S* selectivity and its magnitude from the peak intensity difference in the diastereomeric host-guest complex ions [$(H + G_R)^+$ and $(H + G_{S\text{-}dn})^+$, n: number of deuterium atoms] observed in one FAB mass spectrum. The chiral recognition ability is estimated from the corresponding peak intensity ratio [$I(H + G_R)^+/(H + G_{S\text{-}dn})^+ = I_R/I_{S\text{-}dn}$ value]. Under the competitive conditions, the criteria for chiral recognition properties become quite clear, simple, and direct, as shown below:

Case 1. $I_R/I_{S\text{-}dn} > 1.0$ means that a chiral host binds an *R* guest more strongly. The larger $I_R/I_{S\text{-}dn}$ value means higher chiral recognition ability of the chiral host.
Case 2. $I_R/I_{S\text{-}dn} < 1.0$ means that a chiral host binds an *S* guest more strongly, in an inverse manner.
Case 3. $I_R/I_{S\text{-}dn} = 1.0$ means that a chiral host does not discriminate between the enantiomers of a chiral guest.

The larger $I_R/I_{S\text{-}dn}$ value deviates from unity, the higher the chiral recognition ability of the chiral host becomes. The improved method has other advantages, as follows: (1) no need to use an internal achiral standard, (2) facile and short-time measurements, and (3) excellent agreement with complexation behavior in solution.

The mass spectra are shown in Figure 5.2 as typical examples. Phe-O-$^i$Pr$^+$ (Cl$^-$) was selected as the chiral guest, and the *S* enantiomer of the guest was labeled with deuterium atoms. Hosts **2a** and **2b** are mirror images, i.e., enantiomeric, of each other. The $I_R/I_{S\text{-}dn}$ values of host **2a** and **2b** with Phe-O-$^i$Pr$^+$ (Cl$^-$) were 1.63 and 0.62, respectively.

Hosts **2a** and **2b** are enantiomers. If those host-guest complex ions are highly structured ones, the complex ion of **2a** with *R* guest and the complex ion of **2b** with *S* guest are enantiomers of each other. Therefore, the cross-chiral relationship (CCR) should hold: the degrees of *R* guest preference of **2a** should be equal to the degree of *S* guest preference of **2b**. The CCR is represented by the following equation for each guest: ($I_R/I_{S\text{-}dn}$ for **2a**) × ($I_R/I_{S\text{-}dn}$ for **2b**) = 1. In fact, the CCR of the enantiomeric pair of hosts **2a** and **2b** toward Phe-O-$^i$Pr$^+$ is 1.01 [= 1.63 ($I_R/I_{S\text{-}dn}$ for **2a**) × 0.62 ($I_R/I_{S\text{-}dn}$ for **2b**)], as shown in Figure 5.2b and c. Quantitative CCRs were recognized for several chiral hosts.[21] These experimental findings proved the high reliability of the $I_R/I_{S\text{-}dn}$ values obtained and the highly structured host-guest complex ions detected by the FABMS/EL guest method.[22-27] Therefore, great attention was paid to this method in 1994.[28]

### 5.3.2 Isotope Effects on $I_R/I_{S\text{-}dn}$ Values

In the FABMS/EL guest method, two isotope effects should be considered: (1) overlap between the isotope peak of the host-guest complex ion (including the unlabeled guest) and the main peak of the host-guest complex ion (including the labeled guest), and (2) differences between complexation properties containing a complex structure, binding ability, etc. of the chiral host with the labeled guest and those of the host with the unlabeled guest.

With regard to the former isotope effect, the experimental $I_R/I_{S\text{-dn}}$ values are corrected on the basis of the natural abundance of the overlapped isotope for the complex ion with the unlabeled guest. For example, the composition formula of the complex ion of host **2a** with unlabeled Phe-O-Me$^+$ is $C_{52}H_{58}NO_9$. The natural abundance is as follows: M, 100; (M + 1), 57.62; (M + 2), 18.15; (M + 3), 4.08 (M: exact mass 840.4; the intensities are normalized as M = 100). Using the labeled guest, Phe-O-Me-$d_3^+$, the peak for the exact mass of the complex ion is overlapped by the (M + 3) isotope peak of the complex with the unlabeled guest. In this case, the experimental and corrected $I_R/I_{S\text{-dn}}$ values were 1.78 and 1.92, respectively. Using Phe-O-$^i$Pr-$d_7^+$ or Phe-O-Et-$d_5^+$, the (M + 7) or (M + 5) isotope peak intensity of the complex ion with **2a** (or **2b**) is negligibly small (<0.12).

The later isotope effect was checked in comparison with the $I_{R\text{-dn}}/I_S$ value of the *R* guest labeling case and the $I_R/I_{S\text{-dn}}$ value of the *S* guest labeling case. The agreement of the $I_{R\text{-dn}}/I_S$ value with the $I_R/I_{S\text{-dn}}$ value indicates no isotope effect on chiral recognition. No, or a negligibly small, isotope effect of the deuterium labeling on the FAB mass spectral intensities of the host-guest complex ions was confirmed experimentally in the host-guest complexation systems of chiral hosts such as chiral crown ethers, chiral potands, and permethylated oligosaccharides with the deuterium-labeled alkyl ester of amino acids.[11,21,27]

### 5.3.3 Dependency of $I_R/I_{S\text{-DN}}$ Values on Initial Concentrations of Chiral Host and Chiral Guest Pair

The $I_R/I_{S\text{-dn}}$ value varies depending upon the initial concentration of the chiral host and chiral guest.[22,26] It is therefore important to check if the $I_R/I_{S\text{-dn}}$ values agree with the ratio of the association constants ($K_R/K_S$). The $I_R/I_{S\text{-dn}}$ values reflect mainly the ratio of the concentration of the diastereomeric complex ion in the matrix $[(H + G_R)^+]/[(H + G_S)^+]$. The concentration ratio $[(H + G_R)^+]/[(H + G_S)^+]$ is calculated from the association constants in an organic solvent ($K_R$ and $K_S$), which are determined separately by the NMR or other spectroscopic method, and the initial concentrations of the chiral host and guests ($[H]_0$, $[G_R^+]_0$, and $[G_S^+]_0$). In the case of $[H]_0 \ll [G_R^+]_0$ (= $[G_S^+]_0$), the $[(H + G_R)^+]/[(H + G_S)^+]$ values are approximately equal to $K_R/K_S$.

Indeed, the changes in the $I_R/I_{S\text{-dn}}$ values of several chiral hosts, including permethylated carbohydrates, chiral crown ethers, etc., were experimentally in good agreement with the changes in the $[(H + G_R)^+]/[(H + G_S)^+]$ values.[11,21,24,29–37] For example, the $K_R$ and $K_S$ values of host **2a** with each enantiomer of guest PheOiPr$^+$ (Cl$^-$) were estimated to be 2 $M^{-1}$ and 1 $M^{-1}$ (in chloroform-*d*/dichloromethane-$d_2$ (v/v) = 1/2, at 298 K, by the NMR titration method), respectively.[21] The $[(H + G_R)^+]/[(H + G_S)^+]$ values calculated from $K_R$ and $K_S$ are plotted against $[H]_0$, to give curve (a) in Figure 5.3. The $I_R/I_{S\text{-dn}}$ values in NBA matrix, which are shown as open circles in the figure, are in good agreement with the $[(H + G_R)^+]/[(H + G_S)^+]$ values, despite the fact that these values were determined in different solvents (NBA or chloroform-*d*/dichloromethane-$d_2$ mixture). Certainly, the $K_R$ and $K_S$ values should change, depending on the solvents. However, the effect may be canceled in their ratio ($K_R/K_S$), though there are some exceptions with regard to

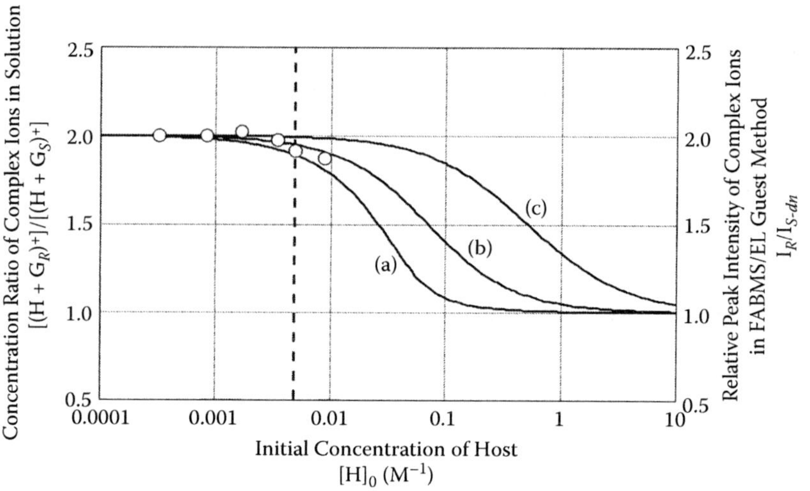


**FIGURE 5.3** Plots of $[(H + G_R)^+]/[(H + G_S)^+]$ against $[H]_0$ in solution and the $I_R/I_{S\text{-dn}}$ values (○) under various $[H]_0$ in FABMS/EL guest method. Host, **2a**, guest, Phe-O $^i$Pr$^+$ (Cl$^-$); matrix NBA. (a) $K_R = 2\ M^{-1}$, $K_S = 1\ M^{-1}$; (b) $K_R = 20\ M^{-1}$, $K_S = 10\ M^{-1}$; (c) $K_R = 200\ M^{-1}$, $K_S = 100\ M^{-1}$. $[G_R^+] = [G_S^+] = 0.025\ M$ is assumed. The dotted line shows $[H]_0$ for the FABMS experiments. (Reproduced by permission of the American Chemical Society, from *J. Am. Chem. Soc.* 117 [1995]: 7726–36.)

the solvent effect. Next, in order to examine the effect of the magnitude of the binding constants ($K_R$ and $K_S$) on the $[(H + G_R)^+]/[(H + G_S)^+]$ values, other curves are plotted against a range of $[H]_0$, as shown in Figure 5.3, assuming (b) $K_R = 20 \times 10^2\ M^{-1}$, $K_S = 10 \times 10^2\ M^{-1}$, and (c) $K_R = 200 \times 10^2\ M^{-1}$, $K_S = 100 \times 10^2\ M^{-1}$. The dotted line shows $[H]_0$ for the FABMS experiments. It is clarified that under the FABMS sampling conditions, the three curves collapse and the $[(H + G_R)^+]/[(H + G_S)^+]$ values of each curve are nearly identical. This means that the chiral recognition ability ($I_R/I_{S\text{-dn}}$) determined by the FABMS/EL guest method is almost identical with that ($K_R/K_S$) evaluated by other methods, such as titration in NMR or in UV-visible spectrometry, and is independent of the magnitude of the binding constants.

Accordingly, under the conditions of $[G_R^+]_0$ ($= [G_S^+]_0) >> [H]_0$), the $I_R/I_{S\text{-dn}}$ values reach the $K_R/K_S$ values, which can be a thermodynamically significant quantity; the $-\Delta\Delta G_{enan}$ values are in units of J mol$^{-1}$:

$$-\Delta\Delta G_{enan} = -\Delta G_R - (-\Delta G_S) = RT \ln(K_R/K_S) \geq RT \ln(I_R/I_{S\text{-dn}}) \tag{5.1}$$

Thus, the $I_R/I_{S\text{-dn}}$ values are reasonably converted into the difference in free energy ($-\Delta\Delta G_{enan}$).

The above correlations between the $I_R/I_{S\text{-dn}}$ values in NBA and $K_R/K_S$ in organic solvents such as chloroform were also experimentally confirmed in the complexation of several chiral host and chiral ammonium ion guest pairs. Some examples are shown in Figure 5.4.[38–40] Also, chiral separation of chiral amine by capillary

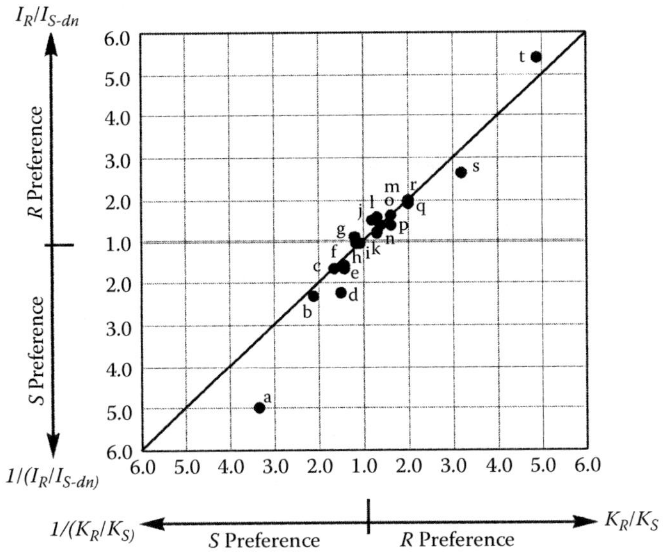


**FIGURE 5.4** Plots of ratio of association constants ($K_R/K_S$) in organic solvents against the $I_R/I_{S\text{-dn}}$ values in the FABMS/EL guest method (NBA matrix). Host, guest, reference number: (a) **3**, Phe-O-$^i$Pr$^+$, 29; (b) **13a**, Trp-O-$^i$Pr$^+$, 35; (c) **3**, Trp-O-$^i$Pr$^+$, 29; (d) **12$_1$a**, Trp-O-$^i$Pr$^+$, 35; (e) **16a**, NEA$^+$, 34; (f) **12$_1$b**, Phe-O-$^i$Pr$^+$, 35; (g) **13b**, Phe-O-$^i$Pr$^+$, 35; (h) **19$_1$**, (*p*-F-)Pgly-O-$^i$Pr$^+$, 32, 38; (i) **12$_1$b**, Pgly-O-$^i$Pr$^+$, 35; (j) **13b**, Trp-O-$^i$Pr$^+$, 35; (k) **11**, NEA$^+$, 39, 41; (l) **9**, Met-O-Me$^+$, 21; (m) **21$_1$**, Trp-O-$^i$Pr$^+$, 32; (n) **11**, Pgly-O-$^i$Pr$^+$, 40, 41; (o) **18b**, NEA$^+$, 34; (p) **18a**, Trp-O-$^i$Pr$^+$, 34; (q) **2a**, Pgly$^+$, 11, 21; (r) **16a**, Trp-O-$^i$Pr$^+$, 34; (s) **4a**, Met-O-Me, 21. Pgly = 1-amino-1-phenyl-acetic acid, NEA = 1-(1-naphthyl)-ethylamine. (Reproduced by permission of the Mass Spectrometry Society of Japan, from *J. Mass Spectrom. Soc. Jpn.* 50 [2002]: 311–29.)

electrophoresis using 18-crown-6-tetracarboxylic acid was examined, and the results were in good agreement with those by the FABMS/EL guest method.[41]

### 5.3.4 Dependency of $I_R/I_{S\text{-dn}}$ Values on Temperature

The $I_R/I_{S\text{-dn}}$ value, which is a new measure of chiral recognition ability determined by the FABMS/EL guest method, is approximately equal to the $K_R/K_S$ values. Therefore, the logarithm of the $I_R/I_{S\text{-dn}}$ values ought to change with a linear relationship to the reciprocal values of absolute temperature ($1/T$). The dependency of the $I_R/I_{S\text{-dn}}$ value on temperature was examined using a thermally controlled FAB probe, which was altered to fix the sample solution on the tip of the probe during measurement.[42] In that case, permethylated oligosaccharide **3** (Chart 5.2) and several ammonium ions of amino acid isopropyl esters (Ala, Val, Tle, Met, and Phe) were used as the chiral host and guests, respectively. The host showed good solubility in NBA matrix under the experimental temperature (8–35°C). The $I_R/I_{S\text{-dn}}$ values of host **3** with each ammonium ion guest approached unity with a rise in temperature. The tendency was recognized in all cases with the given

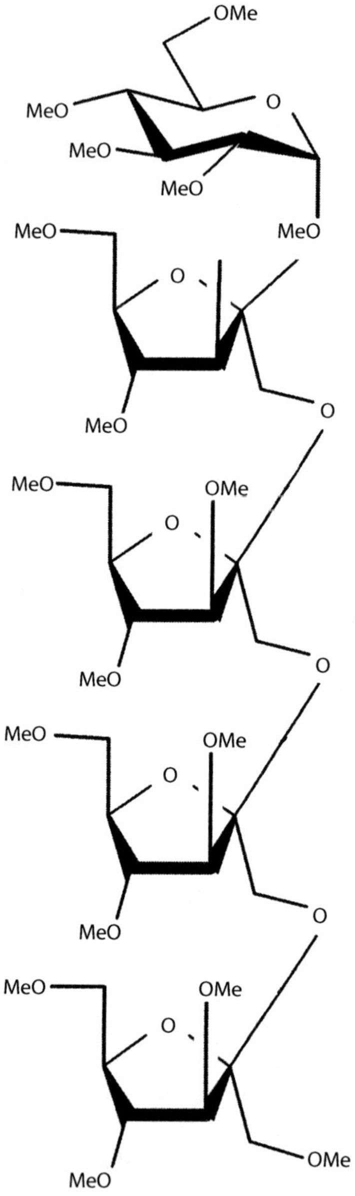


**3**

**CHART 5.2** Permethylated oligosaccharide host (**3**).

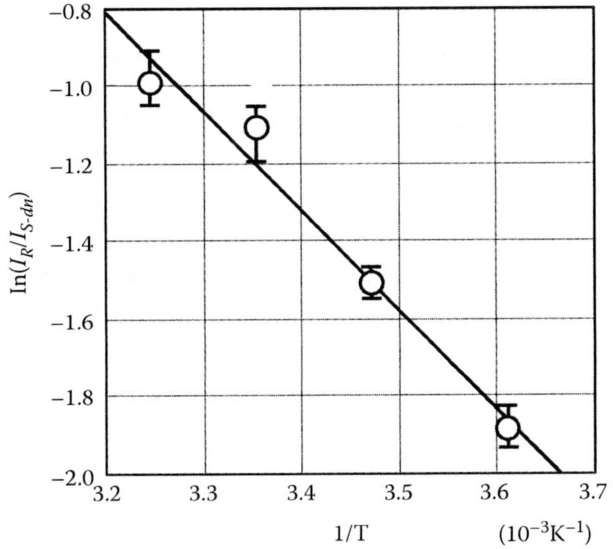


**FIGURE 5.5** Plot of the logarithms of the $I_R/I_{S\text{-dn}}$ values against the reciprocal values of the absolute temperature ($1/T$). Host, **3**; guest, Tle-O-$^i$Pr$^+$ ($Cl^-$) (equimolar mixture of $G_R^+$ and $G_{S\text{-d7}}^+$). Tle = *tert*-leucine.

guests. The decrease in selectivity with an increase in temperature is one of the general behaviors in molecular recognition. When the logarithms of the $I_R/I_{S\text{-dn}}$ values were plotted against the reciprocal values of the absolute temperature, good linear relationships were recognized in all cases with the given guests (correlation coefficient $R^2 = 0.85$–$0.97$). An example is shown in Figure 5.5. This fact supports the induction that the $I_R/I_{S\text{-dn}}$ value corresponds to the $K_R/K_S$ value under competitive conditions.

The difference in free energy for chiral recognition ($-\Delta\Delta G_{enan}$) is related to the thermodynamic parameters, the enthalpy ($-\Delta\Delta II^{o}_{enan}$) and the entropy ($-\Delta\Delta S^{o}_{enan}$), by the following equation, where $R$ is a gas constant.

$$-\Delta\Delta G^{o}_{enan} = -\Delta\Delta H^{o}_{enan} - (-T\Delta\Delta S^{o}_{enan}) \quad (5.2)$$

Then, using Equations 5.1 and 5.2, Equation 5.3 is derived:

$$\ln (I_R/I_{S\text{-dn}}) = -(\Delta\Delta H^{o}_{enan}/R)\,(1/T) + (\Delta\Delta S^{o}_{enan}/R) \quad (5.3)$$

Therefore, $\Delta\Delta H^{o}_{enan}$ and $\Delta\Delta S^{o}_{enan}$ are calculated from the slope and the intercept of the fitted line.[42] In the case of **3** with Tle-O-$^i$Pr$^+$ (Tle: *tert*-leucine), $\Delta\Delta H^{o}_{enan}$ and $\Delta\Delta S^{o}_{enan}$ were calculated at 21 kJ mol$^{-1}$ and 54 J mol$^{-1}$ K$^{-1}$, respectively. Thus, the thermodynamic parameters of the enantioselective host-guest complexation can be estimated by the FABMS/EL guest method.

### 5.3.5 Dependency of $I_R/I_{S\text{-DN}}$ Values on Matrix

Generally, solvents affect a host-guest complexation in solution. The matrix effects on the $I_R/I_{S\text{-dn}}$ values toward Phe-O-$^i$Pr$^+$ (Cl$^-$) were examined using host **3**, which is well dissolved in matrices such as NBA, glycerol (G), α-thioglycerol (TG), 2-nitrophenyl *n*-octyl ether (NPOE), 2,2'-dithidiethanol (DTDE), diethanolamine (DEA), and triethanolamine (TEA).[43]

The peaks of the host-guest complex ions were not observed using the basic matrices, DEA and TEA, because of deprotonation from the ammonium ion guest. The $I_R/I_{S\text{-dn}}$ values changed depending on the matrices, and the order is as follows: 1.00 > G > TG > DTDE ~ NPOE > NBA. The reasons are not clear at the present time. As a hypothesis, the relative concentration of the actual guest for the host may be decreased by the strong solvation of the guest or the complex ion of matrices such as G and TG, and then the host and the guest may form the host-guest complex under noncompetitive conditions.

In the host-guest complexation of crown ether and an alkylated carbohydrate host with cationic guests, NBA is a good matrix. Kim and coworkers evaluated the chiral recognition ability of chiral bis-pyridino-18-crown-6 derivatives using NBA matrix with the FABMS/EL guest method.[44] For other host-guest complexation systems, a suitable matrix for the system should be sought.

### 5.3.6 Application of FABMS/EL Guest Method: Screening of New Chiral Hosts

The advantage of the FABMS/EL guest method is to evaluate the chiral recognition ability of a chiral host in a short measurement time from a single mass spectral chart. Moreover, the sample preparation is very easy, only requiring the mixing of a chiral host and an equimolar mixture of deuterium-labeled/unlabeled enantiomers of a chiral guest in a matrix under competitive conditions. Therefore, this method is suitable for the screening of chiral recognition of hosts.

For example, the chiral discrimination abilities of various permethylated carbohydrates toward several ammonium ions of amino acid isopropyl esters were evaluated combinatorially.[32] By this exhaustive evaluation, it was discovered for the first time that host **3** had remarkable chiral recognition ability toward chiral ammonium ions. The $I_R/I_{S\text{-dn}}$ values of the chiral crown derivatives, permethylated carbohydrates, and the other chiral hosts (Figure 5.6) are summarized in Table 5.1.

The FABMS/EL guest method was applied for a simultaneous estimation of the chiral discrimination abilities of several chiral hosts toward a chiral ammonium ion on the basis of one mass spectrum.[45] In this case, the products by etherification of several chiral alcohols with bistosylate of diethylene glycol were used directly to estimate the chiral discrimination abilities, and it was confirmed experimentally that the $I_R/I_{S\text{-dn}}$ values in the three-component system involved a chiral host and an enantiomeric labeled/unlabeled guest pair that agreed with those in a multi-host/enantiomeric guest pair system. From the results, it seems that the FABMS/EL guest method using the multihost/enantiomeric guest pair system is one of the

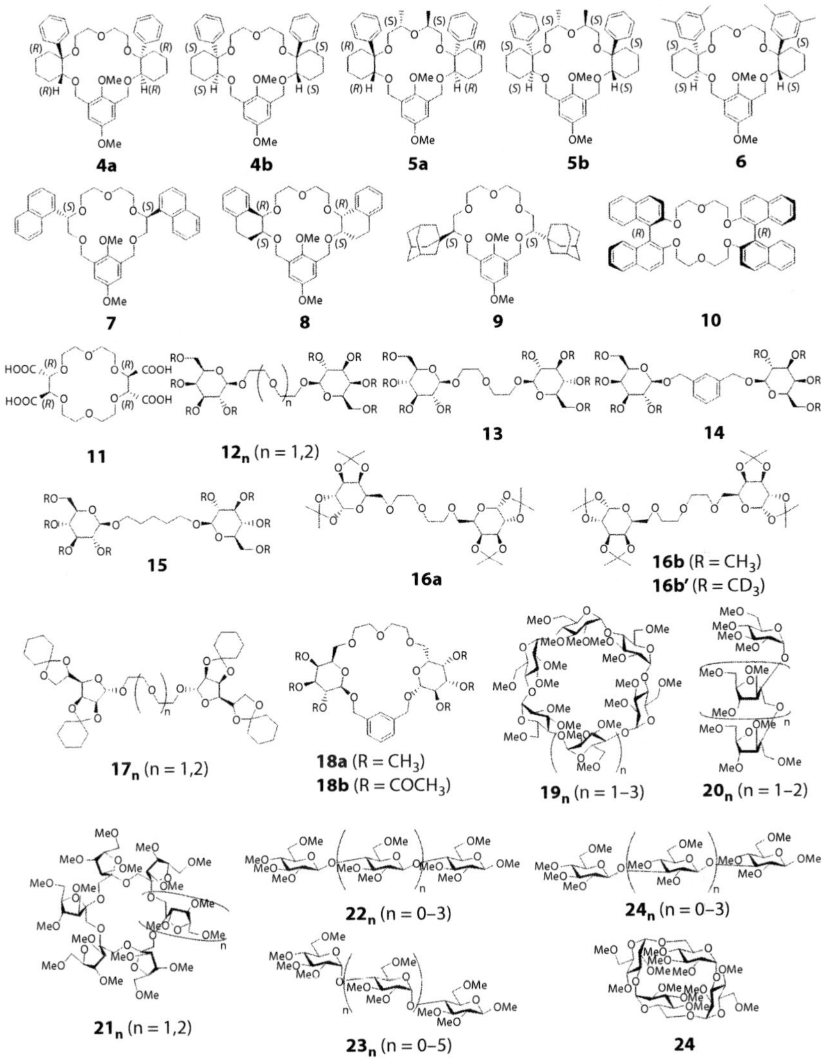


**FIGURE 5.6** Chiral hosts.

good screening methods for evaluating chiral recognition of chiral hosts. However, this system has some disadvantages. In the multihost/enantiomeric guest pair systems, chiral hosts, which have much weaker binding ability than other hosts, are not detected in the mass spectrum simultaneously because of the small relative peak intensity. The chiral discrimination ability of such chiral hosts should be evaluated individually by the FABMS/EL guest method in the three-component system.

**TABLE 5.1**

**$I_R/I_{S\text{-dn}}$ Values of Chiral Hosts with Chiral Guests in FABMS/EL Guest Method (1)[11,21,27,34]**

| | Chiral Guest | | | | | | | | | | | | | | |
|---|---|---|---|---|---|---|---|---|---|---|---|---|---|---|---|
| **Host** | **$AlaOMe^+$** | **$ValOMe^+$** | **$LeuOMe^+$** | **$IleOMe^+$** | **$TleOMe^+$** | **$ProOMe^+$** | **$MetOMe^+$** | **$PheOMe^+$** | **$TrpOMe^+$** | **$PglyOMe^+$** | **$NEA^+$** | **$PheOEt^+$** | **$PglyOEt^+$** | **$PheO^iPr^+$** | **$PglyO^iPr^+$** |
| 2a | 1.58 | 2.69 | 1.27 | 2.07 | 2.77 | 0.89 | 1.57 | 1.93 | 1.53 | 1.94 | 1.2 | 2.25 | 2.09 | 1.67 | |
| 2b | | | | | | | 0.66 | 0.52 | | 0.5 | 0.89 | | 0.48 | | |
| 4a | 4.00 | 5.03 | 3.16 | 3.62 | nd | 0.65 | 5.35 | 4.37 | 3.49 | 1.15 | | 5.03 | | 3.66 | |
| 4b | 0.26 | 0.20 | | | | | 0.22 | 0.22 | | 0.91 | 0.65 | | | | |
| 5 | 0.93 | 6.93 | | 3.36 | | | | 1.87 | | 1.33 | | 2.09 | | 2.45 | 1.36 |
| 6 | 1.27 | 2.17 | | 2.72 | | | | 0.64 | 0.72 | 1.41 | | 0.63 | | 0.71 | 1.41 |
| 7 | 0.51 | 0.5 | | | | | | 0.38 | | 1.25 | 0.63 | | | | |
| 8 | 1.59 | 1.14 | 1.39 | 1.12 | 1.12 | 1.02 | 1.5 | 1.04 | 1.35 | 1.11 | | 1.04 | | 1.16 | |
| 9 | 0.75 | 0.85 | | 0.81 | | nd | | | | 0.54 | 1.17 | 0.71 | | 0.81 | 0.61 |
| 10 | | | | | | | 1.61 | 1.61 | | 1.41 | | | | | |
| 11 | | 0.90 | | | | | | | | | | | | | |
| 12 | 0.96 | 1.12 | 1.2 | | | | 0.96 | 0.97 | 1.02 | 1.39 | 1.23 | | | | |
| 17a | 1.15 | | 1.11 | | | | | 1.27 | 2.30 | 0.62 | 0.63 | | | | |

*Note:* Tle = *tert*-leucine, NEA = 1-(1-naphthyl)ethylamine, Pgly = phenylglycine (1-amino-1-phenyl-acetic acid). The errors of the $I_R/I_{S\text{-dn}}$ values depend on the relative peak intensity of the host-guest complex ions for the base peak in the mass spectra. In the case of the above data, the errors of the $I_R/I_{S\text{-dn}}$ values are ±4%.

## 5.4 FABMS/ENANTIOMER-LABELED (EL) HOST METHOD

### 5.4.1 Basic Concept of FABMS/EL Host Method

The optical purity, i.e., the enantiomeric excess (*ee*), of chiral compounds can be determined by chiral chromatography,[3,4] capillary electrophoresis,[46] and spectrometric methods such as NMR[2] etc. Since the 1990s, the evaluation of *ee* by mass spectrometry has been paid great attention because of the high sensitivity and facility of the measurements.[18,19,47–50] In particular, the optical purity of α-amino acids was determined by Tao and Cooks et al. with high accuracy based on the CID of Cu–(amino acid) complex ions using the ESIMS/MS ion trap.[5]

In *ee* determination by the FABMS/EL method, one enantiomer of a chiral host is labeled with stable isotope atoms such as deuterium. This is called the FABMS/EL host method.[51,52] Here, an *S* enantiomer of a chiral host is labeled with deuterium atoms. For the evaluation of the optical purity of the chiral guest, the relative peak intensity of the diastereomeric host-guest complex ions, which are produced from complexation between an equimolar mixture of deuterium-labeled *S* and unlabeled *R* hosts ($H_{S\text{-}dn}$ and $H_R$) and a chiral guest, is taken as a quantitative measure; n is the number of deuterium labels. The fundamental concept of this methodology is schematically shown in Figure 5.7, where the diastereomeric host-guest complex ion peaks are given. For the conceptual data in the figure, the *R* guest forms a complex with the *R* host by an arbitrary factor of 2.0 ($I_R/I_{S\text{-}dn} = 2.0$) better than the *S* guest. Accordingly, the *S* guest should form a complex with the *S* host by a factor of 2.0 ($I_R/I_{S\text{-}dn} = 0.5$) better than the *R* host because of the mirror image relationship between the host-guest complex ions. Furthermore, the racemic guest should provide a pair of peaks of equal intensity because of the net compensation of a racemic host–racemic guest combination. Therefore, in the case of a given guest with unknown *ee*, one can determine the percent of enantiomeric excess (% *ee*) from the relative peak intensity of the host-guest complex ions.

In the mass spectra by the FABMS/EL host method, the relative peak intensity of two complex ions, (**2a** + G)$^+$ and (**2b**, + G)$^+$, changed depending on the % *ee* of the guests (Figure 5.8a). The intensity excess (*Ie*) values calculated from peak intensities of their complex ions showed a good linear correlation with the actual optical purity (% *ee*), as shown in Figure 5.8b. In the case of Phe-O-$^i$Pr$^+$ (Cl$^-$), the correlation coefficient was $R^2 = 0.9988$ (n = 9). The definition of *Ie* values corresponds to that of enantiomeric excess (*ee*: $|[R] - [S]|/|[R] + [S]|$), as in the following Equation 5.4:

$$Ie = [I(H_R + G)^+ - I(H_{S\text{-}dn} + G)^+]/[I(H_R + G)^+ + I(H_{S\text{-}dn} + G)^+] \quad (5.4)$$

Therefore, the optical purity of an *ee* unknown guest can be determined using the equation $\%\ ee = |Ie|/|Ie_{100}| \times 100$. Sawada et al. mathematically proved that the *Ie* values linearly correspond to the *ee* values.[53] The accuracy of the estimated optical purity strongly depends on the magnitude of the absolute *Ie* value ($|Ie_{100}|$) at 100% *ee*. The $|Ie_{100}|$ values are correlated with the $I_R/I_{S\text{-}dn}$ values, as in the following equation:

$$|Ie_{100}| = (I_R/I_{S\text{-}dn} - 1)/(I_R/I_{S\text{-}dn} + 1) \quad (5.5)$$

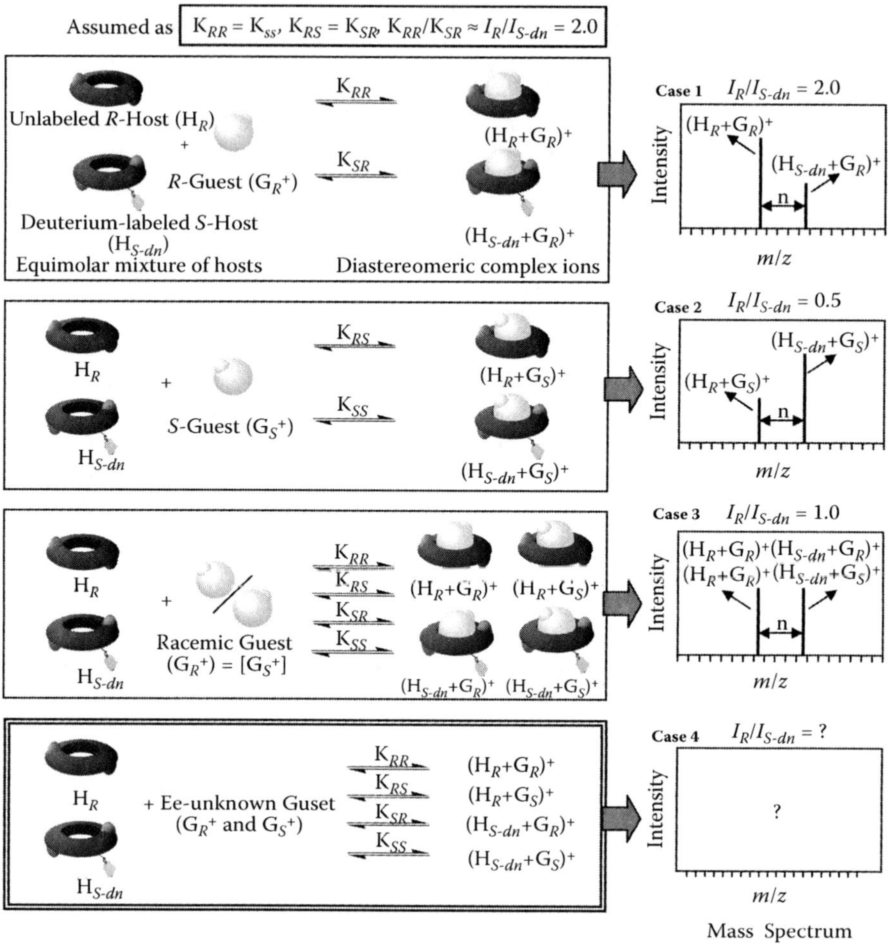


**FIGURE 5.7** Illustration of concept of the FABMS/EL host method.

The $|Ie_{100}|$ values of the host pairs toward various chiral ammonium ions are summarized in Table 5.2.[51,52,54] An $|Ie_{100}|$ value larger than 0.1 is a necessary condition for evaluation of the optical purity of the chiral guests.

When the corresponding deuterium-labeled *S* (or *R*) enantiomer of a target guest was used as the internal standard, the *ee* values had even higher accuracy from a single mass spectrum in the FABMS/EL host method.[55]

### 5.4.2 Relationship between FABMS/EL Host Method and the Guest Method

As described above, the $I_R/I_{S\text{-dn}}$ values evaluated by the FABMS/EL host method are compatible with the *Ie* values, Table 5.3. If the $I_R/I_{S\text{-dn}}$ values reflect the behaviors of host-guest complexation in a matrix, the $I_R/I_{S\text{-dn}}$ values and the $|Ie_{100}|$ values

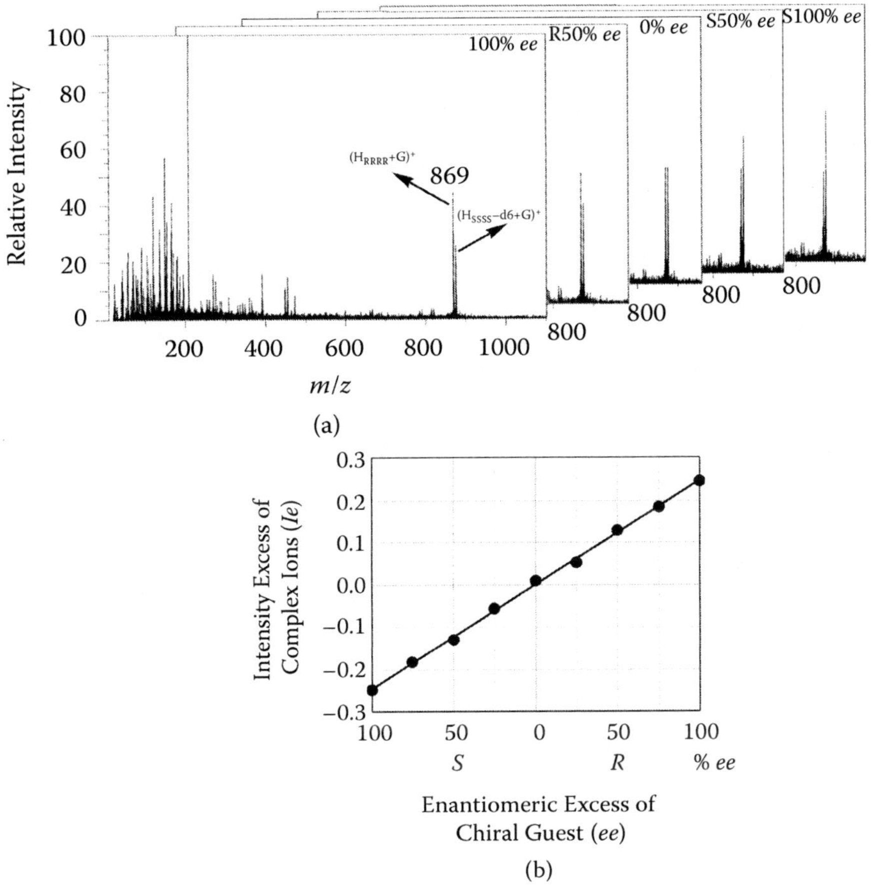


**FIGURE 5.8** (a) FAB mass spectra in the FABMS/EL host method and (b) the correlation between optical purity (% *ee*) of chiral guest and *Ie* values. Hosts, **2a**/**2b**, (equimolar mixture); guest, Phe-O-$^{i}$Pr$^{+}$. The correlation coefficient ($R^2$) of the fitted line is 0.9988 (n = 9). (Dr. M. Shizuma measured the mass spectra for this book.)

determined by the FABMS/EL host method must correspond to those by the FABMS/EL guest method. In Figure 5.9 the $|Ie_{100}|$ values of several chiral hosts evaluated by the FABMS/EL host method are plotted against the $|Ie_{100}|$ values derived from the $I_R/I_{S\text{-dn}}$ values determined by the FABMS/EL guest method. Both $|Ie_{100}|$ values agreed well with each other. For example, the $|Ie_{100}|$ values in the host method and guest method for host **3** and NEA$^{+}$ were 0.220 and 0.227, respectively. The $|Ie_{100}|$ values for host **3** and Trp-O-$^{i}$Pr$^{+}$ were 0.130 and 0.119, respectively. Naturally, the agreement of the $|Ie_{100}|$ values by the FABMS/EL guest method with those by the FABMS/EL host method indicates that the results of the latter method are due to chiral host-guest complexation. The enantioselective complexation of chiral crown ethers with amino sugars and dipeptides was estimated by the FABMS/EL host method.[56,57]

**TABLE 5.2**

**$I_R/I_{S\text{-}dn}$ Values of Chiral Hosts with Chiral Guests in FABMS/EL Guest Method (2)[29–35]**

| | Chiral Guest | | | | | | | | | | |
|---|---|---|---|---|---|---|---|---|---|---|---|
| Host | $AlaOiPr^+$ | $ValOiPr^+$ | $TleOiPr^+$ | $ProOiPr^+$ | $MetOiPr^+$ | $AspOiPr^+$ | $SerOiPr^+$ | $PheOiPr^+$ | $TrpOiPr^+$ | $PglyOiPr^+$ | $NEA^+$ |
| 3 | 0.45 | 0.14 | 0.33 | 1.23 | 0.28 | | | 0.18 | 0.56 | 0.26 | |
| $13_1$a | 0.94 | 0.18 | 1.10 | 0.91 | 0.87 | 1.10 | | 0.58 | 0.45 | 1.00 | 1.40 |
| $13_1$b | 0.84 | 0.18 | 0.98 | 0.91 | 0.93 | 1.10 | | 1.10 | 1.10 | 1.50 | 0.73 |
| $13_2$a | | | | | | | | | | | |
| $13_2$b | 0.83 | 0.98 | 1.10 | 0.74 | 0.87 | 1.3 | | 0.63 | 0.57 | 1.10 | 1.40 |
| 14a | 0.83 | 0.83 | 0.97 | 0.95 | 0.72 | 1.3 | | 0.49 | 0.43 | 0.75 | 1.50 |
| 14b | 1.10 | 0.88 | 1.10 | 1.10 | 0.97 | 1.2 | | 0.95 | 1.2 | 0.84 | 0.96 |
| 15a | 0.93 | 0.95 | 1.10 | 1.00 | 0.98 | 1.1 | | 0.86 | 0.83 | 0.89 | 1.10 |
| 15b | 0.98 | 0.86 | 1.10 | 1.00 | 0.94 | 1.1 | | 0.99 | 0.96 | 0.89 | |
| 16a | 0.92 | 0.93 | 0.94 | 1.00 | 0.92 | 0.99 | | 0.87 | 0.87 | 0.86 | 1.20 |
| 16b | 0.95 | 1.00 | 0.98 | 1.00 | 1.00 | 1.10 | | 1.00 | 0.97 | 1.00 | 1.00 |
| 17a | 1.43 | 0.85 | 1.03 | 1.98 | 1.33 | 1.00 | | 1.78 | 2.64 | 0.76 | 0.63 |
| $18_1$ | | | | | | | | | | | |
| $18_2$ | | | | | | | | | | | |
| 19a | 1.40 | 1.18 | 1.63 | 1.09 | 1.27 | 0.90 | | 0.85 | 1.91 | 0.55 | 0.96 |
| 19b | 1.46 | 1.39 | 1.57 | 1.77 | 1.5 | 0.96 | | 1.88 | 2.09 | 0.96 | 1.61 |
| $20_1$ | 0.99 | 1.00 | 0.95 | 1.07 | 0.91 | | 0.95 | 1.02 | 1.29 | 0.94 | |
| $20_2$ | 1.06 | 1.00 | 0.94 | 1.07 | 0.91 | | 1.15 | 1.01 | 1.23 | 0.91 | |
| $20_3$ | 0.90 | 1.00 | 0.93 | 1.14 | 0.92 | | 0.99 | 1.00 | 1.17 | 0.89 | |
| $21_1$ | 1.11 | 1.19 | 1.23 | 1.16 | 1.18 | | | 1.15 | 1.18 | 0.93 | |
| $21_2$ | 1.16 | 0.87 | 0.85 | 1.19 | 1.54 | | | 1.08 | 1.28 | 1.56 | |
| $22_1$ | 0.94 | 1.28 | 1.18 | 1.08 | 1.04 | | 1.01 | 1.00 | 1.38 | 0.99 | |
| $22_2$ | 0.94 | 0.88 | 1.00 | 1.16 | 0.95 | | 1.18 | 1.01 | 1.38 | 0.76 | |

| | | | | | | | | | | |
|---|---|---|---|---|---|---|---|---|---|---|
| $23_0$ | 0.99 | 1.00 | 0.97 | 0.94 | 1.04 | | 1.02 | 0.97 | 0.62 | 0.9 |
| $23_1$ | 0.97 | 0.97 | 0.98 | 1.09 | 1 | | 1.03 | 1.01 | 0.57 | 0.91 |
| $23_2$ | 0.94 | 0.98 | 0.95 | 1.08 | 0.97 | | 1.03 | 0.98 | 0.59 | 0.89 |
| $23_3$ | 0.94 | 0.93 | 1.05 | 1.06 | 0.98 | | 1.03 | 0.96 | 0.49 | 0.89 |
| $24_0$ | 1.02 | 0.96 | 0.95 | 1.16 | 1.07 | | 1.07 | 1.06 | 1.06 | 0.98 |
| $24_1$ | 1.07 | 0.97 | 0.95 | 1.07 | 1.12 | | 1.10 | 1.05 | 1.12 | 0.91 |
| $24_2$ | 1.05 | 1.01 | 0.99 | 1.08 | 1.08 | | 1.03 | 1.02 | 1.13 | 0.91 |
| $24_3$ | 1.03 | 0.98 | 0.95 | 1.07 | 1.08 | | 1.00 | 1.01 | 1.15 | 0.86 |
| $24_4$ | | 1.02 | 0.92 | 1.17 | 1.08 | | | 1.06 | 1.17 | 0.87 |
| $24_5$ | | 1.03 | 0.97 | 1.08 | 1.15 | | | 1.01 | 1.14 | 0.86 |
| $25_0$ | 1.02 | 0.97 | 0.92 | 1.07 | 1 | | 1.05 | 1.05 | 0.98 | 0.99 |
| $25_1$ | 1.05 | 1.07 | 0.98 | 1.09 | 1.12 | | 1.07 | 1.03 | 0.88 | 1.01 |
| $25_2$ | 1.03 | 1.01 | 0.99 | 1.06 | 1.07 | | 1.04 | 1.02 | 0.78 | 0.98 |
| $25_3$ | 1.04 | 0.97 | 0.91 | 1.16 | 1.01 | | 1.06 | 0.96 | 0.79 | 0.96 |
| 26 | 0.90 | 0.33 | | 3.72 | 0.54 | | | 0.68 | 0.81 | 0.48 |

*Note:* Tle = *tert*-leucine, NEA = 1-(1-naphthyl)ethylamine, Pgly = phenylglycine (1-amino-1-phenyl-acetic acid). The errors of the $I_R/I_{S\text{-dn}}$ values are ±4%.

## TABLE 5.3
## $Ie_{100}$ and $I_R/I_{S\text{-}dn}$ Values in FABMS/EL Host Method

| Host Pair | Guest | $Ie_{100}$ Value | $I_R/I_{S\text{-}dn}$ Value | Ref. |
|---|---|---|---|---|
| **2a/2b'** | Gly-O-Me$^+$ | 0.000 | 1.00 | 51 |
| | (*R*)-Pgly-O-Me$^+$ | 0.296 | 1.84 | 51 |
| | (*S*)-Pgly-O-Me$^+$ | 0.307 | 0.53 | 51 |
| | (*R*)-Pgly-O-Et$^+$ | 0.329 | 1.98 | 51 |
| | (*R*)-Pgly-O-$^i$Pr$^+$ | 0.333 | 2.00 | 51 |
| | (*S*)-Asp-O-Me$^+$ | 0.460 | 0.37 | 51 |
| | (*S*)-Asn-O-Me$^+$ | 0.325 | 0.51 | 51 |
| | (*S*)-Phe-O-Me$^+$ | 0.325 | 0.51 | 51 |
| | (*R*)-Val-O-Me$^+$ | 0.458 | 2.69 | *a* |
| | (*S*)-Val-O-Et$^+$ | 0.481 | 0.35 | 51 |
| | (*R*)-Ala-O-Me$^+$ | 0.225 | 1.58 | *a* |
| | (*R*)-Tle-O-Me$^+$ | 0.469 | 2.77 | *a* |
| | (*R*)-Ile-O-Me$^+$ | 0.349 | 2.07 | *a* |
| | (*R*)-Met-O-Me$^+$ | 0.219 | 1.56 | *a* |
| | (*R*)-Cys-O-Me$^+$ | 0.228 | 1.59 | *a* |
| | (*R*)-His-O-Me$^+$ | 0.167 | ca. 1.4 | *a* |
| | (*R*)-Tys-O-Me$^+$ | 0.283 | 1.79 | *a* |
| | (*R*)-Lys-O-Me$^+$ | 0.048 | 1.10 | *a* |
| | (*R*)-Arg-O-Me$^+$ | 0.091 | ca. 1.2 | *a* |
| | (*R*)-1-phenylethylamine | 0.026 | 0.95 | 51 |
| | (*R*)-1-phenyl-2-hydroxyethylammonium ion | 0.235 | 0.62 | 51 |
| | (*R*)-1-(4-nitrophenyl)ethylammonium ion | 0.015 | 1.03 | 51 |
| **4a/4b'** | (*S*)-Leu-O-Me$^+$ | 0.408 | 0.42 | 52 |
| **17a/17b'** | (*S*)-NEA$^+$ | 0.206 | 1.52 | 34 |
| | (*R*)-NEA$^+$ | 0.220 | 0.64 | 34 |
| | (*R*)-*N*-benzyl-*N*-1-phenylethylammonium ion | 0.149 | 0.74 | 34 |
| | (*R*)-2-butylammonium ion | 0.020 | 0.96 | 34 |
| | (*S*)-2-butylammonium ion | 0.015 | 0.97 | 34 |
| | (*S*)-1-phenylethylammonium ion | 0.005 | 0.99 | 34 |
| | (*S*)-Trp$^+$ | 0.010 | 0.98 | 34 |
| | (*S*)-Phe$^+$ | 0.064 | 0.88 | 34 |
| | (*S*)-Asp$^+$ | 0.020 | 0.96 | 34 |
| | (*R*)-Trp-O-Me$^+$ | 0.217 | 1.55 | 34 |
| | (*S*)-Trp-O-$^i$Pr$^+$ | 0.408 | 0.42 | 34 |
| | (*S*)-Pro-O-$^i$Pr$^+$ | 0.322 | 1.95 | 54 |
| | (*S*)-Phe-O-Me$^+$ | 0.130 | 0.77 | 34 |
| | (*S*)-Val-O-Me$^+$ | 0.197 | 1.49 | 34 |
| | (*S*)-Pro-O-Bn$^+$ | 0.190 | 0.68 | 34 |

*Note:* *a* = unpulished data, Pgly = phenylglycine (1-amino-1-phenyl-acetic acid), Tle = *tert*-leucine, NEA = 1-(1-naphthyl)ethylamine, Bn = benzyl.

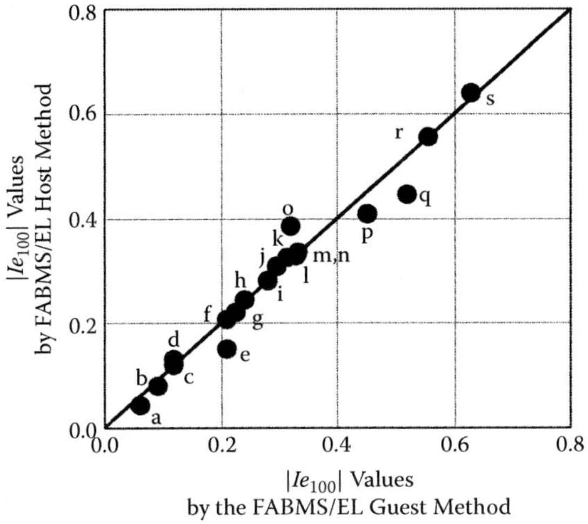


**FIGURE 5.9** Correlation between $|Ie_{100}|$ values by the FABMS/EL guest method and those by the FABMS/EL host method. Host, guest: (a) **16**, Ala-O-$^{i}$Pr$^{+}$; (b) **2**, NEA$^{+}$; (c) **4**, Leu-O-Me$^{+}$; (d) **16**, Phe-O-Me$^{+}$; (e) **4**, NEA$^{+}$; (f) **2**, Trp-O-Me$^{+}$; (g) **16**, NEA$^{+}$; (h) **2**, Phe-O-$^{i}$Pr$^{+}$; (i) **16**, Phe-O-$^{i}$Pr$^{+}$; (j) **2**, Pgly-O-Me$^{+}$; (k) **2**, Phe-O-Me$^{+}$; (l) **2**, Pgly-O-Et$^{+}$; (m) **16**, Pro-O-$^{i}$Pr$^{+}$; (n) **2**, Pgly-O-$^{i}$Pr$^{+}$; (o) **2**, Phe-O-Et$^{+}$; (p) **16**, Trp-O-$^{i}$Pr$^{+}$; (q) **4**, Leu-O-Me$^{+}$; (r) **4**, Trp-O-Me$^{+}$; (s) **4**, Phe-O-Me$^{+}$. References: 21, 51, 52 (for host **2** and **4**); 31, 54 (for host **16**). NEA = 1-(1-naphthyl)-ethylamine, Pgly = 1-amino-1-phenyl-acetic acid.

## 5.5 ESI AND MALDI MS/ENANTIOMER-LABELED (EL) METHOD

Chiral recognition of the chiral host was examined by the EL guest method using other ionization of mass spectrometry, such as ESI and MALDI. In ESIMS, the diastereomeric host-guest complex ion peaks were detected with good sensitivity. However, the $I_R/I_{S\text{-}dn}$ values in ESIMS were depressed in comparison with those in FABMS.[58,59] Moreover, the magnitude of depression of the $I_R/I_{S\text{-}dn}$ values changed depending on the mechanical structure of the ion source, the structure of the chiral host, and the chiral guest.[60] Thus, the chiral recognition ability of the chiral host can be estimated only qualitatively using the ESIMS/EL guest method. The reproducibility of the $I_R/I_{S\text{-}dn}$ in ESIMS is very high under the same instrumental and sample concentration conditions.[58–60] Therefore, the optical purity of the chiral guest was evaluated with high accuracy by the ESIMS/EL host method. The *ee* values of the chiral carboxylate guest were estimated successfully by the ESIMS/EL host method using a host pair, which was a lanthanum complex ion with a chiral ligand containing nitrogen atoms, as shown in Figure 5.10.[61,62]

In the MALDIMS/EL guest method, all $I_R/I_{S\text{-}dn}$ values were unity because of interruption of the competitive host-guest complexation by crystallization processes.[63]

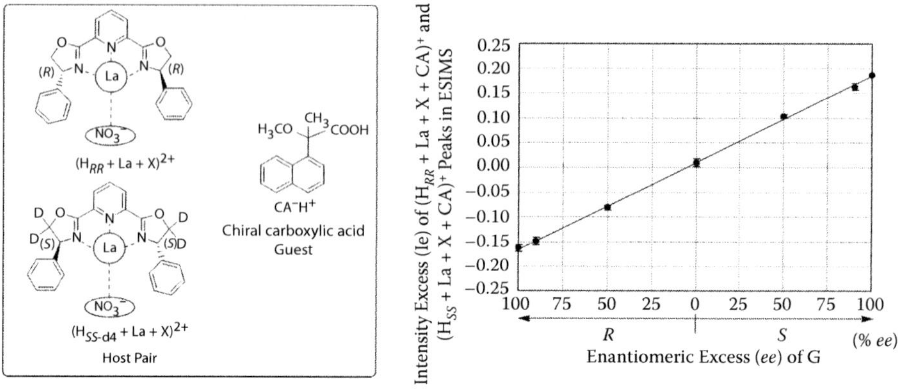


**FIGURE 5.10** Deuterium-labeled *S,S* and unlabeled *R,R* host pair for evaluating the optical purity of carboxylic acid, and correlation between the *Ie* values of the complex ion peaks and the actual *ee* of the guest. (Reproduced by permission of John Wiley & Sons, Ltd., from *J. Mass Spectrom.* 41 [2006]: 266–68.)

## 5.6 CONCLUSION

Chiral recognition of the chiral host was evaluated quantitatively using the FABMS/EL guest method. In particular, this method is a powerful tool for the screening of the chiral recognition ability of new natural/synthetic host compounds because of the single-stage detection in MS, while the optical purity of the chiral guest is estimated easily by the FABMS/EL host method. In both methods, chiral recognition ability of the hosts is the key point for the accuracy of quantitative analyses and coverage of the target chiral compounds.

## REFERENCES


1. Cram, D. J., Cram, J. M. 1974. Host-guest chemistry. Complexes between organic compounds simulate the substrate selectivity of enzymes. *Science* 183:803–9.
2. Parker, D. 1991. NMR determination of enantiomeric purity. *Chem. Rev.* 91:1441–57.
3. König, W. A. 1992. *Gas chromatographic enantiomer separation with modified cyclodextrin*. Heidelberg: Hüthig.
4. Hara, S., Cazes, J., eds. 1986. Optical resolution by liquid chromatography [Special issue]. *J. Liq. Chromatogr.* 9:241–94.
5. Tao, W. A., Zhang, D., Wang, F., Thomas, P. D., Cooks, R. G. 1999. Kinetic resolution of D, L-amino acid based on gas-phase dissociation copper (II) complexes. *Anal. Chem.* 71:4427–29.
6. Barber, M., Bordoll, R. S., Elliot, G., Sedgwick, R. D., Tyler, A. N. 1982. Fast atom bombardment mass spectrometry. *Anal. Chem.* 54:645A–57A.
7. Fenn, J. B., Mann, M., Meng, C. K., Wong, S. F., Whitehouse, C. M. 1989. Electrospray ionization for mass spectrometry of large biomolecules. *Science* 246:64–71.
8. Karas, M., Bachman, D., Bahr, U., Hillenkamp, F. 1987. Matrix-assisted ultraviolet laser desorption of non-volatile compounds. *Int. J. Mass Spectrom. Ion Proc.* 78:53–68.
9. Baldwin, M. A., Howell, S. A., Welham, K. J., Winker, F. J. 1988. Identification of chiral isomers by fast atom bombardment/mass spectrometry: dialkyl tartrates. *Biomed. Environ. Mass Spectrom.* 16:357–60.

10. Sawada, M., Shizuma, M., Takai, Y., Yamada, H., Kaneda, T., Hanafusa, T. 1992. Enantioselectivity in FAB mass spectrometry. *J. Am. Chem. Soc.* 114:4005–6.
11. Sawada, M., Takai, Y., Yamada, H., Kaneda, K., Mizooku, T., Hirose, K., Tobe, Y., Naemura, K. 1994. Chiral recognition in molecular complexation for the crown ether-amino ester system. A facile FAB mass spectrometric approach. *J. Chem. Soc. Chem. Commun.* 2497–98.
12. Sawada, M., Okumura, Y., Shizuma, M., Takai, Y., Hidaka, Y., Yamada, H., Tanaka, T., Kaneda, T., Hirose, K., Misumi, S., Takahashi, S. 1993. Enantioselective complexation of carbohydrate or crown ether hosts with organic ammonium ion guests detected by FAB mass spectrometry. *J. Am. Chem. Soc.* 115:7381–88.
13. Sawada, M., Okumura, Y., Yamada, H., Takai, Y., Takahashi, S., Kaneda, T., Hirose, K., Misumi, S. 1993. Cross-chiral examinations of molecular enantioselective recognition by fast atom bombardment mass spectrometry: Host-guest complexations between chiral crown ethers and chiral organic ammonium ions. *Org. Mass Spectrom.* 28:1525–28.
14. Sawada, M. 1994. In *Biological mass spectrometry, present and future*, ed. T. Matsuo, R. M. Caprioli, M. L. Gross, Y. Seyama, 639–46. New York: John Wiley & Sons.
15. Sawada, M. 1996. Advances in the application of FAB mass spectrometry to studies on chiral recognition of chiral crown ethers. [In Japanese]. *Dojin News* 78:10–17.
16. Pócsfalvi, G., Lipták, M., Huszthy, P., Bradshaw, J. S., Izatt, R. M., Vékey, K. 1996. Characterization of chiral host–guest complexation in fast atom bombardment mass spectrometry. *Anal. Chem.* 68:792–95.
17. Dobó, A., Lipták, M., Huszthy, P., Vékey, K. 1997. Chiral recognition via host-guest interactions detected by fast-atom bombardment mass spectrometry: Principles and limitations. *Rapid Commun. Mass Spectrom.* 11:889–96.
18. Davey, S. N., Leigh, D. A., Smart, J. P., Tetler, L. W., Truscello, A. M. 1996. Fast atom bombardment mass spectrometry as a tool for the rapid determination of enantioselective binding of methylated cyclodextrins. *Carbohydr. Res.* 290:117–23.
19. Liang, Y., Bradshaw, J. S., Izatt, R. M., Pope, R. M., Dearden, D. V. 1999. Analysis of enantiomeric excess using mass spectrometry: Fast atom bombardment/sector and electrospray ionization/Fourier transform mass spectrometric approaches. *Int. J. Mass Spectrom.* 185:977–88.
20. Krishna, P., Prabhaker, S., Vairamani, M., Monoharan, M., Jemmis, E. D. 1999. Chiral recognition and the determination of optical purity of a-phenylethylamine using monosaccharide as a chiral selector under liquid secondary ion mass spectral conditions. *Eur. J. Mass Spectrom.* 5:485–88.
21. Sawada, M., Takai, Y., Yamada, H., Hirayama, S., Kaneda, T., Tanaka, T., Kamada, K., Mizooku, T., Takeuchi, S., Ueno, K., Hirose, K., Tobe, Y., Naemura, K. 1995. Chiral recognition in host-guest complexation determined by the enantiomer-labeled guest method using fast atom bombardment mass spectrometry. *J. Am. Chem. Soc.* 117:7726–36.
22. Sawada, M. 1997. Chiral recognition detected by fast atom bombardment mass spectrometry. *Mass Spectrom. Rev.* 16:73–90.
23. Sawada, M. 1997. Chiral recognition in host-guest complexation determined by FAB mass spectrometry [in Japanese]. *J. Mass Spectrom. Soc. Jpn.* 45:439–58.
24. Shizuma, M. 1998. Detection of chiral discrimination ability of chiral crown ether by FAB mass spectrometry [in Japanese]. *J. Mass Spectrom. Soc. Jpn.* 46:211–18.
25. Sawada, M. 2002. Development of quantitative chiral recognition mass spectrometry [in Japanese]. *J. Mass Spectrom. Soc. Jpn.* 50:311–29.
26. Sawada, M. 2005. In *The encyclopedia of mass spectrometry*, ed. M. L. Gross, R. M. Caprioli, N. M. M. Nibbering, 740–47. Vol. 4. Amsterdam: Elsevier.

27. Sawada, M., Hagita, K., Imamura, H., Tabuchi, H., Yodoya, S., Umeda, M., Takai, Y., Yamada, H., Yamaoka, H., Hirose, K., Tobe, Y., Tanaka, T., Takahashi, S. 2000. Chiral recognition ability of crown ethers toward organic amine compounds: FAB mass spectrometry coupled with the enantiomer-labeled guest method [in Japanese]. *J. Mass Spectrom. Soc. Jpn.* 48:323–32.
28. 1995. Analytical currents. *Anal. Chem.* 67:586A.
29. Shizuma, M., Adachi, H., Kawamura, M., Takai, Y., Takeda, T., Sawada, M. 2001. Chiral discrimination of fructo-oligosaccharides toward amino acid derivatives by induced-fitting chiral recognition. *J. Chem. Soc. Perkin Trans.* 2:592–601.
30. Sawada, M., Shizuma, M., Takai, Y., Adachi, H., Takeda, T., Uchiyama, T. 1998. Measurement of chiral amino acid discrimination by cyclic oligosaccharides: A direct FAB mass spectrometric approach. *Chem. Commun.* 1453–54.
31. Shizuma, M., Adachi, H., Amemura, A., Takai, Y., Takeda, T., Sawada, M. 2001. Chiral discrimination of permethylated gluco-oligosaccharide toward amino acid ester salts. *Tetrahedron* 57:4567–78.
32. Shizuma, M., Adachi, H., Takai, Y., Hayashi, M., Tanaka, T., Takeda, T., Sawada, M. 2001. Combinatorial evaluation of the chiral discrimination of permethylated carbohydrates using fast-atom bombardment mass spectrometry. *Carbohydr. Res.* 335:275–81.
33. Shizuma, M., Kiso, T., Terauchi, H., Takai, Y., Yamada, H., Nishimoto, T., Ono, D., Shimomura, O., Nomura, R., Miwa, Y., Nakamura, M., Nakano, H. 2008. Evaluation of chiral amino acid discrimination by a permethylated cyclic tetrasaccharide, cyclo-{→6)-α-D-Glcp-(1→3)-α-D-Glcp-(1→6)-α-D-Glcp-(1→3)-α-D-Glcp-(1→}, using FAB mass spectrometry. *Chem. Lett.* 37:1054–55.
34. Shizuma, M., Kadoya, Y., Takai, Y., Imamura, H., Yamada, H., Takeda, T., Arakawa, R., Takahashi, S., Sawada, M. 2002. New artificial host compounds containing galactose end groups for binding chiral organic amine guests: Chiral discrimination and their complex structures. *J. Org. Chem.* 67:4795–807.
35. Shizuma, M., Ohta, M., Yamada, H., Takai, Y., Nakaoki, T., Takeda, T., Sawada, M. 2002. Enantioselective complexation of chiral linear hosts containing monosaccharide moieties with chiral organic amines. *Tetrahedron* 58:4319–30.
36. Shizuma, M., Ono, D., Nakamura, M., Yamada, H., Takai, Y., Sawada, M. 2005. Chiral discrimination ability of a simple chiral linear host toward chiral amino acid derivatives [in Japanese]. *Kagaku To Kogyo* (Osaka) 79:397–402.
37. Sato, H., Shizuma, M. 2008. Triazole-linked host compounds for chiral-discrimination toward amino acid ester guests. *J. Oleo Sci.* 57:503–11.
38. Easton, C. J., Lincoln, S. F. 1996. Chiral discrimination by modified cyclodextrins. *Chem. Soc. Rev.* 25:163–70.
39. Bang, E., Jung, J.-W., Lee, W., Lee, D. W., Lee, W. 2001. Chiral recognition of (18-crown-6)-tetracarboxylic acid as a chiral selector determined by NMR spectroscopy. *J. Chem. Soc. Perkin Trans.* 2:1685–92.
40. Machida, Y., Nishi, H., Nakamura, K. 1998. Nuclear magnetic resonance studies for the chiral recognition of the novel chiral stationary phase derived from 18-crown-6 tetracarboxylic acid. *J. Chromatogr. A* 810:33–41.
41. Sawada, M., Yamauchi, Y., Shizuma, M., Takai, Y., Nakano, K., Kuroda, M., Arakawa, R. 2000. Chiral recognition of 18-crown-6-tetracarboxylic acid toward amino acids and organic amines by fast atom bombardment mass spectrometry. A comparison with capillary electrophoresis [in English]. *J. Mass Spectrom. Soc. Jpn.* 48:380–86.
42. Shizuma, M., Yamada, H., Takai, Y., Sawada, M. 2005. Estimation of thermodynamic parameters on enantioselective complexation equilibria by temperature-dependent FAB mass spectrometry. *Chem. Lett.* 34:1182–83.

43. Shizuma, M., Yamada, H., Adachi, H., Takai, Y., Takeda, T., Sawada, M. 2002. Matrix effects on the chiral recognition determined by the relative peak intensity of diastereomeric host-guest complex ions using the FAB mass spectrometry/enantiomer labeled method [in English]. *J. Mass Spectrom. Soc. Jpn.* 50:290–94.
44. Kim, J.-K., Lee, J. G., Lee, S., Seo, J. J., Hong, J., Suh, H. 2002. Chiral molecular recognition in fast atom bombardment (FAB-MS) enantiomer labeled (EL) guest method using new chiral bis-piridino-18-crown-6. *Bull. Korean Chem. Soc.* 23:543–44.
45. Shizuma, M., Adachi, H., Ono, D., Sato, H., Nakamura, M. 2009. Direct screening of chiral discrimination abilities of chiral hosts using mass spectrometry. *Chirality* 21:324–30.
46. Fanali, S. 1997. Controlling enantioselectivity in chiral capillary electrophoresis with inclusion—Complexation. *J. Chromatogr. A* 792:227–67.
47. Guo, J., Wu, J., Siuzdak, G., Finn, M. G. 1999. Measurement of enantiomeric excess by kinetic resolution and mass spectrometry. *Angew. Chem. Int. Ed.* 38:1755–58.
48. Wu, Y.-N., Tu, Y.-P., Pan, Y.-J., Chen, Y.-Z., Cui, M., Song, F.-R., Liu, S.-Y. 1997. Stereochemical effects in mass spectrometry—Determination of the optical purity of enantiomers by mass spectrometry. *Anal. Lett.* 30:1399–406.
49. Grigorean, G., Ramirez, J., Ahn, S. H., Lebrilla, C. B. 2000. A mass spectrometry method for the determination of enantiomeric excess in mixtures of d,l-amino acids. *Anal. Chem.* 72:4275–81.
50. Fago, G., Filippi, A., Giardini, A., Laganà, A., Paladini, A., Speranza, M. 2001. Chiral recognition of O-phosphoserine by mass spectrometry. *Angew. Chem. Int. Ed.* 40:4051–54.
51. Sawada, M., Yamaoka, H., Takai, Y., Kawai, Y., Yamada, H., Azuma, T., Fujioka, T., Tanaka, T. 1998. Determination of enantiomeric excess for amino acid ester using FAB mass spectrometry. *Chem. Commun.* 1569–79.
52. Sawada, M., Yamaoka, H., Takai, Y., Kawai, Y., Yamada, H., Azuma, T., Fujioka, T., Tanaka, T. 1999. Determination of enantiomeric excess for organic primary amine compounds by chiral recognition fast-atom bombardment mass spectrometry. *Int. J. Mass Spectrom.* (Molecular Recognition Special Issue) 193:123–30.
53. Sawada, M., Takai, Y., Imamura, H., Yamada, H., Takahashi, S., Yamaoka, H., Hirose, K., Tobe, Y., Tanaka, J. 2001. Chiral recognizable host-guest interactions detected by fast-atom bombardment mass spectrometry: Application to the enantiomeric excess determination of primary amines. *Eur. J. Mass Spectrom.* 7:447–59.
54. Shizuma, M., Imamura, H., Takai, Y., Yamada, H., Takeda, T., Takahashi, S., Sawada, M. 2000. A new reagent to evaluate optical purity of organic amines by FAB mass spectrometry. *Chem. Lett.* 29:1292–93.
55. Shizuma, M., Imamura, H., Takai, Y., Yamada, H., Takeda, T., Takahashi, S., Sawada, M. 2001. Facile ee-determination from a single measurement by fast atom bombardment mass spectrometry: A double labeling method. *Int. J. Mass Spectrom.* 210/211:585–90.
56. Sawada, M., Nishiwaki, T., Yamaoka, H., Yamada, H., Takai, Y., Arakawa, R. 2000. Stereoisomer discrimination of some amino sugars: Chiral recognition FAB mass spectrometry coupled with the enantiomer-labeled host method [in Japanese]. *J. Mass Spectrom. Soc. Jpn.* 48:231–37.
57. Sawada, M., Kamei, A., Ueno, H., Yamada, H., Takai, Y., Shizuma, M., Yamaoka, H., Tanaka, J., Arakawa, R. 2004. Enantiomer excess determination of amines and dipeptides by ESI and FAB mass spectrometry coupled with the enantiomer-labeled host method [in Japanese]. *J. Mass Spectrom. Soc. Jpn.* 52:289–94.
58. Sawada, M., Takai, Y., Kaneda, T., Arakawa, R., Okamoito, M., Doe, H., Matsuo, T., Naemura, K., Hirose, K., Tobe, Y. 1996. Chiral molecular recognition in electrospray ionization mass spectrometry. *Chem. Commun.* 1735–36.

59. Sawada, M., Takai, Y., Yamada, H., Nishida, J., Kaneda, T., Arakawa, R., Okamoto, M., Hirose, K., Tanaka, T., Naemura, K. 1998. *J. Chem. Soc. Perkin Trans.* 2:701–10.
60. Sawada, M., Takai, Y., Yamada, H., Yoshikawa, M., Arakawa, R., Tabuchi, H., Takada, M., Tanaka, J., Shizuma, M., Yamaoka, H., Hirose, K., Fukuda, K., Tobe, Y. 2004. Depression of the apparent chiral recognition ability obtained in the host-guest complexation systems by electrospray and nano-electrospray ionization mass spectrometry. *Eur. J. Mass Spectrom.* 10:27–37.
61. Sawada, M., Nomura, S., Miyamoto, Y., Egawa, N., Shizuma, M., Yamada, H., Takai, Y., Tanaka, J., Yamaoka, H. 2004. Chiral recognition of carboxylic acids by ESI mass spectrometry [in Japanese]. *J. Mass Spectrom. Soc. Jpn.* 52:154–57.
62. Takai, Y., Iguchi, K., Yamada, H., Shizuma, M., Arakawa, R., Sawada, M. 2006. Enantiomeric excess determination of a chiral carboxylic acid using the enantiomer-labeled host method by electrospray ionization mass spectrometry. *J. Mass Spectrom.* 41:266–68.
63. Sawada, M., Harada, M., Takai, Y., Nakano, K., Kuroda, M., Arakawa, R. 2000. Measurement of chiral recognition properties of crown ethers using matrix-assisted laser desorption ionization mass spectrometry [in English]. *J. Mass Spectrom. Soc. Jpn.* 48:141–44.

# 6 Enantioselectivity in Gas-Phase Ion-Molecule Reactions

*Maurizio Speranza*

## CONTENTS

6.1 Introduction ........ 87
6.2 Experimental Methodologies ........ 89
6.3 Chiral Ion Recognition ........ 90
6.3.1 Enantioselective Reactions in Collisionally Activated Metal-Bound Complexes ........ 91
6.3.2 Enantioselective Reactions in Thermally Activated Proton-Bound Complexes ........ 95
6.3.2.1 Oligosaccharides as Chiral Hosts ........ 95
6.3.2.2 Cytochrome *c* as Chiral Host ........ 104
6.3.2.3 Resorcinarenes as Chiral Hosts ........ 104
6.3.2.4 Tetra-Amide Macrocycles as Chiral Hosts ........ 120
6.4 Conclusions and Outlook ........ 127
Acknowledgments ........ 128
References ........ 128

## 6.1 INTRODUCTION

A process is defined as *enantioselective* when one enantiomer of a chiral compound is formed or destroyed in preference to the other enantiomer. Since enantiomers have the same physicochemical properties in isotropic conditions, enantioselectivity shows up only in reactions occurring in anisotropic conditions, i.e., when the chiral compound interacts either with a dissymetric selector to form a diastereomeric pair or with circularly polarized electromagnetic radiations. The ability of the dissymetric probe to differentiate between two enantiomers of a chiral molecule is particularly important in biochemistry[1,2] and organic synthesis.[3] An enantioselective reaction yields an optically active product from achiral starting materials, using either a chiral catalyst, an enzyme, or a chiral reagent. The degree of selectivity is measured by the enantiomeric excess. An important variant is kinetic resolution, in which a preexisting chiral center undergoes reaction with a chiral catalyst, an enzyme, or a chiral reagent such that one enantiomer reacts faster than the other and leaves behind the less reactive enantiomer, or in which a preexisting chiral center influences the reactivity of a reaction center elsewhere in the same molecule.

In this context, several procedures can be followed to evaluate the enantioselectivity of a dissymetric selector M toward the enantiomeric forms $A^R$ and $A^S$ of a chiral molecule. One possibility is to let M react with the $A^R/A^S$ racemate under conditions allowing equilibration between the relevant diastereomeric [M•$A^R$] and [M•$A^S$] encounter complexes (Figure 6.1). In this case, if $\Delta\Delta G = \Delta\Delta G^{\ddagger}$, the enantioselectivity reflects exclusively their different thermodynamic stability, and therefore we speak of *thermodynamic enantioselectivity*. If instead $\Delta\Delta G = 0$ and $\Delta\Delta G^{\ddagger} \neq 0$, the enantioselectivity reflects exclusively the relative stability of the diastereomeric transition structures during evolution of the diastereomeric encounter complexes to the product. In this case, we speak of *kinetic enantioselectivity*. In the most frequent cases where $\Delta\Delta G \neq 0$ and $\Delta\Delta G^{\ddagger} \neq 0$, the process exhibits an enantioselectivity that is governed by both thermodynamic and kinetic factors. Finally, if $\Delta\Delta G \neq 0$ and $\Delta\Delta G^{\ddagger} = 0$, the process appears unselective even though involving diastereomeric encounter complexes of different stability.

Evaluation of the enantioselectivity can be also carried out by generating separately the [M•$A^R$] and [M•$A^S$] complexes from combination of M with the pure $A^R$ or $A^S$ enantiomers and by measuring their reaction kinetics. Since here equilibration between [M•$A^R$] and [M•$A^S$] is prevented, we can speak of *thermodynamic enantioselectivity* when $\Delta\Delta G \neq 0$ and $\Delta\Delta G^{\ddagger} = 0$, and of *kinetic enantioselectivity* when $\Delta\Delta G = 0$ and $\Delta\Delta G^{\ddagger} \neq 0$. Obviously, if $\Delta\Delta G \neq 0$ and $\Delta\Delta G^{\ddagger} \neq 0$, the kinetic measurements reflect an enantioselectivity that depends on both thermodynamic and kinetic factors. Finally, if $\Delta\Delta G = \Delta\Delta G^{\ddagger}$, differently stable [M•$A^R$] and [M•$A^S$] complexes will react at the same rate. The enantioselectivity of the processes described in this chapter has been mainly determined in this latter context.

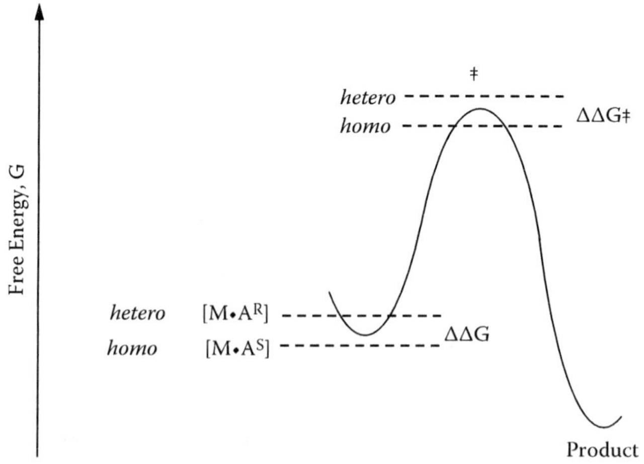


**FIGURE 6.1** Schematic free energy profile for the reaction between reactants M and A. The full line stands for a hypothetical $[M•A]^+$ encounter complex where either A or M is achiral, whereas the dotted lines refer to diastereomeric $[M•A]^+$ encounter complexes where A and M are both chiral and have the same (*homo*) or opposite (*hetero*) configuration.

The role of the solvent in asymmetric catalysis, in general, and in enzymatic catalysis, in particular, is hardly understood. Therefore, any piece of information concerning (1) the nature and the dynamics of shape-specific intermolecular interactions between functionalities located on the enzyme/molecule complementary surfaces and (2) the rate of acceleration due to partial or total desolvation of the functionalities themselves in the enzyme cavity[4,5] would be of great interest.

Such an ambitious task has been pursued through the use of many spectroscopic and mass spectrometric (MS) techniques, and this book is aimed at presenting the state of the art in this field. Although MS is traditionally regarded as a "blind" tool for stereochemical analysis, a body of evidence is currently available witnessing the potential of such a technique for structural and stereochemical studies.[6–18] At the same time, the ability of MS to characterize diastereomeric complexes in the absence of perturbing environmental factors and to measure ion abundance differences with high sensitivity and reproducibility makes it particularly attractive when small differences in the energetics and reactivity of diastereomeric species have to be determined.

Chiral recognition by MS is usually based on the measurement of (1) the relative abundance of noncovalent diastereomeric adducts between a chiral host and the two enantiomers (one isotopically labeled) of a guest,[8,12,19] (2) the relative stability of diastereomeric adducts by equilibrium measurements[13,15,20] or by collision-induced dissociation (CID) experiments (Cooks' kinetic method),[9,21–28] and (3) the rates of ion-molecule reactions between diastereomeric adducts and suitable chiral or achiral reactants.[14,29–33] Many of the chapters of this book illustrate the state of the art of the research on themes 1 and 2. Most important studies on theme 3 are described in this chapter. The first part of the chapter provides a brief description of the MS techniques employed. Readers willing to go deeper into the instrumental details are urged to refer to more specialized books.[34] Above all, this book is addressed to scientists interested in chiral recognition processes who are willing to implement their current studies with gas-phase investigations.

## 6.2 EXPERIMENTAL METHODOLOGIES

An ever-growing array of techniques are being implemented to generate chiral ionic clusters and measure their properties, including electrospray ionization (ESI), matrix-assisted laser desorption and ionization (MALDI), and fast atom bombardment (FAB) sources, coupled with tandem mass spectrometers. ESI refers to a sequence of complex processes by which an intense electric field disperses a sample liquid into a bath gas as a fine spray of highly charged droplets. Evaporation of those charged droplets produces gas-phase ions by mechanisms that remain the subject of much argument and debate.[35–40] Electrospray is a very gentle, nonfragmenting ionization technique that may leave unsolvated ions with a memory of their solution phase structure. Weakly bound complexes can be easily studied by electrospray. An uncharged bonus is that ions formed by this method are often multiply charged by the addition of protons. This multiple charging allows detection of species that would normally be well beyond the mass range of the instrument.

Multiple stages of mass analysis separation can be accomplished with individual mass spectrometer elements separated in space, or in a single mass spectrometer with the mass spectrometric steps separated in time. In tandem mass spectrometers separated in space, such as EBE,[41] BEBE,[41] QHQ,[42] or Q-TOF,[43] the separation elements are physically separated and distinct, although there is a connection between them for ion transmission under high-vacuum conditions. In tandem mass spectrometers separated in time, the separation is accomplished with ions trapped in the same place, with multiple separation steps taking place over time. Quadrupole ion trap (ITMS),[42,44,45] Fourier-transform ion cyclotron resonance (FT-ICR-MS),[46–48] and other mass spectrometric variants can be used for such an analysis. Trapping instruments can perform multiple steps of analysis, which is sometimes referred to as $MS^n$ (for example, $MS^3$ indicates three stages of separation).

The most distinctive features of all these multistage mass selectors are their extensive ion manipulation capabilities, useful for measuring the stability and reactivity of ions and for probing their structure. Targeted ions can be selectively isolated from unwanted accompanying species. After this isolation step, a number of experiments can be carried out. For instance, selected ions can be accelerated into a neutral inert gas (e.g., Ar) to produce fragment ions. This process, called collision-induced dissociation (CID),[49,50] can give information upon the structure of a covalently bound ion or upon the relative stability of ionic fragments arising from decomposition of an ion-neutral cluster (see the following section). As an alternative to CID processes, various methods can be adopted for fragmenting ions in tandem MS, including electron capture dissociation (ECD), electron transfer dissociation (ETD), infrared multiphoton dissociation (IRMPD), and blackbody infrared radiative dissociation (BIRD). The architecture of the Fourier-transform ion cyclotron resonance (FT-ICR) mass spectrometer allows storage of trapped ions for long periods of time (up to several hours), provided that a high vacuum is maintained to reduce the number of destabilizing collisions between ions and residual neutral molecules. Under these conditions, the reactivity and the stability of the trapped ions can be conveniently probed by measuring the kinetics and the equilibrium constant of their reaction with suitable neutral reactants.

## 6.3 CHIRAL ION RECOGNITION

The first example of chiral ion recognition with a mass spectrometer was reported in 1977 by Fales and Wright.[51] Their study showed that the chirality of diisopropyltartrates strongly influences the stability of their diastereomeric proton-bound dimers, generated by isobutane chemical ionization (CIMS) of their racemic mixtures.[52,53] In order to differentiate the protonated species of the homochiral self-dimers ("*homo*") from the heterochiral one ("*hetero*"), the CIMS experiments were carried out on an equimolar mixture of one enantiomer, deuterium labeled at the estereal function, with the other unlabeled enantiomer. The significant deviation of the relative abundances of the three protonated diastereomeric dimers ([$d_2$-*homo*]:[*d*-*hetero*]:[*homo*] = 1:1.3:1) from the expected 1:2:1 statistical ratio indicated that the homochiral $d_2$-*homo* and *homo* dimers are more stable than the heterochiral *d-hetero* one. The lower stability of the heterochiral dimer is ascribed to steric repulsion between the estereal functions of the two monomers in the hydrogen-bonded basket-type structure of the complex. Similar chirality effects have been measured for the same systems using FAB as the ionization mode.[53,54]

After Fales and Wright's pioneering work, a number of papers appeared in the literature along the same stream,[55–71] which are illustrated in Chapter 5. The relative stability of charged diastereomeric complexes has been measured using the so-called kinetic method, first introduced by Cooks and coworkers over thirty years ago for the determination of proton affinities.[72]

### 6.3.1 Enantioselective Reactions in Collisionally Activated Metal-Bound Complexes

The first attempt to perform enantioselective reactions in collisionally activated metal-bound diastereomeric complexes was reported by Schröder and Schwarz in 2004.[73] In the ESI-QHQ-CID of the complexes of Scheme 6.1, the stereochemical constraints in the (*R*)-1,1'-bi-2-naphtholate ((*R*)-BINOLate) ligand affect the activation of the other chiral 2-alkanol ($RCH(OH)CH_3$) ligand. Ions **1** were mass selected in the first quadrupole, subjected to CID with xenon in the hexapole at variable collision energies ($E_{lab}$ = 0–10 eV), and the products analyzed using the second quadrupole. CID of **1** (R = $CH_3$) leads to loss of neutral water (6.1a) as a minor route and expulsion of the entire $RCH(OH)CH_3$ ligand (6.1b) as the major one. Analysis of the $E_{lab}$ dependence of the relative abundance of the residual ionic fragments indicates

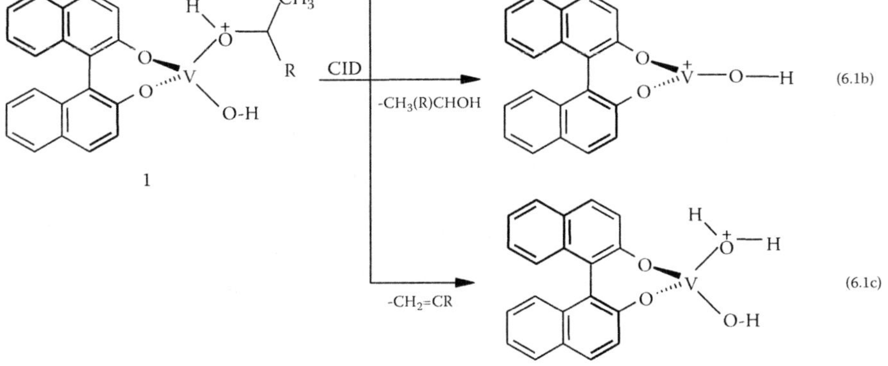


**SCHEME 6.1**

that (6.1a) is thermochemically favored, but kinetically hindered compared to the endothermic pathway (6.1b). This conclusion is consistent with the suggested structure **1**. No significant differences in the CID fragmentation patterns of the homo- and the heterochiral complexes **1** (R = $C_2H_5$) were appreciated. Further enlargement of the R group (R = $C_6H_{13}$; $C_6H_5$) leads to significant stereochemical effects in the CID spectra. The major difference between the diastereomers with 2-octanol is related to the extension of the fragmentation channel (6.1c), which instead is absent with smaller alcoholic ligands. It is concluded that the magnitude of the observed effects is much too small to draw any decisive conclusions about the enantioselectivity of the reactive processes observed in the activated diastereomeric adducts **1**.

Later on, Schwarz and coworkers investigated the enantioselective CID of Ni(II) complexes of chiral secondary alcohols and (*R*)-BINOLate ligands (Scheme 6.2).[74,75] CID of the ESI-formed ion **2** leads to the simple loss of the coordinated alkanol (path 6.2a), in competition with loss of the corresponding ketone (path 6.2b) through an oxidation process that must involve C-H bond activation. No enantioselectivity was observed in the fragmentation patterns of ions **2** (R = $C_2H_5$, n-$C_3H_7$, n-$C_4H_9$, and n-$C_5H_{11}$). A small enantioselectivity is detectable with R = n-$C_6H_{13}$. A more pronounced stereochemical effect is observed with R = $C_6H_5$ (SE = 1.39 ± 0.07), indicating a disfavored bond activation in the homochiral couple, or a more strongly bound alkanol ligand in the case of the heterochiral complex.

Although ESI is widely recognized as one of the most convenient tools for putting in the gas phase a variety of nonvolatile analytes, nevertheless the processes and mechanisms operating in ESI are still inadequately understood.[76–80] The problem arises from the intrinsic difficulties in observing the formation and the behavior of submicron charged droplets from solution and, more importantly, in evaluating whether and to what extent the nature and the relative concentration of the analytes in these droplets change during the time evolution of the electrospray plume.

Some insights into this problem were provided by ESI-QHQ-CID of complexes generated from methanolic solutions containing $Co(NO_3)_2$ together with variable concentrations of either pure (1*S*,2*S*)-(+)-N-methylpseudoephedrine ($^{(+)}$**3**) or its mixtures

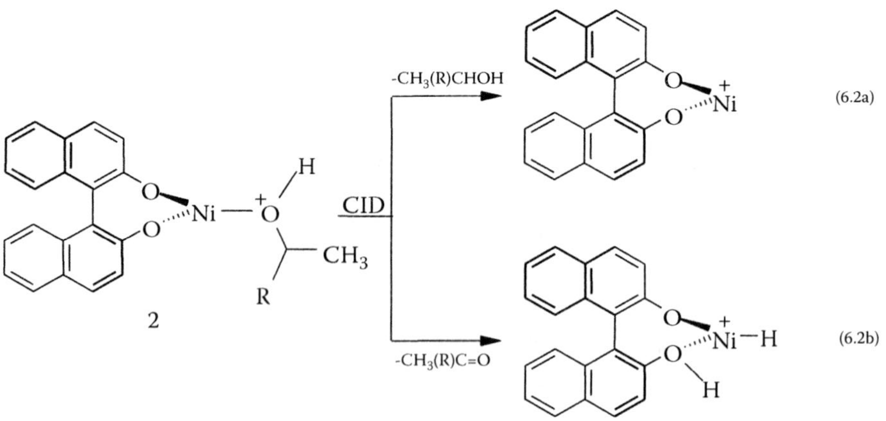


**SCHEME 6.2**

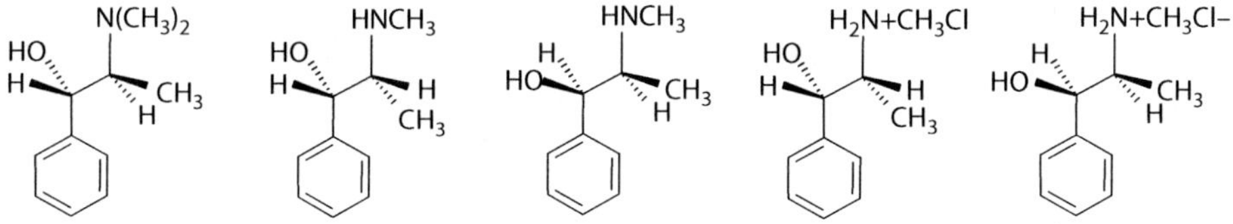


(+)**3** (+)**4** (–)**4** (+)**4*** (–)**4***

**CHART 6.1**

with the enantiomers of ephedrine **4** (1*S*,2*R*)-(+)-ephedrine ($^{(+)}$**4**), (1*R*,2*S*)-(–)-ephedrine ($^{(-)}$**4**), or their hydrochlorides ($^{(+)}$**4***; $^{(-)}$**4***) (Chart 6.1).[81] CID of the isolated *m/z* 479 ion (Scheme 6.3), formally corresponding to [($^{(+)}$**3**)$_2$•CoNO$_3$]$^+$, reflects not only the expected bond connectivity with formation of [($^{(+)}$**3**-H)•$^{(+)}$**3**•Co]$^+$ (*m/z* 416), but also that of the isomeric [$^{(+)}$**3**•**4**•CoCH$_2$ONO$_2$]$^+$ and [($^{(+)}$**3**-H)•**4**•CoCH$_3$ONO$_2$]$^+$ structures. In particular, appreciable formation of *m/z* 403 and 402 fragments (and their dehydrogenated derivatives), and of the *m/z* 166 one (and its dehydrated derivative) were observed, attributed to [$^{(+)}$**3**•**4**•Co]$^+$, [($^{(+)}$**3**-H)•**4**•Co]$^+$, and [**4**•H]$^+$, respectively (Scheme 6.3). The CID pattern of *m/z* 479 is strongly sensitive to (1) the specific form of ephedrine, whether as a neutral molecule **4** or as the hydrochloride salt **4***, (2) the [$^{(+)}$**3**]/[**4**] (or [$^{(+)}$**3**]/[**4***]) concentration ratio, and (3) the specific configuration of ephedrine **4**. In the presence of $^{(-)}$**4** (or $^{(-)}$**4***), the relative abundance of [($^{(+)}$**3**-H)•$^{(+)}$**3**•Co]$^+$ (*m/z* 416) increases by increasing the [$^{(+)}$**3**]/[**4**] (or [$^{(+)}$**3**]/[**4***]) concentration ratio, whereas the reverse is true for the [$^{(+)}$**3**•**4**•Co]$^+$ (*m/z* 403) and [($^{(+)}$**3**-H)•**4**•Co]$^+$ (*m/z* 402) fragments. This opposite behavior can be accounted for by fast **4**$\leftrightarrows$$^{(+)}$**3** (or **4***$\leftrightarrows$$^{(+)}$**3**) ligand exchange in the ESI-formed high-order aggregate precursors of the *m/z* 479 ion. At the highest [**4**] (or [**4***]), the **4**$\rightarrow$$^{(+)}$**3** (or **4***$\rightarrow$$^{(+)}$**3**) displacement in these high-order aggregates is favored over the reverse **4**$\leftarrow$$^{(+)}$**3** (or **4***$\leftarrow$$^{(+)}$**3**) one.

As illustrated in Scheme 6.4, this enhances the contribution of the [$^{(+)}$**3**•**4**•CoCH$_2$ONO$_2$]$^+$ and [($^{(+)}$**3**-H)•**4**•CoCH$_3$ONO$_2$]$^+$ isomeric structures to the

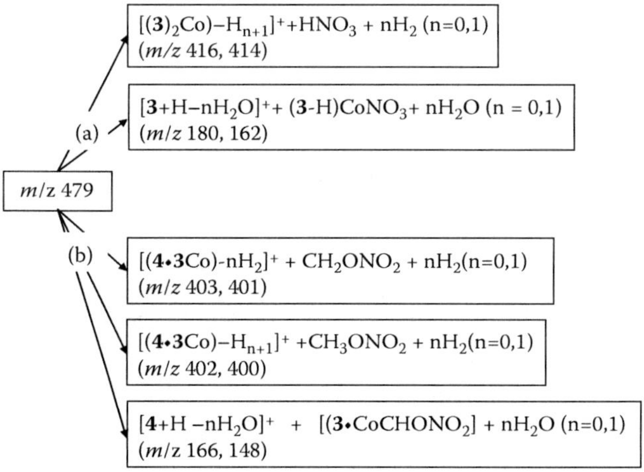


**SCHEME 6.3**

*m/z* 479 signal to the expenses of the $[(^{(+)}\mathbf{3})_2{\bullet}CoNO_3]^+$ and $[(^{(+)}\mathbf{3}\text{-H}){\bullet}^{(+)}\mathbf{3}{\bullet}CoNO_3H]^+$ ones. The consequence is a decrease of the *m/z* 479 → *m/z* 416 fragmentation channels and a parallel increase of the *m/z* 479 → *m/z* 403; *m/z* 402 ones. At the lowest [**4**], the **4**←$^{(+)}$**3** displacement is favored and the opposite trend is observed. The fact that these effects are much less evident in the presence of $^{(+)}$**4** or $^{(+)}$**4*** components denotes a marked stereoselectivity of the **4**⇆$^{(+)}$**3** (or **4***⇆$^{(+)}$**3**) ligand exchange in the precursors of the *m/z* 479 ion. On the grounds of the above evidence, it is thought that the *m/z* 479 is generated in the ESI droplets by decomposition of higher-order aggregates of N-methylpseudoephedrine and ephedrine around the $CoNO_3^+$ center, wherein extensive structural reorganization takes place before decomposition to the isomeric intermediates of Scheme 6.4. The nature and the fate of these higher-order aggregates markedly depend on the presence, charge state, relative concentration, and configuration of the ephedrine molecules in the ESI droplets.

This conclusion is reinforced by the isolation of an ion at *m/z* 493 among the ionic products from ESI of $^{(+)}\mathbf{3}/Co(NO_3)_2/CH_3OH$ solutions, formally corresponding to $[(^{(+)}\mathbf{3}\text{-H}){\bullet}^{(+)}\mathbf{3}{\bullet}CoCH_3ONO_2]^+$ and $[(^{(+)}\mathbf{3})_2{\bullet}CoCH_2ONO_2]^+$ structures (Scheme 6.4). CID of this ion is characterized by the predominant formation of *m/z* 417 and 416 fragments, corresponding to $[(^{(+)}\mathbf{3})_2{\bullet}Co]^+$ and $[(^{(+)}\mathbf{3}\text{-H}){\bullet}^{(+)}\mathbf{3}{\bullet}Co]^+$. Formation of the $[(^{(+)}\mathbf{3}\text{-H}){\bullet}^{(+)}\mathbf{3}{\bullet}CoCH_3ONO_2]^+$ and $[(^{(+)}\mathbf{3})_2{\bullet}CoCH_2ONO_2]^+$ structures necessarily requires, respectively, the formal methylene and methyl group transfer from $^{(+)}$**3** to the $NO_3$ moiety of higher-order $^{(+)}\mathbf{3}/CoNO_3^+$ aggregates during the ESI droplet evaporation.

As a final remark, it should be pointed out that the effect of the configuration, charge state, and concentration of ephedrine on the structure of the Co(II) complexes ESI formed from $^{(+)}\mathbf{3}/\mathbf{4}(Co(NO_3)_2$ methanolic mixtures have some connections with

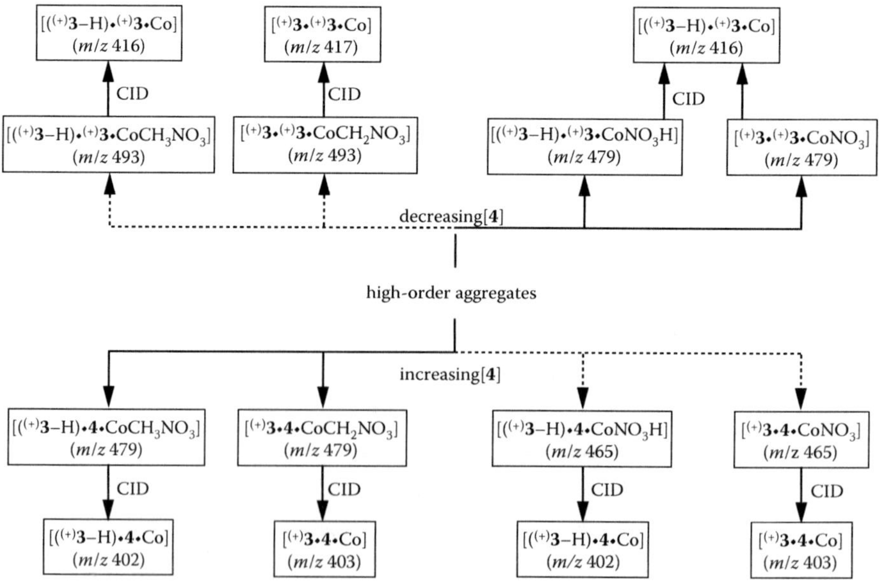


**SCHEME 6.4**

the mass spectrometric observation of different structures for the $[Co(III)(acac)_2$/diisopropyl-D-tartrate$]^+$ and $[Co(III)(acac)_2$/diisopropyl-L-tartrate$]^+$ complexes induced by the presence in the relevant solutions of the (*R*,*R*)-(+)- or (*S*,*S*)-(−)-hydrobenzoin "spectator" molecules, which are actively and selectively involved in their formation.[82] It should be stressed that here, differently from the above Co(II) complexes, the spectator molecules do not induce any reaction in the ESI-formed complexes, but simply influence the structural landscape of the $[Co(III)(acac)_2$/tartrate$]^+$ complexes by participating to their formation.

## 6.3.2 Enantioselective Reactions in Thermally Activated Proton-Bound Complexes

### 6.3.2.1 Oligosaccharides as Chiral Hosts

The methodology employed for measuring the gas-phase kinetic enantioselectivity of a chiral selector M toward a chiral molecule A is based on the generation of the corresponding proton-bound diastereomeric complexes ($[M{\bullet}H{\bullet}A]^+$) by electrospray ionization (ESI) of suitable M/A mixtures. The proton-bound $[M{\bullet}H{\bullet}A]^+$ complex is transferred into the resonance cell of the FT-ICR-MS by a system of potentials and lenses and isolated by broad-band ejection of the accompanying ions. Then, the complex is quenched by collisions with an inert gas pulsed into the FT-ICR-MS cell through a magnetic valve, isolated by broad-band ejection of the accompanying ions, and allowed to react in the presence of an externally introduced chiral or achiral reagent B present in the cell at a fixed pressure (Equation 6.1):

$$[M{\bullet}H{\bullet}A]^+ + B \rightarrow A + [M{\bullet}H{\bullet}B]^+ \qquad (6.1)$$

The rate constant of the guest exchange reaction (Equation 6.1) is extracted based on the decay of the isolated ion $[M{\bullet}H{\bullet}A]^+$ as a function of time $t$. If $I$ is the intensity of complex $[M{\bullet}H{\bullet}A]^+$ at the delay time $t$ and $I_0$ is the sum of the intensities of $[M{\bullet}H{\bullet}A]^+$ and $[M{\bullet}H{\bullet}B]^+$, a monoexponential $\ln(I/I_0)$ vs. $t$ plot is often obtained whose slope provides the pseudo-first-order rate constant $k'$ for Equation 6.1. The corresponding second-order rate constants $k$ are calculated from the ratio between the slope of the first-order plots and the B concentration ($k = k'/[B]$). In some instances, the $\ln(I/I_0)$ vs. $t$ plot is not linear and presents a curvature that denotes the occurrence of biexponential kinetics (open circles in Figure 6.2). This is due to the coexistence of at least two stable isomeric structures for $[M{\bullet}H{\bullet}A]^+$, one less reactive (denoted with the *slow* subscript) and the other more reactive (denoted with the *fast* subscript).[14,29–33] The time dependence of $[M{\bullet}H{\bullet}A]^+_{fast}$ (full circles in Figure 6.2) can be inferred from the overall $[M{\bullet}H{\bullet}A]^+$ decay (open circles in Figure 6.2) after subtracting the first-order decay of $[M{\bullet}H{\bullet}A]^+_{slow}$ (broken line in Figure 6.2). The Y intercepts of the first-order decay of $[M{\bullet}H{\bullet}A]^+_{slow}$ and $[M{\bullet}H{\bullet}A]^+_{fast}$ provide an estimate of their starting relative distribution. In contrast, any monoexponential kinetics is generally attributed to the occurrence of a single structure or, alternatively, of several stable isomers, but with comparable reactivity toward B.

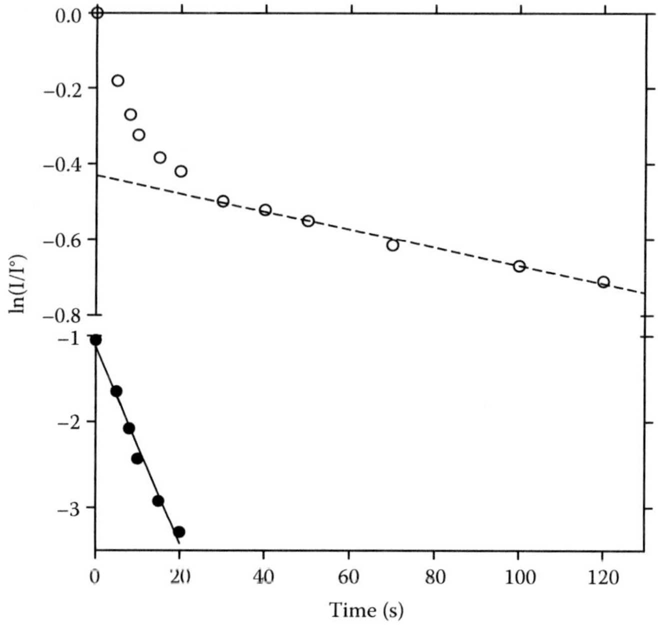


**FIGURE 6.2** Plot of a typical biexponential kinetics.

Kinetic enantioselectivity of Equation 6.1 is obtained by comparing the second-order rate constants $k$ for the same reaction involving the diastereomeric $[M{\bullet}H{\bullet}A^R]^+$ and $[M{\bullet}H{\bullet}A^S]^+$ complexes. When the host and the guest in the complex have the same configuration, the rate constant is denoted as $k_{homo}$; when instead they have opposite configurations, the rate constant is denoted as $k_{hetero}$. The enantioselectivity factor ρ is expressed by the $k_{homo}/k_{hetero}$ ratio. A $\rho > 1$ value indicates that the reactant B displaces the guest from the homochiral complex faster than the guest from the heterochiral one. The opposite is true when $\rho < 1$. A $\rho = 1$ value corresponds to equal displacement rates.

In 1998, Lebrilla and coworkers observed that the rate of Equation 6.1 is found to be sensitive to the *D* or *L* configuration of the guest amino acid (A) when using permethylated β-cyclodextrin (**5**) as M host, and 1-aminopropane as the B reactant.[29–31] The magnitude of the reaction enantioselectivity is as large as the measured $k_D/k_L$ ratio is far from unity. The measured enantioselectivity increases from alanine (**ala**) ($k_D/k_L = 0.62$) to valine (**val**) ($k_D/k_L = 0.32$), leucine (**leu**) ($k_D/k_L = 0.28$), isoleucine (**ile**) ($k_D/k_L = 0.26$), and *allo*-isoleucine (**aile**) ($k_D/k_L = 0.24$). Even when the amino acid side chain is hydroxylated, as in **thr** and **ser**, which would make more favored its interaction with the cavity of **5**, the enantioselectivity increases with the size of the side chain. Proline (**pro**) ($k_D/k_L = 0.67$) and *cis*-4-hydroxyproline (**hpro**) ($k_D/k_L = 0.71$) display low selectivities because these molecules are rigid and compact. Aromatic amino acids, like phenylalanine (**phe**) ($k_D/k_L = 1.22$) and tyrosine (**tyr**) ($k_D/k_L = 1.49$), exhibit significantly smaller and opposite enantioselectivities. Similar trends are obtained when alkylamines more basic than 1-aminopropane are used, i.e., 2-aminobutane and 1-amino-2-propanol.[32] These results indicate that increasing the size of the side chain of the amino acidic guest A increases the enantioselectivity of Equation 6.1 to a certain extent, beyond which it tends to decrease (Figure 6.3).

According to molecular modeling (MM) calculations,[30,31] the different enantioselectivities of Figure 6.3 can be accounted for by the structure of the relevant $[M{\bullet}H{\bullet}A]^+$ complexes. **Leu**, **ile**, and **aile** with four carbons in the side chain have the optimal size to fit into the cavity of **5**, while **phe** and **tyr** with seven carbons are too large to fit into the same cavity. Noticeable differences in the interactions of the two enantiomers of A with host **5** are observed. Chiral differentiation occurs when the access to the protonated amino group of A is limited either by its alkyl side chain or by the methoxy groups of the host that are drawn in by hydrogen-bonding interactions. These differences in binding translate to differences in reaction rates.

Phenylalanine (**phe**) behaves differently from **val** in **5**. Both *L*- and *D*-**phe** interact in the same way. In fact, the predominant interaction of both the ammonium and the carboxylic acid group of **phe** forces its phenyl group to remain inside the cavity. The similarity in the binding of the two enantiomers is responsible for the observed small enantioselectivity ($k_D/k_L = 1.22$). In contrast, both the ammonium and the carboxylic acid group of *L*-**val** interact preferentially with the narrow rim of **5**, whereas only the ammonium of *D*-**val** interacts in the same way, and its carboxylic acid group interacts preferentially with the wider rim of the host. Thereby, the high (and reverse) enantioselectivity ($k_D/k_L = 0.32$) exhibited by the two enantiomers.

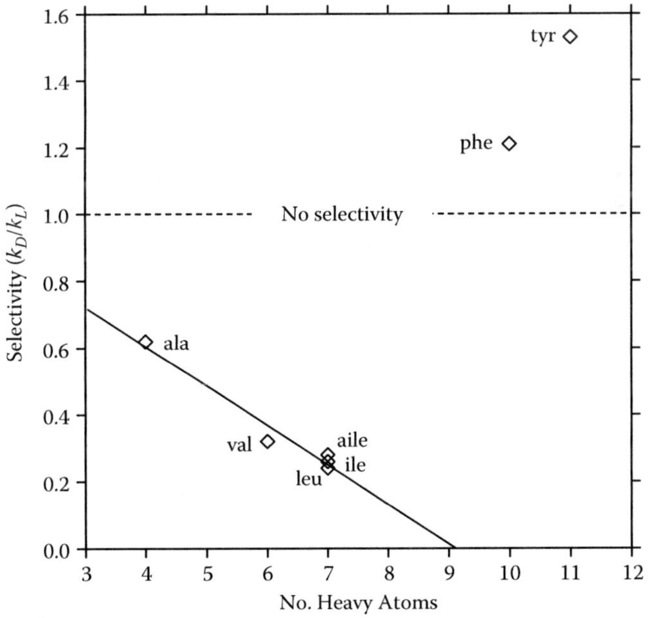


**FIGURE 6.3** Chiral selectivity ($k_D/k_L$) as a function of the number of nonhydrogen atoms on the amino acid. Host: permethylated β-cyclodextrin **5**; reagent base: 1-aminopropane. Chiral selectivity tends to increase with the size of the amino acids; **phe** and **tyr**, however, do not follow this trend.

Additional experiments were performed to examine the effect of the host size on the enantioselectivity. Since the methyl groups in methylated β-CD orient themselves toward the center of the cavity, it is expected that decreasing the number of $CH_3O$ groups in β-CD from 21 (as in trimethyl-β-CD (**5**)) to 14 (as in the dimethyl-β-CD (**6**)) increases the effective size of the cavity.[30] In fact, the enantioselectivity of **val** ($k_D/k_L$ = 0.32 with M = **5**) increases to $k_D/k_L$ = 0.18 with M = **6**. No significant effect of the cavity size is observed with the smaller **ala**.

Another way to increase the host cavity is by using permethylated γ-CD **7** (diameter of the cavity from 7.5 to 8.3 Å) instead of permethylated β-CD **5** (diameter of the cavity from 6.0 to 6.5 Å).[30] The larger cavity size of **7** decreases (and inverts) the enantioselectivity of **val** from $k_D/k_L$ = 0.32 to $k_D/k_L$ = 1.41, and that of **ile** from $k_D/k_L$ = 0.26 to $k_D/k_L$ = 2.28. This observation indicates that these amino acids have optimal enantioselectivity with **5**. Conversely, **phe** increases in selectivity from $k_D/k_L$ = 1.2 (with **5**) to $k_D/k_L$ = 1.8 (with **7**), suggesting that the wider cavity of **7** allows each **phe** enantiomer to find more distinct interactions with the host.

The gas-phase guest exchange reaction (Equation 6.1) has been employed to probe the enantioselectivity of permethylated β-CD **5** (M) toward pharmacologically important compounds, such as A = DOPA, amphetamine, ephedrine, and penicillamine (Chart 6.2).[33] A variety of alkyl amines B, including 1-aminopropane, 1,2-diaminoethane, 1,3-diaminopropane, and (*R*)-1-amino-2-propanol, have been used as reactants. The guest exchange kinetic results are reported in Table 6.1. The presence of more than one reacting $[M{\bullet}H{\bullet}A]^+$ structure is observed with A = DOPA and penicillamine. The results have been rationalized in terms of specific interactions in the relevant inclusion complexes, which determine their structure and relative stability.

The maltose-based oligomers are exact linear analogues of CDs. For example, maltoheptaose **8** (Chart 6.3) is composed of seven α(1-4)-linked glucose units such as β-CD. Its linear chain is sufficiently flexible to wrap around guest molecules and form "quasi-inclusion" complexes in the gas phase. The $k_D/k_L$ values for the exchange reactions between 1-aminopropane and the complexes of some amino acids with

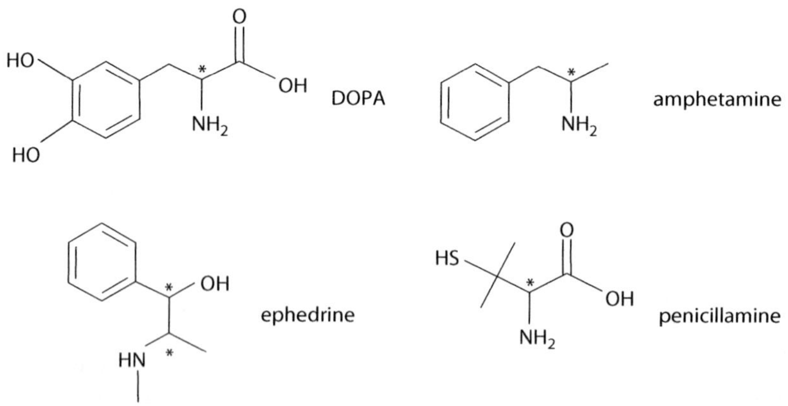


**CHART 6.2**

**TABLE 6.1**
**Reaction Selectivity ($k_D/k_L$) of Various Proton-Bound Trimethyl-β-Cyclodextrin (5)/Drug Complexes with Several Bases ($k \times 10^{11}$ cm[1] mol[-1] s[-1])**

| Guest | | 1-aminopropane | (R)-1-amino-2-propanol | 1,2-diaminoethane | 1,3-diaminopropane |
|---|---|---|---|---|---|
| DOPA | $k_D$ | $<10^{-4}$–15 | $<10^{-4}$–15 | 0.0024 | fast: 0.131; slow: 0.014 |
| | $k_L$ | $<10^{-4}$–15 | 0.0047 | 0.0121 | fast: 0.122; slow: 0.0131 |
| | $k_D/k_L$ | – | – | 0.20 | fast: 1.07; slow: 0.46 |
| amphetamine | $k_D$ | 0.27 | 1.34 | – | – |
| | $k_L$ | 0.40 | 1.78 | – | – |
| | $k_D/k_L$ | 0.68 | 0.75 | – | – |
| ephedrine | $k(+)$ | 0.031 | 0.53 | – | – |
| | $k(-)$ | – | 0.64 | – | – |
| | $k(+)/k(-)$ | – | 0.83 | – | – |
| penicillamide | $k_D$ | fast: 1.80; slow: 0.55 | – | – | – |
| | $k_L$ | 3.4 | – | – | – |
| | $k_D/k_L$ | fast: 0.53; slow: 0.16 | – | – | – |

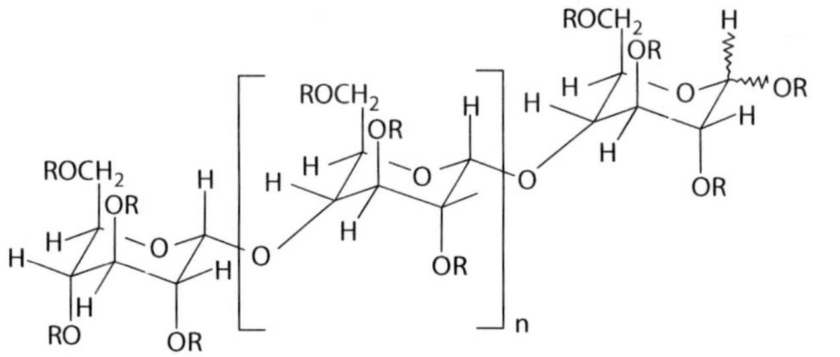


R = H (n = 5 (**8**))
R = $CH_3$ (n = 5 (**9**); 4 (**10**), 3 (**11**))

**CHART 6.3**

permethylated maltoheptaoses **9–11** have been determined.[32] The measured enantioselectivities are slightly less than those of the complexes between **5** and the same amino acids. The same trend is observed as a function of the side chain size of the guests. Remarkably, the reaction on the maltoheptaose complexes with **tyr** and **phe** exhibits a significantly greater enantioselectivity relative to those with **5**. For example, the

$k_D/k_L$ ratio for **phe** decreases from 1.21 with **5** to 0.25 with maltoheptaose **9**. Comparative analysis of the enantioselectivity of amino acids toward **5** and **9** reveals the large effect of the cavity size of **5**, which instead disappears when the flexible **9** is used as host (Figure 6.4).

MM calculations give important insight into the origin of such large enantioselectivity differences. Unlike the rigid **5**, the linear maltoheptaose **9** allows the enantiomers of **phe** to find the most favorable conformation. The phenyl group of *L*-**phe** is oriented toward the C6 center of the hosts, while that of *D*-**phe** is oriented toward the C2 and C3 centers of the hosts. Smaller linear hosts, such as maltohexaose **10** and maltopentaose **11**, show lower enantioselectivities. Molecular modeling calculations predict that these hosts are too small to fully envelop the guest and create an environment for high enantioselectivity.

More recently, the investigation has been extended to complexes with peptides as guests.[83] The selected peptides contain either the **phe** or the **val** moieties, since these amino acids displayed the lowest and highest enantioselectivity with **5**. The **phe**, **gly-phe**, **gly-gly-phe** series was examined to observe the effects of lengthening the distance between the site of protonation (the N terminus) and the chiral center (the **phe** residue). The **gly-val** and **phe-gly** dipeptides were included for comparison. The same reactions on the complexes with permethylated maltoheptaose **9** (Chart 6.3), the linear analogue of **5**, have been investigated as well.

A summary of the selectivities for **5** and **9** as hosts with the selected peptides is presented in Table 6.2.[83] The most notable trend was the increase in selectivities

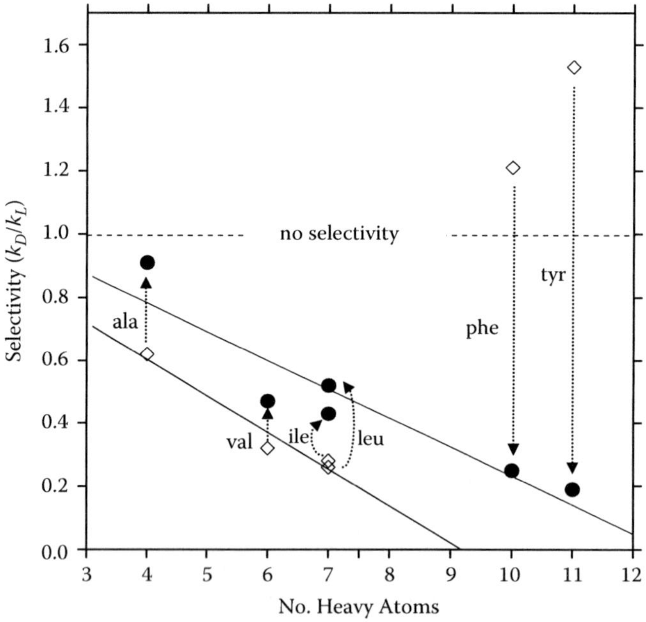


**FIGURE 6.4** Enantioselectivity ($k_D/k_L$) as a function of the number of nonhydrogen atoms on the amino acid. Hosts: permethylated β-cyclodextrin **5** (open diamonds), permethylated maltoheptaose **9** (full circles); reagent base: 1-aminopropane.

**TABLE 6.2**
**Reaction 6.3 Selectivities ($\rho = k_D/k_L$) of Selected Amino Acids with Trlmethyl-β-cyclodextrin (5) and Linear Permethylated Maltoheptaose (9) (Chart 6.3)**

| Guest (A) | GBa (kcal mol$^{-1}$) | Host (M) | Amine (B) | GB[a] (kcal mol$^{-1}$) | Enantioselectivity ($\rho = k_D/k_L$) |
|---|---|---|---|---|---|
| **phe** | 217 | 5 | n-$C_3H_7NH_2$ | 206 | 1.25 |
| | | 9 | n-$C_3H_7NH_2$ | | 0.22 |
| **gly-phe** | 215 | 5 | $H_2NC_2H_4NH_2$ | 215 | 0.37 |
| | | 9 | $H_2NC_2H_4NH_2$ | | 0.15 |
| **gly-phe** | 215 | 5 | $H_2NC_3H_6NH_2$ | 221 | 0.48 |
| **gly-gly-phe** | 217 | 5 | $H_2NC_3H_6NH_2$ | 221 | 0.50 |
| | | 9 | $H_2NC_3H_6NH_2$ | | 0.61 |
| **phe-gly** | | 5 | $H_2NC_3H_6NH_2$ | 221 | 3.30 |
| | | 9 | $H_2NC_3H_6NH_2$ | | 0.46 |
| **val** | | 5 | n-$C_3H_7NH_2$ | 206 | 0.32 |
| | | 9 | n-$C_3H_7NH_2$ | | 0.48 |
| **gly-val** | | 5 | $H_2NC_2H_4NH_2$ | 215 | 1.15 |
| | | 9 | $H_2NC_2H_4NH_2$ | | 0.14 |

[a] GB = Gas-Phase Basicity.

between **phe** [ρ = 1.25 (M = 5); 0.22 (M = **9**)] and **gly-phe** [ρ = 0.37 (M = **5**; 0.15 (M = **9**)]. With **gly-gly-phe**, the selectivity decreases but remains quite substantial [ρ = 0.50 (M = **5**); 0.61 (M = **9**)]. Selectivity appears to decrease with more basic amines but not to a large extent {for [**5**•H•**gly-phe**]$^+$: ρ = 0.37 (B = $H_2NC_2H_4NH_2$); 0.48 (B = $H_2NC_3H_6NH_2$)}. The enantioselectivity of the [M•H•**phe-gly**]$^+$ complexes is more similar to that of [M•H•**phe**]$^+$ than to that of [M•H•**gly-phe**]$^+$. The behavior of [**5**•H•**val**]$^+$ (ρ = 0.32) and [**5**•H•**gly-val**]$^+$ (ρ = 1.15) is opposite of that of [**5**•H•**phe**]$^+$ (ρ = 1.25) and [**5**•H•**gly-phe**]$^+$ (ρ = 0.37). The highest selectivities are observed with M = **9** [ρ = 0.15 (**gly-phe**); 0.14 (**gly-val**)].

As mentioned before, the poor selectivity of [**5**•H•**phe**]$^+$ is accounted for by the relatively rigid cavity of the host, which constrains the phenyl group of the guest in such a way that its ammonium and the carboxyl groups can interact with the host rims only to a limited extent. With [**5**•H•**gly-phe**]$^+$ and [**5**•H•**gly-gly-phe**]$^+$, an increased selectivity is observed. Molecular modeling calculations indicate that lengthening the guest effectively pushes the phenyl group out of the cavity to maintain the interaction between the protonated N terminus with the narrow rim. With the phenyl group outside the host cavity, the guest enantiomers are not as strongly constrained and interact with the cyclodextrin host in more unique ways, resulting in a greater selectivity. Selectivity is markedly increased when M = **9**. Molecular modeling calculations indicate that **9** wraps around the guest in a helical fashion, producing intimate interactions with the guest. The adjustable cavity of **9** allows the phenyl group of the guest to find unique orientations for both enantiomers that are well distinguishable from one another by the exchange reaction (Equation 6.1).

Besides Equation 6.1, an additional reaction channel is observed corresponding to the formation of the ternary complex $[M•H•A•B]^+$ when $[M•H•A]^+$ (M = underivatized β-CD (**12**); A = amino acid) reacts with alkyl amine B. The positive charge of the protonated B is suspected to coordinate with the carboxyl group of A. The zwitterionic amino acid is then stabilized through hydrogen bonding to the CD lower rim (Figure 6.5).[84] Formation of zwitterionic structure in **12** is favored for amino acids with high-gas-phase basicities as well as with structural compatibility with the host. Indeed, with underivatized α-CD (**13**), only **pro** forms the ternary complex $[M•H•A•B]^+$ (B = n-$C_3H_7NH_2$), and therefore, it is thought to be included in the host cavity in the zwitterion form.[85] The low-energy [**13**•H•**pro**]$^+$ structure from consistent valence force field (CVFF) calculations indicates that **pro** is encapsulated in **13** in such a way that its ammonium and carboxylic groups can interact with the narrow rim of the host while allowing the narrow rim to further interact with B = n-$C_3H_7NH_2$ (Figure 6.5). No other amino acid is able to interact in the same way. Indeed, the most favored [**13**•H•A]$^+$ structures with these guests place the side chain of the amino acid outside the cavity while still maintaining the interaction between the narrow rim with both amine and the carboxylic groups. The side chain prevents B from simultaneously interacting with the nascent carboxylate and the narrow rim. In the larger cavity of **12**, the amino acids have more freedom to rearrange and interact favorably to yield zwitterions. Of the twenty amino acids tested, only the least basic five (**gly**, **ala**, **cys**, **asp**, and **val**) plus **ser** and **thr** do not form the ternary complex $[M•H•A•B]^+$ (B = n-$C_3H_7NH_2$), and therefore the corresponding zwitterionic protomers. In the even larger cavity of the underivatized γ-CD (**14**), all twenty amino acids tested, but the least basic **gly** and **ala**, do form the ternary complex $[M•H•A•B]^+$ (B = n-$C_3H_7NH_2$), and therefore the corresponding zwitterionic protomers. Maltoheptose (**8** in Chart 6.3), the linear analogue of **12**, was also investigated as a host molecule. It is more flexible than the closed cyclic structure, and its

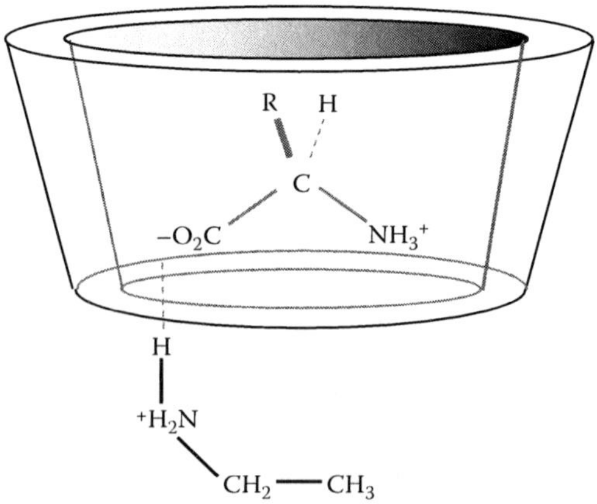


**FIGURE 6.5** Proposed structure of gas-phase zwitterionic amino acids in the cavity of underivatized β-cyclodextrin **12** (represented as a toroid cavity).

"pseudo" cavity provides a structure that can orient itself to better interact with the guest molecule. With this host, only the guest exchange reaction (Equation 6.1) was observed, thus excluding any zwitterionic form for guest A.

Table 6.3 reports the rate constants of both the exchange reaction (Equation 6.3) and the competing B addition pathways on diastereomeric [**12**•H•A]$^+$ complexes. With A = **ala**, **asp**, **val**, and **thr**, and B = n-$C_3H_7NH_2$, the ln($I/I°$) vs. time plots are monoexponential. This indicates that only a single [**12**•H•A]$^+$ structure is formed whose reaction with B leads exclusively to the exchanged product [**12**•H•B]$^+$. With A = **glu** and **gln** (B = n-$C_3H_7NH_2$) or with A = **his**, **lys**, and **arg** (B = $H_2NC_2H_4NH_2$), the ln($I/I°$) vs. time plots are monoexponential as well. This implies that, also in

**TABLE 6.3**
**Rate Constants (Host: M = Underivatized β-Cyclodextrin 12) (*k* × 10" $cm^3$ $mol^{-1}$ $s^{-1}$)**

| Guest (A) | Amine (B) | [M•H•A]+B→[M•H•B]+A | | [M•H•A]+B→[M•H•A•B] | |
|---|---|---|---|---|---|
| | | *k* | *ρ* | *k* | *ρ* |
| **ala**$^D$ | n-$C_3H_7NH_2$ | 0.54 | | no reaction | |
| **ala**$^L$ | | 1.9 | | no reaction | |
| **asp**$^D$ | n-$C_3H_7NH_2$ | 0.0052 | | no reaction | |
| **asp**$^L$ | | 0.014 | | no reaction | |
| **val**$^D$ | n-$C_3H_7NH_2$ | 0.61 | | no reaction | |
| **val**$^L$ | | 2.3 | | no reaction | |
| **glu**$^D$ | n-$C_3H_7NH_2$ | no reaction | | 0.0017 | 0.29 |
| **glu**$^L$ | | no reaction | | 0.0058 | |
| **leu**$^D$ | n-$C_3H_7NH_2$ | 0.082 | 0.28 | 0.091 | 0.24 |
| **leu**$^L$ | | 0.32 | | 0.38 | |
| **ile**$^D$ | n-$C_3H_7NH_2$ | 0.13 | 0.37 | 0.14 | 0.25 |
| **ile**$^L$ | | 0.53 | | 0.57 | |
| **pro**$^D$ | n-$C_3H_7NH_2$ | 0.3 | 0.26 | 0.46 | 0.21 |
| **pro**$^L$ | | 1.6 | | 2.2 | |
| **thr**$^D$ | n-$C_3H_7NH_2$ | 0.1 | | no reaction | |
| **thr**$^L$ | | 0.26 | | no reaction | |
| **phe**$^D$ | n-$C_3H_7NH_2$ | 0.48 | 0.26 | 0.94 | 0.94 |
| **phe**$^L$ | | 1 | | 1.7 | |
| **tyr**$^D$ | n-$C_3H_7NH_2$ | 0.0062 | 0.24 | 0.01 | 0.91 |
| **tyr**$^L$ | | 0.0073 | | 0.011 | |
| **gln**$^D$ | n-$C_3H_7NH_2$ | no reaction | 0.19 | 0.049 | 0.27 |
| **gln**$^L$ | | no reaction | | 0.18 | |
| **his**$^D$ | $H_2NC_2H_4NH_2$ | no reaction | 0.38 | 0.0025 | 0.58 |
| **his**$^L$ | | no reaction | | 0.0043 | |
| **lys**$^D$ | $H_2NC_2H_4NH_2$ | no reaction | 0.48 | 0.0031 | 0.53 |
| **lys**$^L$ | | no reaction | | 0.0058 | |
| **arg**$^D$ | $H_2NC_2H_4NH_2$ | no reaction | 0.85 | 0.0039 | 0.42 |
| **arg**$^L$ | | no reaction | | 0.0092 | |

this case, only a single [**12**•H•A]$^+$ isomer is formed, but now yielding the addition product [**12**•H•A•B]$^+$ by reaction with B. With A = **leu**, **ile**, **pro**, **phe**, and **tyr**, and B = n-$C_3H_7NH_2$, the ln($I/I°$) vs. time decay curves are biexponential. This suggests the presence of two stable [**12**•H•A]$^+$ structural isomers, the more reactive yielding the addition product [**12**•H•A•B]$^+$ and the less reactive the exchanged one, [**12**•H•B]$^+$. The occurrence of two distinct structures for these latter complexes was corroborated by molecular dynamic (MD) simulations, which could recognize low-energy [**12**•H•**leu**]$^+$ isomers. In one structure, denoted as the "threaded through" structure, the ammonium group of **leu** interacts with the narrow rim and the carboxyl group with the wider rim of the host. In the other, denoted as the "narrow only" structure, both the carboxyl and the ammonium groups of **leu** interact with the narrow rim of the host. The first structure accounts for the formation of the exchanged [**12**•H•B]$^+$ product and the latter for the formation of the addition [**12**•H•A•B]$^+$ one.

The ρ enantioselectivity factors of Table 6.3 indicate that the selectivity of aromatic amino acids is much lower than those with alkyl side chain. For instance, **phe** and **tyr** show ρ = 0.48 (A = **phe**) and 0.85 (A = **tyr**) for the exchange reaction and ρ = 0.55 (A = **phe**) and 0.91 (A = **tyr**) for the B addition pathway. This reduced enantioselectivity was explained by the molecular modeling of **phe**, which illustrates that, inside the **12** cavity, both the N and the C terminus interact with the lower rim of the host. The inflexible phenyl ring is forced into similar, sterically acceptable conformations for both enantiomers that do not contribute to stereoselectivity.

#### 6.3.2.2 Cytochrome *c* as Chiral Host

In another set of ESI-FT-ICR-MS experiments, Lebrilla and coworkers demonstrated that proton transfer from multiply charged [cytochrome *c*]$^{+n}$ (n = 7–9) to (*R*)- ($B^R$) and (*S*)-2-aminobutane ($B^S$) shows a significant enantioselectivity (Table 6.4).[86] Proton transfer to $B^R$ is invariably faster than that to $B^S$, irrespective of the charge state of cytochrome *c*. The decay of the [cytochrome *c*]$^{+9}$ ions with reaction time is best represented by a single rate constant, while that of the [cytochrome *c*]$^{+n}$ (n = 7, 8) ions is best represented by two rate constants (*fast* and *slow* in Table 6.4). This is indicative of a single conformer for [cytochrome *c*]$^{+9}$ and of two conformers for [cytochrome *c*]$^{+n}$ (n = 7, 8).[87] The relative amounts of these conformers (percent in Table 6.4) are the same with both $B^R$ and $B^S$. Enantiomer $B^S$ exhibits a reactivity that decreases by decreasing n, as expected from purely Coulombic considerations. In this regard, both 1-aminopropane and *tert*-butylamine behave in the same manner. With $B^R$, the rate constants are approximately equal when n = 9, 8 (*fast* channel), and 7 (*fast* channel), suggesting a similar reactive site for the three charged states. Toward [cytochrome *c*]$^{+n}$ (n = 8, 9), *tert*-butylamine is one order of magnitude less reactive than 1-aminopropane, despite the 2.9-kcal mol$^{-1}$ higher basicity. These findings are interpreted in terms of the strong influential role of steric effects in the highly specific arrangement of the alkylamine toward the multiply charged protein.

#### 6.3.2.3 Resorcinarenes as Chiral Hosts

Chiral calixarenes and their analogues, resorcinarenes, are conformationally more or less flexible macrocyclic compounds with chiral substituents attached to their wider upper rim or narrower lower rim. They may exist in a highly symmetric

**TABLE 6.4**
**Rate Constants ($k \times 10^{12}$ $cm^3$ $mol^{-1}$ $s^{-1}$) for the Proton Transfer from [Cytochrome c] (n = 7–9) to Aklylamines**

| [cytochrome c]$^{+n}$ | | (R)-(-)-2-aminobutane (GB, 211.7)[a] | | (S)-(+)-2-aminobutane (GB, 211.7)[a] | | 1-aminopropane (GB, 210.1)[a] | tert-butylamine (GB, 213.0)[a] |
|---|---|---|---|---|---|---|---|
| n | type | $k_R$ | % | $k_S$ | % | k | k |
| 9 | | 15±3 | | 2.5±0.2 | | 2.2 | 0.61 |
| 8[b] | | 2.3±0.5 | | 0.46±0.11 | | 0.29 | 0.038 |
| 8 | fast | 10±3 | 45 | 1.9±0.4 | 46 | 0.31 | 0.37 |
| 8 | slow | 1.4±0.1 | 55 | 0.37±0.10 | 54 | | |
| 7[b] | | 0.23±0.01 | | 0.084±0.036 | | 0.072 | |
| 7 | fast | 11±1 | 21 | 1.4±0.3 | 30 | 0.14 | 0.051 |
| 7 | slow | 0.13±0.11 | 79 | 0.14±0.19 | 70 | | |

[a] GB, Gas Phase Basicity (kcal $mol^{-1}$);
[b] Further reaction of the lower charged state arising from proton transfer.

bowl-shaped, so-called cone conformation, or in several other conformations that do not exhibit as perfect a cavity as does the cone conformation.

Chiral recognition by calixarenes in the gas phase was virtually unknown before 2002, when Speranza and coworkers started a comprehensive ESI-FT-ICR-MS study on the displacement of several amino acids A from the chiral amido[4]resorcinarene host M = **15**$^S$ (Figure 6.6) using the 2-aminobutane enantiomers ($B^R$ and $B^S$) as reactants (Equation 6.1).[88–90] The experimental procedure is the same described in the previous section, with the only variant that, since here chiral B is employed, another enantioselectivity factor must be introduced besides the ρ term. Indeed, when the guest exchange 4 involves the same [M•H•A]$^+$ complex and a chiral reactant B [either $B^S$ ($k_S$) or $B^R$ ($k_R$)], the corresponding rate constant ratio $\xi = k_R/k_S$ expresses the enantioselectivity of the reaction relative to the configuration of the base. A $\xi > 1$ value indicates that the displacement of the A guest from a given [M•H•A]$^+$ diastereomer is faster with $B^R$ than with $B^S$. The opposite is true when $\xi < 1$. A $\xi = 1$ value corresponds to equal displacement rates.

The molecular asymmetry of **15**$^S$ is due to the four axial pendants containing the chiral *L*-valine group. The efficiency of the gas-phase exchange reaction 6.3 is expressed by eff = $k/k_C$, i.e., by the ratio between the second-order rate constants *k* of the process and the relevant collision rate constants ($k_C$), estimated according to Su's trajectory calculation method.[91] The results of Table 6.5 indicate that the efficiency of the gas-phase exchange reaction (Equation 6.1) is affected by the nature and the configuration of both the amino acid guest (when $\rho = k_{homo}/k_{hetero} \neq 1$) and the amine B (when $\xi = k_R/k_S \neq 1$). Besides, the presence of more than one reacting [M•H•A]$^+$ structure (i.e., [M•H•A]$^+$*fast* and [M•H•A]$^+$*slow*) is observed only with A = tyrosine methyl ester (**tyrOMe**), 3,4-dihydroxyphenylalanine (**DOPA**), and tryptophan (**trp**).

The emerging selectivity picture, discussed in the light of MM and MD calculations, points to chiral recognition by **15**$^S$ as determined by the effects of the host asymmetric frame upon the structure, stability, and rearrangement dynamics of the diastereomeric [**15**$^S$•H•A]$^+$ complexes and the orientation of the amine reactant B

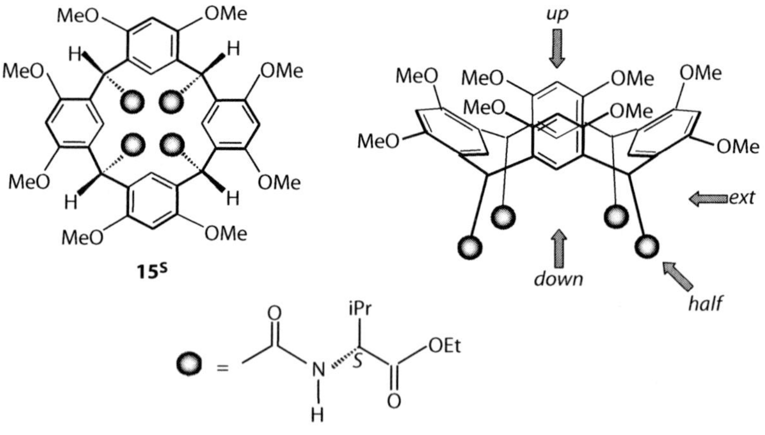


**FIGURE 6.6** Formula of 2,8,14,20-tetrakis(*L*-valinamido)[4]resorcinarene (**15**$^S$) in the flattened cone conformation and possible interaction regions (*up*, *ext*, *half*, and *down*) with chiral amino acids.

**TABLE 6.5**
**Exchange Rate Constants (Host: M = $15^{S}$) ($k \times 10^{10}$ $cm^3$ $mol^{-1}$ $s^{-1}$)**

| Guest (A) | (R)-(-)-$C_4H_9NH_2$ | | | (S)-(+)-$C_4H_9NH_2$ | | | |
|---|---|---|---|---|---|---|---|
| | $k$ | eff | $\rho$ | $k$ | eff | $\rho$ | $\xi$ |
| $\mathbf{ala}^{D}$ | 7.69±0.25 | 0.69 | | 7.06±0.08 | 0.63 | | 1.09±0.26 |
| $\mathbf{ala}^{L}$ | 6.87±0.15 | 0.45 | 0.89±0.04 | 5.89±0.08 | 0.53 | 0.83±0.02 | 0.86±0.17 |
| $\mathbf{leu}^{D}$ | 3.76±0.02 | 0.34 | | 4.68±0.05 | 0.42 | | 0.80±0.02 |
| $\mathbf{leu}^{L}$ | 2.82±0.02 | 0.25 | 0.75±0.01 | 3.64±0.10 | 0.32 | 0.78±0.03 | 0.77±0.03 |
| $\mathbf{ser}^{D}$ | 4.59±0.06 | 0.41 | | 3.70±0.06 | 0.34 | | 1.24±0.05 |
| $\mathbf{ser}^{L}$ | 5.05±0.05 | 0.62 | 1.10±0.03 | 7.56±0.06 | 0.68 | 2.04±0.05 | 0.91±0.02 |
| $\mathbf{thr}^{D}$ | 2.89±0.04 | 0.25 | | 2.79±0.05 | 0.24 | | 1.03±0.04 |
| $\mathbf{thr}^{L}$ | 3.56±0.08 | 0.30 | 1.23±0.04 | 3.65±0.13 | 0.31 | 1.31±0.07 | 0.97±0.06 |
| $\mathbf{athr}^{D}$ | 1.75±0.07 | 0.15 | | 2.49±0.01 | 0.21 | | 0.70±0.03 |
| $\mathbf{athr}^{L}$ | 4.35±0.09 | 0.37 | 2.49±0.15 | 4.38±0.08 | 0.37 | 1.76±0.04 | 0.99±0.04 |
| $\mathbf{phg}^{D}$ | 4.98±0.12 | 0.43 | | 5.19±0.06 | 0.45 | | 0.96±0.03 |
| $\mathbf{phg}^{L}$ | 4.86±0.10 | 0.42 | 0.98±0.04 | 4.24±0.06 | 0.37 | 0.82±0.02 | 1.14±0.05 |
| $\mathbf{phe}^{D}$ | 3.60±0.03 | 0.32 | | 3.56±0.04 | 0.32 | | 1.01±0.02 |
| $\mathbf{phe}^{L}$ | 2.20±0.03 | 0.20 | 0.61±0.01 | 2.28±0.04 | 0.21 | 0.64±0.02 | 0.96±0.03 |
| $\mathbf{tyr}^{D}$ | 0.86±0.02 | 0.08 | | 1.42±0.02 | 0.13 | | 0.61±0.02 |
| $\mathbf{tyr}^{L}$ | 1.07±0.02 | 0.10 | 1.24±0.06 | 1.36±0.02 | 0.13 | 0.96±0.03 | 0.79±0.02 |
| $(\mathbf{tyrOMe}^{D})_{fast}$ | 1.63±0.19 | 0.15 | | 1.30±0.04 | 0.12 | | 1.25±0.19 |
| $(\mathbf{tyrOMe}^{L})_{fast}$ | 2.61±0.30 | 0.23 | 1.60±0.42 | 1.80±0.09 | 0.16 | 1.38±0.12 | 1.45±0.25 |
| $(\mathbf{tyrOMe}^{D})_{slow}$ | 0.12±0.01 | 0.01 | | 0.08±0.01 | 0.01 | | 1.43±0.15 |
| $(\mathbf{tyrOMe}^{L})_{slow}$ | 0.18±0.01 | 0.02 | 1.50±0.23 | 0.16±0.01 | 0.01 | 2.00±0.43 | 1.17±0.11 |
| $(\mathbf{DOPA}^{D})_{fast}$ | 2.28±0.08 | 0.20 | | 1.26±0.05 | 0.11 | | 1.81±0.12 |
| $(\mathbf{DOPA}^{L})_{fast}$ | 3.00±0.09 | 0.27 | 1.32±0.08 | 1.82±0.20 | 0.16 | 1.44±0.23 | 1.65±0.24 |
| $(\mathbf{DOPA}^{D})_{slow}$ | 0.07±0.01 | 0.01 | | 0.06±0.01 | 0.01 | | 1.14±0.28 |
| $(\mathbf{DOPA}^{L})_{slow}$ | 0.10±0.01 | 0.02 | 1.43±0.40 | 0.08±0.01 | 0.01 | 1.23±0.34 | 1.27±0.25 |
| $(\mathbf{trp}^{D})_{fast}$ | 1.82±0.04 | 0.16 | | 1.92±0.04 | 0.12 | | 0.95±0.04 |
| $(\mathbf{trp}^{L})_{fast}$ | 1.14±0.04 | 0.10 | 0.63±0.04 | 2.33±0.20 | 0.16 | 1.21±0.12 | 0.49±0.06 |
| $(\mathbf{trp}^{D})_{slow}$ | 0.15±0.01 | 0.01 | | 0.23±0.01 | 0.02 | | 0.65±0.04 |
| $(\mathbf{trp}^{L})_{slow}$ | 0.14±0.01 | 0.01 | 0.93±0.14 | 0.37±0.07 | 0.03 | 1.61±0.36 | 0.37±0.09 |
| $\mathbf{pro}^{D}$ | 1.51±0.01 | 0.13 | | 1.38±0.02 | 0.12 | | 1.09±0.03 |
| $\mathbf{pro}^{L}$ | 1.64±0.01 | 0.15 | 1.09±0.01 | 1.50±0.03 | 0.14 | 1.09±0.03 | 1.09±0.03 |
| $\mathbf{pip}^{D}$ | 0.15±0.01 | 0.01 | | 0.12±0.01 | 0.010 | | 1.26±0.04 |
| $\mathbf{pip}^{L}$ | 0.16±0.01 | 0.01 | 1.09±0.03 | 0.16±0.01 | 0.014 | 1.34±0.02 | 1.02±0.04 |

in the encounters with $[\mathbf{15}^{S}\bullet H\bullet A]^+$. Three regions of protonated $\mathbf{15}^{S}$ are found to be most suited for hosting the amino acid A = **ala**, **ser**, **DOPA**, **tyr**, **tyrOMe**, and **trp**: (1) inside the achiral upper-rim cavity (*up*), (2) among the four chiral pendants in correspondence to its chiral lower-rim cavity (*down*), and (3) in the external position in proximity of two adjacent pendants (*ext*) (Figure 6.6).[88,89] MD simulations on low-energy $[\mathbf{15}^{S}\bullet H\bullet \mathbf{ala}]^+$ geometries point to *ext* as the thermodynamically most favored structures at room temperature. The relevant diastereomeric structures are almost equally stable. Therefore, the observed enantioselectivity has to be attributed to specific stabilization of the exchange transition structures. Concerning the [**15**$^{S}$•H•**ser**]$^{+}$ complexes, MM and MD calculations point to *down* as the most favored hosting region for both *L*- (**ser**$^{L}$) and *D*-serine (**ser**$^{D}$), although the complex with **ser**$^{L}$ is ca. 6 kcal mol$^{-1}$ less stable than with **ser**$^{D}$. Accordingly, the pronounced effect of the serine configuration on the exchange reaction (Equation 6.1; ρ = 1.10 ± 0.03 (with B$^{R}$); 2.04±0.05 (with B$^{S}$); Table 6.5) is accounted for by the greater stability of [**15**$^{S}$•H•**ser**$^{D}$]$^{+}_{down}$ relative to [**15**$^{S}$•H•**ser**$^{L}$]$^{+}_{ext}$. The [**15**$^{S}$•H•A]$^{+}$ (A = **thr**, **athr**) systems react with a kinetics and an enantioselectivity similar to those exhibited by [**15**$^{S}$•H•**ser**]$^{+}$.[92]

MM/MD simulations starting from the global minimum of both diastereomeric [**15**$^{S}$•H•**DOPA**]$^{+}$ complexes suggest *down* as the most favorable hosting region of the amino acidic guest, irrespective of its configuration.[93] In order to optimize the interaction with the guest, the pendants of [**15**$^{S}$•H]$^{+}$ adopt a preorganized canyon-shaped architecture due to the formation of a pair of stable H-bonds between adjacent host pendants. The guest in [**15**$^{S}$•H•**DOPA**]$^{+}$ can be permanently trapped not only into the chiral *down* canyon-shaped cleft of the host, but also onto its achiral *up* region. The relevant diastereomeric structures are almost equally stable, with the *down* structures ca. 5 kcal mol$^{-1}$ more stable than the *up* ones. This computational result is consistent with the formation of more than one reacting [**15**$^{S}$•H•**DOPA**]$^{+}$ structure, the most reactive corresponding to *up* and the less reactive to *down*. The *up* stucture displays the highest effect of the B configuration [ξ = 1.81 ± 0.12 (**DOPA**$^{D}$); 1.65 ± 0.24 (**DOPA**$^{L}$)]. This implies that the B amine must get completely inside the empty chiral cavity of [**15**$^{S}$•H•**DOPA**]$^{+}_{up}$ to displace the guest (a "backside" displacement). In contrast, the slower exchange reaction with [15$^{S}$•H•**DOPA**]$^{+}_{down}$ exhibits a much smaller effect of the B configuration [ξ = 1.14 ± 0.28 (**DOPA**$^{D}$); 1.27 ± 0.25 (**DOPA**$^{L}$)], which is similar to that accompanying the reaction with [**15**$^{S}$•H•**ser**$^{D}$]$^{+}_{down}$ (ξ = 1.24 ± 0.05). This observation indicates the involvement of a congested, high-energy transition structure with B not fully inserted into the host chiral cavity, which therefore may exert only in part the effects of its asymmetry toward it.

MM calculations for the diastereomeric [**15**$^{S}$•H•**tyr**]$^{+}$ and [**15**$^{S}$•H•**tyrOMe**]$^{+}$ complexes indicates that, at 0 K, **tyr** preferentially occupies two specific regions of [**15**$^{S}$•H]$^{+}$, namely, the *up* region and a region of the host with the guest placed halfway between the *down* and the *ext* positions.[92] This position has been denoted as the *half* region (Figure 6.6). It should be noted that, relative to **tyr**, the **tyrOMe** guest is located slightly more inward in the host cavity, and more surrounded by its chiral pendants. MD simulations on [**15**$^{S}$•H•**tyr**]$^{+}$ suggest that, at 300 K, the amino acidic guests can maintain permanently both the *half* and the *up* positions of [**15**$^{S}$•H]$^{+}$. A similar behavior has been observed with the *up* and *half* structures of the diastereomeric [**15**$^{S}$•H•**tyrOMe**]$^{+}$ complexes as well. The *half* and *up* regioisomers of either [**15**$^{S}$•H•**tyrOMe**$^{L}$]$^{+}$ and [**15**$^{S}$•H•**tyrOMe**$^{D}$]$^{+}$ exhibit almost the same average potential energy (within 0.2 kcal mol$^{-1}$), when scaled to 300 K. In contrast, the average potential energy of the *up* regioisomer of [**15**$^{S}$•H•**tyr**]$^{+}$ exceeds that of the *half* regioisomer by 2 kcal mol$^{-1}$, in agreement with the largely predominant occurrence of the latter structure under the ESI-FT-ICR-MS conditions.

The concomitant occurrence of stable *half* and *up* regioisomers of the [**15**$^S$•H•**tyrOMe**]$^+$ complexes is consistent with their biexponential reaction kinetics. In analogy with the [**15**$^S$•H•**DOPA**]$^+$ systems and in qualitative agreement with the computed enthalpy gap between the *up* and *half* [**15**$^S$•H•**tyrOMe**]$^+$ structures, the most stable *half* regioisomer is identified as the less reactive [**15**$^S$•H•**tyrOMe**]$^+_{slow}$ component. The less stable *up* regioisomer is obviously identified as the most reactive [**15**$^S$•H•**tyrOMe**]$^+_{fast}$ component (Table 6.5). In contrast, the large stability gap between the *half* and *up* regioisomers of the [**15**$^S$•H•**tyr**]$^+$ complexes (2 kcal mol$^{-1}$) prevents the formation of the latter one, and therefore, only the stable *half* regioisomer is formed under ESI-FT-ICR-MS conditions. This accounts for the monoexponential kinetics followed by the [**15**$^S$•H•**tyr**]$^+$ complexes (Table 6.5).

The comparable ρ values, measured for the reaction with [**15**$^S$•H•**tyr**]$^+$ [$\rho = 1.24 \pm 0.06$ (with B$^R$); $0.96 \pm 0.03$ (with B$^S$)] and [**15**$^S$•H•**tyrOMe**]$^+$ [$\rho = 1.32 \pm 0.08$ (*fast* channel with B$^R$); $1.44 \pm 0.23$ (*fast* channel with B$^S$); $1.43 \pm 0.40$ (*slow* channel with B$^R$); $1.23 \pm 0.34$ (*slow* channel with B$^S$)], find an explanation in the remarkable similarity of the corresponding *half* structures, both characterized by intense *C=O*···H-N-guest and $C_{arom}OH$···O=C intermolecular interactions (the groups belonging to the guest are in italic). The significant difference in the measured ξ terms ($\xi < 1$ for [**15**$^S$•H•**tyr**]$^+$ and $\xi > 1$ for [**15**$^S$•H•**tyrOMe**]$^+$; Table 6.5) is due to the fact that, as pointed out before, **tyrOMe** is located more inward in the host cavity, and surrounded by its chiral pendants, than **tyr**. As a consequence, relative to **tyr**, displacement of **tyrOMe** requires that the amine B enters more in depth into the host cavity, which therefore may exert more the effects of its asymmetry toward it.

MM calculations on the diastereomeric [**15**$^S$•H•**trp**]$^+$ complexes point to the guest enantiomers as residing preferentially at the *up* and *down* regions of the [**15**$^S$•H]$^+$ host.[90,92] MD simulations starting from the global [**15**$^S$•H•**trp**]$^+_{down}$ minima suggest that the guest can reside permanently in the *down* region nearby its chiral lower rim, as well as in its apparently achiral upper rim (the *up* region) (Figure 6.7). The occurrence of stable *down* and *up* structures for the diastereomeric [**15**$^S$•H•**trp**]$^+$ complexes accounts for their biexponential reaction kinetics (Table 6.5). The average potential energy gaps, scaled to 300 K, between the *down* and *up* regioisomers of [**15**$^S$•H•**trp**$^D$]$^+$

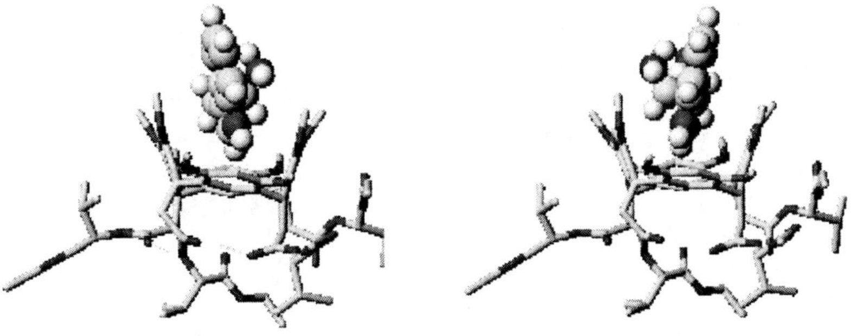

(a) (b)

**FIGURE 6.7** Side view of the fully minimized *up* structures of (a) [**15**$^S$•H•**trp**$^D$]$^+$ and (b) [**15**$^S$•H•**trp**$^L$]$^+$ complexes.

and [$\mathbf{15^S}$•H•$\mathbf{trp^L}$]$^+$ amount, respectively, to 0.7 and 1.9 kcal mol$^{-1}$ in favor of the *down* structures. The average enthalpies of complexation of the diastereomeric [$\mathbf{15^S}$•H•**trp**]$^+$ structures ($\Delta\Delta H_{av} = (\Delta H_{av})_{homo} - (\Delta H_{av})_{hetero}$), calculated from the difference between the average combined enthalpy of the two isolated components and that of the complexes, are estimated to amount to –1.7 and –0.5 kcal mol$^{-1}$ for the *down* and *up* structures, respectively.

In contrast to **DOPA**, the *ext* region of [$\mathbf{15^S}$•H]$^+$ is not suited for hosting **trp**. This different behavior is probably caused by the absence of appropriate OH "hooks" on the **trp** aromatic rings. Indeed, their presence in **DOPA** allows the formation of high-energy *ext* structures. The observation that, like **trp**, both **tyr** and **tyrOMe** never reside at the *ext* region of [$\mathbf{15^S}$•H]$^+$ suggests that the presence of only one OH hook on their aromatic ring is not enough for the formation of the *ext* structure. In spite of the evident differences in the complexation modes, chiral recognition of **DOPA** and trp by [$\mathbf{15^S}$•H]$^+$ exhibits close similarities. The steric energy difference between the relevant *up* and *down* structures is indeed rather close and unaffected by the configuration of the amino acidic guest (the global minima of the diastereomeric complexes are quasi-degenerate). In analogy with **DOPA**, the [$\mathbf{15^S}$•H]$^+$ host adopts a preorganized canyon-shaped architecture by H-bondings between adjacent pendants to embody **trp**. Such H-bonds are maintained during the MD runs, and therefore, the indole moiety of **trp** is firmly accommodated in the canyon-shaped cleft, where it finds sterically complementary and electronically appropriate surfaces of the host. Here, the amino acidic groups of $\mathbf{trp^D}$ interact with two different chiral pendants of the host (four hydrogen bonds), while in the case of $\mathbf{trp^L}$, the indolic NH and the amino acidic groups form three hydrogen bonds with two different host pendants. These different interactions are responsible for the quasi-orthogonal orientation of the **trp** enantiomers in the *down* cavity of [$\mathbf{15^S}$•H]$^+$. The consequence is a different host/guest packing for the relevant *down* [$\mathbf{15^S}$•H•**trp**]$^+$ structures, which may account for the opposite ρ (≥1 with $B^R$; <1 with $B^S$) and the largely different ξ enantioselectivity factors, measured for [$\mathbf{15^S}$•H•**trp**]$^+_{slow}$ (Table 6.5).

The reasons of the *up*-trapping of **trp** can be certainly ascribed to both the low polarity of the guest and the structure of its side chain, which allow the guest to establish a very effective π-π stacking interaction with the electron-rich cavity at the upper rim of the host, strengthened by a dipolar interaction of the NH group, properly located between the methoxy groups of two facing aromatic rings (Figure 6.7a and b). Such a favorable spatial orientation shows close analogies with that of the complex with **DOPA**, wherein the dipolar interaction of the indole NH functionality is replaced by that of the *meta* OH group. In this connection, the similar ρ and ξ values, measured for the *up* and *down* [$\mathbf{15^S}$•H•**trp**]$^+$ structures, are rather surprising since, differently from *down* regioisomers, the *up* ones display a lower-rim cavity that is not appreciably perturbed by the presence of the guest (Figure 6.7a and b). Thus, no appreciable enantioselectivity (ρ ≈ 1) would be expected in the reaction of the [$\mathbf{15^S}$•H•**trp**]$^+$ *up* regioisomers with both amine enantiomers. In contrast, ρ = 1.60 ± 0.11 (with $B^R$) and ρ = 0.82 ± 0.10 (with $B^S$) (Table 6.5). However, a closer inspection of Figure 6.7a and b reveals that the [$\mathbf{15^S}$•H•**trp**]$^+$ *up* structures do have a supramolecular chirality due to the bent position of the guest in the upper-rim cavity of the host.[92] Therefore, despite the chiral lower-rim cavity of the diastereomeric

$[15^S{\bullet}H{\bullet}trp]^+$ *up* complexes being unaffected by the guest, its presence at the *up* region of the host nevertheless generates two supramolecular diastereomeric forms that may be responsible for the observed enantioselectivity. In other words, the enantioselective behavior of the diastereomeric $[15^S{\bullet}H{\bullet}trp]^+$ *up* complexes toward the amine enantiomers is attributed to the operation of a heterotropic allosteric effect in these supramolecular systems.[92]

A peculiar feature of calixarene macrocycles is their nonplanar cyclophane structure, which may give rise to relatively stable stereoisomeric and conformational forms. Besides the flattened cone structure **15**$^S$, the 2,8,14,20-tetrakis(*L*-valinamido) [4]resorcinarene can display several other stable stereoisomeric structures, such as 1,2-alternate (**16**$^S$), chair 1 (**17**$^S$), 1,3-alternate, and chair 2 (Figure 6.8).[93] Its stereoisomerism allows the tailoring of variously shaped cavities that can be probed to evaluate their effect on ion or molecular recognition.[94]

The intrinsic selectivity of the **16**$^S$ and **17**$^S$ isomers toward the pure *D*- and *L*-enantiomers of **phe**, **tyr**, and **DOPA** was measured through Equation 6.1 and compared with the results obtained with **15**$^S$ as host.[94] Linear rate plots were invariably observed with all their $[M{\bullet}H{\bullet}A]^+$ complexes, except those with M/A = **17**$^S$**/tyr** and **17**$^S$**/DOPA**. Therefore, at least two stable isomeric $[M{\bullet}H{\bullet}A]^+$ structures, i.e., $[M{\bullet}H{\bullet}A]^+_{fast}$ and $[M{\bullet}H{\bullet}A]^+_{slow}$, are formed in these systems.

Table 6.6 reports the relevant kinetic enantioselectivity factors. Its analysis shows the strict relationship between the structural features of the selected stereoisomeric hosts and their kinetic enantioselectivity toward a given chiral guest. The relative abundance of the isomeric $[15^S{\bullet}H{\bullet}DOPA]^+_{fast}$ and $[15^S{\bullet}H{\bullet}DOPA]^+_{slow}$ structures (20 ± 4 vs. 80 ± 4%) qualitatively conforms to their computed stability difference (5 kcal mol$^{-1}$).[89] As discussed above, the loss of **DOPA** from the *up* $[15^S{\bullet}H{\bullet}DOPA]^+_{fast}$ regioisomer is promoted by the uptake of B into the chiral hydrophilic rim of the host. A similar uptake is prevented in the *down* $[15^S{\bullet}H{\bullet}DOPA]^+_{slow}$ regioisomer

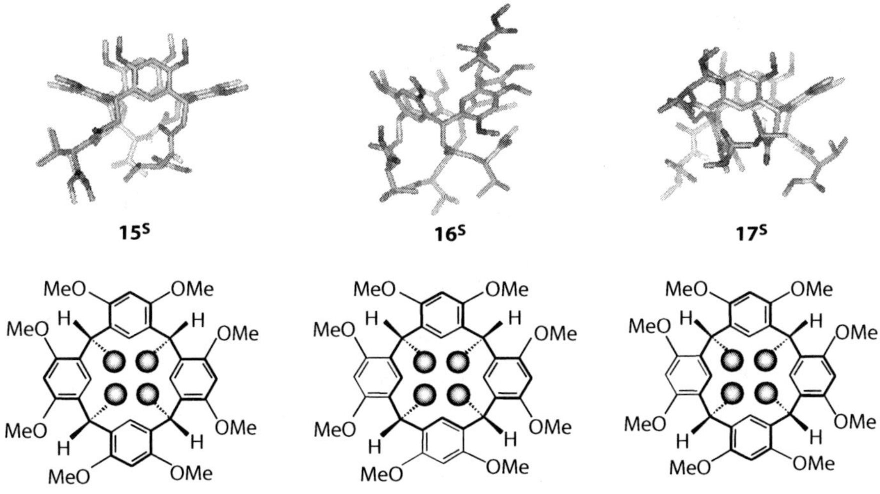


**FIGURE 6.8** Side view of local minimum geometry of 2,8,14,20-tetrakis(*L*-valinamido) [4] resorcinarene in the flattened-cone (**15**$^S$), 1,2-alternate (**16**$^S$), and chair 1 (**17**$^S$) structures.

by the presence of the guest in the hydrophilic cavity of the host. Along this line, the comparatively more leveled distribution of the isomeric $[\mathbf{17^S}\text{•H•}\mathbf{DOPA}]^+_{fast}$ and $[\mathbf{17^S}\text{•H•}\mathbf{DOPA}]^+_{slow}$ structures [48 ± 2 vs. 52 ± 2% (**DOPA**$^\mathbf{D}$); 39 ± 3 vs. 61 ± 3% (**DOPA**$^\mathbf{L}$)] indicates that their stability difference is somewhat reduced by the orientation of the chiral hydrophilic pendants not only along the *down*-like cavity of the host but also nearby its *up*-like region. This effect is mirrored by the reduced reactivity difference of $[\mathbf{17^S}\text{•H•}\mathbf{DOPA}]^+_{fast}$ and $[\mathbf{17^S}\text{•H•}\mathbf{DOPA}]^+_{slow}$ relative to that of the $[\mathbf{15^S}\text{•H•}\mathbf{DOPA}]^+_{fast}$ and $[\mathbf{15^S}\text{•H•}\mathbf{DOPA}]^+_{slow}$ pair (Table 6.6). The size and physical properties of the *up*-like and *down*-like regions of the chair stereoisomer $\mathbf{17^S}$ are markedly different from the *up* and *down* regions of the flattened cone stereoisomer $\mathbf{15^S}$. This may account for their opposite enantioselectivities toward **DOPA** (ρ factors in Table 6.6). A diverse orientation of the incoming base B toward [M•H•**DOPA**]$^+$ (M = $\mathbf{15^S}$; $\mathbf{17^S}$) may play a role as well, as suggested by the appreciable differences in the corresponding ξ terms.

The apparently linear rate plots associated with the B attack on the diastereomeric $[\mathbf{16^S}\text{•H•}\mathbf{DOPA}]^+$ complexes can be interpreted not only with the occurrence of a single complex structure, but also with the occurrence of different structures with comparable reactivity toward the base B. Although no clear-cut evidence is available, the reaction patterns shown by their [M•H•**DOPA**]$^+$ (M = $\mathbf{15^S}$; $\mathbf{17^S}$) stereoisomers lend support to the latter hypothesis. This view well conforms to the ρ and ξ enantioselectivity values of the $[\mathbf{16^S}\text{•H•}\mathbf{DOPA}]^+$ complexes, which are amid those of their [M•H•**DOPA**]$^+$ (M = $\mathbf{15^S}$; $\mathbf{17^S}$) analogues (Table 6.6).

The general enantioselectivity picture shown by the [M•H•**DOPA**]$^+$ complexes found some confirmation in their [M•H•**tyr**]$^+$ analogues. Indeed, although here the unequivocal assignment of two stable isomeric structures is limited only to the $[\mathbf{17^S}\text{•H•}\mathbf{tyr}]^+$ complex, nevertheless the same ρ value trend is observed for both [M•H•**tyr**]$^+$ and [M•H•**DOPA**]$^+$ series, which increases in the M order: $\mathbf{17^S} \leq \mathbf{16^S} < \mathbf{15^S}$ (Table 6.6). Instead, the complete absence of phenolic OH groups in phe has a remarkable effect on the enantioselectivity of the reaction between B and the corresponding [M•H•**phe**]$^+$ complexes. Differently from their [M•H•**tyr**]$^+$ analogues, the [M•H•**phe**]$^+$ complexes invariably exhibited ρ < 1 values, which increase in the reverse M order: $\mathbf{15^S} < \mathbf{16^S} \sim \mathbf{17^S}$.

The effects of the physical environment on the stability of diastereomeric $[\mathbf{15^S}\text{•H•}\mathbf{A}]^+$ (**A**= **DOPA** methyl ester (**DOPAOMe**), **DOPA** ethyl ester (**DOPAOEt**), and **trp** ethyl ester (**trpOEt**)) complexes have been investigated in the gas phase by ESI-FT-ICR-MS and in CDCl3 solutions by $^1$H and $^{13}$C-NMR spectroscopy.[95] It was found that the noncovalent $[\mathbf{15^S}\text{•H•}\mathbf{DOPAOMe}]^+$ and $[\mathbf{15^S}\text{•H•}\mathbf{DOPAOEt}]^+$ are stable in the gas phase. Their exchange reaction 6.1 with either (S)-(+)- ($B^S$) or (R)-(-)-2-aminobutane ($B^R$) follows a different kinetic law. Indeed, $[\mathbf{15^S}\text{•H•}\mathbf{DOPAOMe}]^+$ follows a bi-exponential kinetics, much like $[\mathbf{15^S}\text{•H•}\mathbf{DOPA}]^+$, which implies the occurrence of at least two stable isomeric structures, most likely *up* and *down*. In contrast, $[\mathbf{15^S}\text{•H•}\mathbf{DOPAOEt}]^+$ follows a mono-exponential kinetics, which suggests the occurrence of a single stable *down* structure.

A stable $[\mathbf{15^S}\text{•H•}\mathbf{DOPAOEt}]^+$ complex has been observed in $CDCl_3$ solutions as well by $^1$H and $^{13}$C-NMR spectroscopy. No evidence for a similarly stable $[\mathbf{15^S}\text{•H•}\mathbf{trpOEt}]^+$ complex has been obtained under the same conditions. The

**TABLE 6.6**
**Exchange Rate Constants ($k \times 10^{10}$ cm$^3$ mol$^{-1}$ s$^{-1}$)**

| Host (M) | Guest (A) | (%) | (R)-(-)-$C_4H_9NH_2$ | | | (S)-(+)-$C_4H_9NH_2$ | | | |
|---|---|---|---|---|---|---|---|---|---|
| | | | $k$ | eff | $\rho$ | $k$ | eff | $\rho$ | $\xi$ |
| $15^S$ | $phe^D$ | | 3.60±0.03 | 0.32 | | 3.56±0.04 | 0.32 | | 1.01±0.02 |
| | $phe^L$ | | 2.20±0.03 | 0.20 | 0.61±0.01 | 2.28±0.04 | 0.21 | 0.64±0.02 | 0.96±0.03 |
| | $tyr^D$ | | 0.86±0.02 | 0.08 | | 1.42±0.02 | 0.13 | | 0.61±0.02 |
| | $tyr^L$ | | 1.07±0.02 | 0.10 | 1.24±0.06 | 1.36±0.02 | 0.13 | 0.96±0.03 | 0.79±0.02 |
| | $(DOPA^D)_{fast}$ | 20±4 | 2.28±0.08 | 0.20 | | 1.26±0.05 | 0.11 | | 1.81±0.12 |
| | $(DOPA^L)_{fast}$ | 19±3 | 3.00±0.09 | 0.27 | 1.32±0.08 | 1.82±0.20 | 0.16 | 1.44±0.23 | 1.65±0.24 |
| | $(DOPA^D)_{slow}$ | 80±4 | 0.07±0.01 | 0.01 | | 0.06±0.01 | 0.01 | | 1.14±0.28 |
| | $(DOPA^L)_{slow}$ | 81±3 | 0.10±0.01 | 0.02 | 1.43±0.40 | 0.08±0.01 | 0.01 | 1.23±0.34 | 1.27±0.25 |
| $16^S$ | $phe^D$ | | 4.06±0.09 | 0.36 | | 5.15±0.06 | 0.46 | | 0.79±0.02 |
| | $phe^L$ | | 3.78±0.07 | 0.34 | 0.93±0.06 | 4.93±0.06 | 0.44 | 0.96±0.02 | 0.77±0.02 |
| | $tyr^D$ | | 2.29±0.04 | 0.21 | | 2.93±0.03 | 0.26 | | 0.78±0.02 |
| | $tyr^L$ | | 3.06±0.06 | 0.27 | 1.34±0.05 | 3.81±0.06 | 0.34 | 1.30±0.03 | 0.80±0.03 |
| | $DOPA^D$ | | 1.57±0.06 | 0.14 | | 1.91±0.06 | 0.17 | | 0.82±0.02 |
| | $DOPA^L$ | | 1.47±0.05 | 0.13 | 0.94±0.07 | 1.55±0.05 | 0.14 | 0.81±0.05 | 0.95±0.03 |
| $17^S$ | $phe^D$ | | 3.67±0.04 | 0.33 | | 3.61±0.04 | 0.32 | | 1.02±0.02 |
| | $phe^L$ | | 3.20±0.05 | 0.29 | 0.87±0.01 | 3.25±0.04 | 0.29 | 0.90±0.02 | 0.98±0.04 |
| | $(tyr^D)_{fast}$ | 42±3 | 3.39±0.08 | 0.30 | | 3.95±0.11 | 0.35 | | 0.86±0.04 |
| | $(tyr^L)_{fast}$ | 44±2 | 3.24±0.06 | 0.29 | 0.96±0.02 | 3.00±0.13 | 0.27 | 0.76±0.05 | 1.08±0.07 |
| | $(tyr^D)_{slow}$ | 58±3 | 0.90±0.03 | 0.08 | | 0.97±0.04 | 0.09 | | 0.93±0.06 |
| | $(tyr^L)_{slow}$ | 56±2 | 0.58±0.03 | 0.05 | 0.64±0.06 | 0.61±0.03 | 0.06 | 0.63±0.06 | 0.95±0.09 |
| | $(DOPA^D)_{fast}$ | 48±2 | 1.27±0.03 | 0.11 | | 0.87±0.10 | 0.08 | | 1.45±0.23 |
| | $(DOPA^L)_{fast}$ | 39±3 | 0.47±0.04 | 0.04 | 0.37±0.04 | 1.12±0.10 | 0.10 | 1.44±0.23 | 0.42±0.08 |
| | $(DOPA^D)_{slow}$ | 52±2 | 0.09±0.01 | 0.008 | | 0.09±0.03 | 0.009 | | 0.99±0.64 |
| | $(DOPA^L)_{slow}$ | 61±3 | 0.04±0.01 | 0.004 | 0.44±0.08 | 0.05±0.01 | 0.004 | 0.48±0.40 | 0.86±0.35 |

formation of the stable [**15^S•H•DOPAOEt**]$^+$ complex in $CDCl_3$ is not affected by the presence of traces of additives, like $D_2O$, DCl, or ethyl acetate, or by absorption on silica. APT-$^{13}$C-NMR analysis of [**15^S•H•DOPAOEt**]$^+$ indicated that the amino ester is preferentially located inside the **15**$^S$ chiral cavity, in conformity with the most stable down structure predicted by MM/MD simulations.[89,92]

N-Linked peptidoresorc[4]arenes **18**$^R$ and **19**$^R$ (Figure 6.9) exhibit a pronounced enantioselectivity towards the homologue dipeptides **leu-val** and **val-leu**, respectively, both in solution and in the gas phase.[96] The gas-phase structure and enantioselectivity of the enantiomers of the more rigid bis(diamido)-bridged basket resorcin[4] arenes **20**$^{R/S}$ and **21**$^{R/S}$ (Figure 6.9) have been investigated as well.[96] The dimethoxybenzene rings of **20**$^{R/S}$ and **21**$^{R/S}$ assume the expected flattened cone arrangement, whereas, as shown in Figure 6.10, three different conformations are accessible as regard to the position of the bridged 1,2-diamino substituents, designated as "open wings," "mixed wings," and "folded wings" structures. The exchange reaction 6.1 between diastereomeric [M•H•A]$^+$ (M=**21**$^{R/S}$; A=*L*-**tyrOMe**, L-amphetamine (*L*-**amph**)) complexes and (*R*)-(-)-2-aminobutane ($B^R$) or (*S*)-(+)-2-aminobutane ($B^S$) has been investigated by ESI-FT-ICR-MS.

The results of Table 6.7, in particular the ρ and ξ terms diverging from unity, indicate that the kinetics of the gas phase displacement depend on the nature and the configuration of the guest A and on the configuration of the B amine. Thus, the

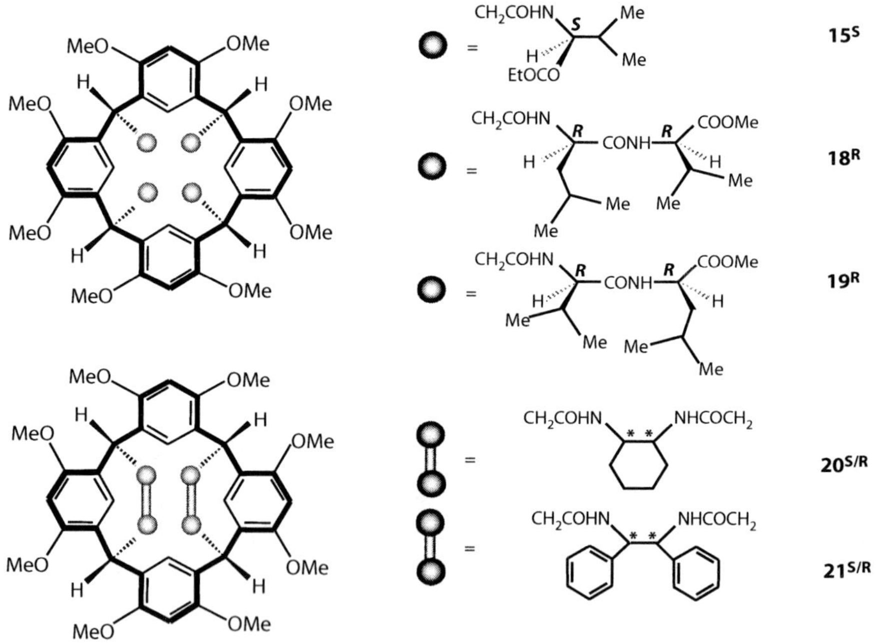


**FIGURE 6.9** Formulas of 2,8,14,20-tetrakis(L-valinamido)[4]resorcinarene (**15**$^S$), 2,8,14,20-tetrakis (*D*-leucyl-*D*-valinamido)[4]resorcinarene (**18**$^R$), 2,8,14,20-tetrakis(*D*-valyl-*D*-leucinamido) [4] resorcinarene (**19**$^R$), basket bis(diamidocyclohexane)resorcin[4]arenes (**20**$^{S/R}$), and basket bis(diamidodiphenylethylene)resorcin[4]arenes (**21**$^{S/R}$).

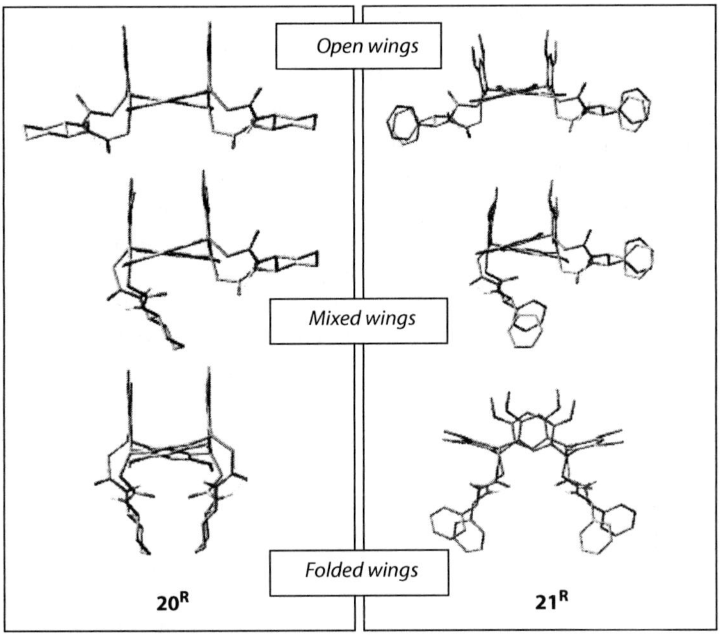


**FIGURE 6.10** Relevant minimum energy structures of the **$20^R$** and **$21^R$** hosts found by conformational search.

**TABLE 6.7**
**Exchange Rate Constants ($k \times 10^{-10}$ cm$^3$ mol$^{-1}$ s$^{-1}$).**

| Host (M) | Guest (A) | (R)-(−)-$C_4H_9NH_2$ | | | (S)-(+)-$C_4H_9NH_2$ | | | |
|---|---|---|---|---|---|---|---|---|
| | | *k* | eff | *ρ* | *k* | eff | *ρ* | ξ |
| **$21^R$** | **L-tyr$^{OMe}$** | 1.20±0.01 | 0.10 | | 1.35±0.03 | 0.11 | | 0.89±0.03 |
| **$21^S$** | **L-tyr$^{OMe}$** | 1.12±0.02 | 0.09 | 0.93±0.03 | 1.06±0.02 | 0.09 | 0.78±0.04 | 1.06±0.04 |
| **$21^R$** | **L-amph** | 1.34±0.05 | 0.11 | | 1.07±0.03 | 0.09 | | 1.25±0.09 |
| **$21^S$** | **L-amph** | 1.06±0.02 | 0.09 | 1.26±0.09 | 1.17±0.04 | 0.10 | 0.91±0.06 | 0.91±0.05 |

diastereomeric [**$21^{R/S}$•H•L-tyrOMe**]$^+$ complexes invariably exhibit ρ<1 values, in qualitative agreement with the *homo>hetero* stability trend of the same complexes determined by Cooks' kinetic method. In contrast, the diastereomeric [**$21^{R/S}$•H•L-amph**]$^+$ complexes do not follow an univocal trend, as they display ρ>1 factors in the reaction with $B^R$ and ρ<1 factors in that with BS. This opposite enantioselectivity is reflected in the corresponding ξ terms, which exhibit a ξ<1 value with the heterochiral complex and a ξ>1 value with the homochiral one. This picture confirms the view that the enantioselectivity of reaction 6.1 with [**$21^{R/S}$•H•L-amph**]$^+$ is essentially kinetic. It is governed by the effects of the resorcin[4]arene frame upon the transition structures involved in the displacement reaction, whereas the relative stability of the diastereomeric [$\mathbf{21}^{R/S}$•H•**L-amph**]$^+$ complexes plays only a minor role.

The enantioselectivity of the M hosts of Figure 6.9 towards the enantiomers of amphetamine ($A^R$=*L*-**amph**; $A^S$=*D*-**amph**) was checked by measuring the rate constants of their exchange reaction 6.1 with the enantiomers of 2-aminobutane $B^R$ or $B^S$.[97] The proton-bound complexes [M•H•A]$^+$ were generated in the FT-ICR-MS by electrospraying M/A methanolic solutions. The same complexes are formed together with proton-bound [M•H•$(A)_2$•HCl]$^+$ aggregates from ESI of M/$AH^+$•$Cl^-$ methanolic solutions. The reaction between [M•H•A]$^+$ and amines B leads to the exclusive formation of the guest-exchange product [M•H•B]$^+$ [eq. 6.1], while that with [M•H•$(A)_2$•HCl]$^+$ proceeds through a consecutive B-to-A displacement sequence with formation of the [M•H•A•B•HCl]$^+$ and [M•H•$(B)_2$•HCl]$^+$ products [eqs. 6.2 and 6.3].

$$[M•H•(A)_2•HCl]^+ + B \rightarrow [M•H•A•B•HCl]^+ + A \quad (6.2)$$

$$[M•H•A•B•HCl]^+ + B \rightarrow [M•H•(B)_2•HCl]^+ + A \quad (6.3)$$

As mentioned before, the pseudo-first-order rate constants ($k'$) of Equations 6.2 and 6.3 were obtained from the slopes of the relevant $\ln(I/I_0)$ vs. $t$ plots, where $I$ is the signal intensity of the corresponding starting complex at the delay time $t$, and $I_0$ is the sum of the signal intensities of the starting complex and its products. The pseudo-first-order rate constant of step 6.3 was derived by best-fitting the relative abundance of the [M•H•$(A)_{2\text{-}n}$•$B_n$•HCl]$^+$ (n = 0–2) ions as a function of the delay time $t$ using as the only constraint the $k'$ value obtained for step 6.2 from the relevant ln $(I/I_0)$ vs. $t$ plot. The second-order rate constants ($k = k'/[B]$) are denoted according to the configurations of the A, M, and B molecules.

Irrespective of the starting methanolic solution, whether containing the free A base or its $AH^+$•$Cl^-$ hydrochloride, the ESI-formed [M•H•A]$^+$ complex always exhibits the same exchange rate constant. This suggests that the same [M•H•A]$^+$ structure is formed both from the free A and from the $AH^+$•$Cl^-$ precursors. Linear rate plots are invariably observed in both Equations 6.2 and 6.3. This common kinetic behavior points to a single isomeric structure for both the [M•H•A]$^+$ and [M•H•$(A)_2$•HCl]$^+$ complexes. The second-order rate constants ($k$) for all the displacement reactions (Equations 6.1–6.3) investigated are listed in Table 6.8. With regard to the exchange reaction (Equation 6.1), the [M•H•A]$^+$ complexes with M = $\mathbf{15}^S$ and $\mathbf{19}^R$ show the greatest enantioselectivity. In general, the effect of the configuration of B on the reaction kinetics is not very pronounced ($0.82 < \xi < 1.25$). Base-induced loss of the first A molecule from the [M•H•$(A)_2$•HCl]$^+$ adducts (Equation 6.2) is up to five times faster than the loss of the same molecule from the corresponding [M•H•A]$^+$ complexes (Equation 6.1). Equation 6.2 displays significant enantioselectivity only when the diastereomeric [$\mathbf{19}^R$•H•$(A)_2$•HCl]$^+$ complexes are involved. Base-induced loss of the residual A molecule from the [M•H•A•B•HCl]$^+$ adducts (Equation 6.3) is normally up to five times slower than the loss of the first A molecule from the [M•H•$(A)_2$•HCl]$^+$

adducts (Equation 6.2). However, Equation 6.3 displays an enantioselectivity that can be qualitatively and quantitatively different from that exhibited by the preceding step (step 6.4). For instance, the ρ values measured with the diastereomeric [**20**•H•$A^R$•B•HCl]$^+$ complexes amount to ca. 1.3, whereas those obtained for their [**20**•H•$(A^R)_2$•HCl]$^+$ precursors do not exceed 1.1. In addition, $\rho < 1$ values are invariably measured with the diastereomeric [**18$^R$**•H•A•B•HCl]$^+$ complexes, whereas $\rho > 1$ values were obtained with their [**18$^R$**•H•$(A)_2$•HCl]$^+$ precursors.

The relatively high enantioselectivity, displayed by the [15$^S$•H•**amph**]$^+$ complexes, can be accounted for by MM and MD calculations, which suggest that the homochiral [**15$^S$**•H•*D*-**amph**]$^+$ complex has higher density of rotational states accessible to the guest than does the heterochiral [**15$^S$**•H•*L*-**amph**]$^+$ structure. The guest in both cases is located in the external position, but only in the homochiral complex does it have the possibility to rotate with more frequent and complete oscillations, maintaining the hydrogen bond with the host's pendant. This rotational disorder of homochiral species is reflected into the higher entropy that makes this structure more stable and less reactive in the displacement reaction. The results obtained on the structures of [**15$^S$**•H•$(A)_{n+1}$•$(HCl)_n$]$^+$ (n = 0, 1) point to the chloride ion placed inside the host cavity and pushing the two molecules of guests far from the chiral frame. This result explains the lower or even absent effect of the chirality in this set of experiments.

Analysis of Table 6.8 reveals that lengthening of the chiral pendants of the host from *L*-valine ethyl ester (M = **15$^S$**) to the *D*-leucyl-*D*-valine (M = **18$^R$**) and *D*-valyl-*D*-leucine (M = **19$^R$**) methyl esters has a significant effect on the kinetics and enantioselectivities of the B-to-A displacements in the diastereomeric [M•H•A]$^+$ complexes. In particular, while the ρ enantioselectivity factors for the B-induced displacement in [**15$^S$**•H•A]$^+$ [$\rho = 0.39 \pm 0.02$ ($B^R$); $0.45 \pm 0.02$ ($B^S$)] and in [**19$^R$**•H•A]$^+$ [$\rho = 0.55 \pm 0.02$ ($B^R$); $0.51 \pm 0.01$($B^S$)] are wholly comparable, those for the same reaction with [**18$^R$**•H•A]$^+$ are significantly different and close to unity [$\rho = 1.05 \pm 0.05$ ($B^R$); $1.26 \pm 0.04$ ($B^S$)]. This observation suggests that the most populated [**19$^R$**•H•A]$^+$ structure strictly resembles the corresponding *ext*-[**15$^S$**•H•A]$^+$ one, in which the A guest is proton bonded to one of the valinamido carbonyls of the host (Figure 6.11). In [**18$^R$**•H•A]$^+$, the same interaction instead connects the A moiety with the leucinamido carbonyls of the host. The different physical environment of the departing A in the B-to-A displacement may be responsible for the different ρ values between *ext*-[**18$^R$**•H•A]$^+$ and the *ext*-[**15$^S$**•H•A]$^+$/*ext*-[**19$^R$**•H•A]$^+$ pair. The effects of the lengths of the chiral pendants of the host also extend to the enantioselectivities of the B-to-A displacements in the diastereomeric [M•H•$(A)_2$•HCl]$^+$ complexes. In particular, the sequential base-induced losses of the A molecule from the [**19$^R$**•H•$(A)_2$•HCl]$^+$ adducts (Equations 6.2 and 6.3) are both appreciably faster and more selective [$\rho = 0.68 \pm 0.02$ (Equation 6.2), 0.71 (Equation 6.3) ($B^R$); $0.66 \pm 0.04$ (Equation 6.2), 0.60 (Equation 6.3) ($B^S$)] than the same processes with [**15$^S$**•H•$(A)_2$•HCl]$^+$ and [**18$^R$**•H•$(A)_2$•HCl]$^+$ (Table 6.8). Such a significant enantioselectivity can be accounted for by the fact that in [M•H•$(A)_2$•HCl]$^+$ the chloride ion occupies a position nearby the inner NH group of the host (Figure 6.11) and therefore pushes the **amph**H$^+$ moiety down to the host pendants. Unlike the diastereomeric [**15$^S$**•H•$(A)_2$•HCl]$^+$ structures, the proton-bonded A guests in the [**19$^R$**•H•$(A)_2$•HCl]$^+$ congeners are still close to the $Cl^-$ "spacer," but now they are surrounded by the chiral leucine methyl ester tails. If they are instead surrounded by the less encumbered

**TABLE 6.8**
**Exchange Rate Constants: A = Amphetamine; B = 2-Aminobutane ($k \times 10^{10}$ cm$^3$ mol$^{-1}$ s$^{-1}$)**

| Host | Complex | (R)-(-)-$C_4H_9NH_2$ | $\rho$ | (S)-(+)-$C_4H_9NH_2$ | $\rho$ | $\xi$ | Reaction Efficiency[a] | |
|---|---|---|---|---|---|---|---|---|
| M | | $k_R$ | | $k_S$ | | | $k_R/k_C$ | $k_S/k_C$ |
| $15^S$ | $[M{\bullet}H{\bullet}A]^+_{hetero}$ | 1.03±0.03 | | 0.83±0.03 | | 1.23±0.09 | 0.09 | 0.07 |
| | $[M{\bullet}H{\bullet}A]^+_{homo}$ | 0.40±0.01 | 0.39±0.02 | 0.37±0.01 | 0.45±0.02 | 1.09±0.03 | 0.03 | 0.03 |
| | $[M{\bullet}H{\bullet}(A)_2{\bullet}HCl]^+_{hetero}$ | 1.84±0.01 | | 1.42±0.06 | | 1.30±0.03 | 0.16 | 0.12 |
| | $[M{\bullet}H{\bullet}(A)_2{\bullet}HCl]^+_{homo}$ | 1.99±0.04 | 1.08±0.03 | 1.70±0.06 | 1.20±0.09 | 1.17±0.07 | 0.17 | 0.15 |
| | $[M{\bullet}H{\bullet}A{\bullet}B{\bullet}HCl]^+_{hetero}$ | 0.97 | | 0.62 | | 1.56 | 0.08 | 0.05 |
| | $[M{\bullet}H{\bullet}A{\bullet}B{\bullet}HCl]^+_{homo}$ | 0.96 | 0.99 | 0.84 | 1.35 | 1.14 | 0.08 | 0.07 |
| $16^R$ | $[M{\bullet}H{\bullet}A]^+_{hetero}$ | 1.11±0.03 | | 1.14±0.01 | | 0.97±0.04 | 0.09 | 0.10 |
| | $[M{\bullet}H{\bullet}A]^+_{homo}$ | 1.17±0.02 | 1.05±0.05 | 1.42±0.02 | 1.26±0.04 | 0.82±0.05 | 0.10 | 0.12 |
| | $[M{\bullet}H{\bullet}(A)_2{\bullet}HCl]^+_{hetero}$ | 1.81±0.03 | | 1.07±0.05 | | 1.69±0.11 | 0.15 | 0.09 |
| | $[M{\bullet}H{\bullet}(A)_2{\bullet}HCl]^+_{homo}$ | 1.91±0.02 | 1.05±0.02 | 2.00±0.09 | 1.87±0.18 | 0.95±0.06 | 0.16 | 0.17 |
| | $[M{\bullet}H{\bullet}A{\bullet}B{\bullet}HCl]^+_{hetero}$ | 0.96 | | 0.80 | | 1.20 | 0.09 | 0.07 |
| | $[M{\bullet}H{\bullet}A{\bullet}B{\bullet}HCl]^+_{homo}$ | 0.70 | 0.73 | 0.79 | 0.99 | 0.88 | 0.06 | 0.07 |
| $17^R$ | $[M{\bullet}H{\bullet}A]^+_{hetero}$ | 1.50±0.01 | | 1.54±0.02 | | 0.97±0.02 | 0.13 | 0.13 |
| | $[M{\bullet}H{\bullet}A]^+_{homo}$ | 0.82±0.02 | 0.55±0.02 | 0.78±0.01 | 0.51±0.01 | 1.05±0.04 | 0.07 | 0.07 |
| | $[M{\bullet}H{\bullet}(A)2{\bullet}HCl]^+_{hetero}$ | 4.14±0.08 | | 4.48±0.13 | | 0.92±0.05 | 0.36 | 0.38 |
| | $[M{\bullet}H{\bullet}(A)2{\bullet}HCl]^+_{homo}$ | 2.82±0.02 | 0.68±0.02 | 2.94±0.09 | 0.66±0.04 | 0.96±0.04 | 0.24 | 0.25 |
| | $[M{\bullet}H{\bullet}A{\bullet}B{\bullet}HCl]^+_{hetero}$ | 1.61 | | 1.73 | | 0.93 | 0.14 | 0.15 |
| | $[M{\bullet}H{\bullet}A{\bullet}B{\bullet}HCl]^+_{homo}$ | 1.14 | 0.71 | 1.02 | 0.60 | 1.12 | 0.10 | 0.09 |
| $18^{R/S}$ | $[M{\bullet}H{\bullet}A^R]^+_{hetero}$ | 1.70±0.07 | | 1.62±0.07 | | 1.05±0.07 | 0.14 | 0.14 |
| | $[M{\bullet}H{\bullet}A^R]^+_{homo}$ | 1.90±0.07 | 1.12±0.09 | 1.85±0.06 | 1.14±0.09 | 1.03±0.04 | 0.16 | 0.16 |

| | | | | | | | | |
|---|---|---|---|---|---|---|---|---|
| | $[M{\bullet}H{\bullet}(A^R)2{\bullet}HCl]^+_{hetero}$ | 3.87±0.07 | | 3.79±0.05 | | 1.02±0.03 | 0.33 | 0.32 |
| | $[M{\bullet}H{\bullet}(A^R)2{\bullet}HCl]^+_{homo}$ | 4.30±0.08 | 1.11±0.04 | 3.79±0.06 | 1.00±0.03 | 1.13±0.04 | 0.37 | 0.32 |
| | $[M{\bullet}H{\bullet}A^R{\bullet}B{\bullet}HCl]^+_{hetero}$ | 1.04 | | 1.15 | | 0.90 | 0.09 | 0.10 |
| | $[M{\bullet}H{\bullet}A^R{\bullet}B{\bullet}HCl]^+_{homo}$ | 1.31 | 1.26 | 1.46 | 1.27 | 0.90 | 0.11 | 0.12 |
| **19**$^{R/S}$ | $[M{\bullet}H{\bullet}A^R]^+_{hetero}$ | 1.06±0.02 | | 1.17±0.04 | | 0.91±0.05 | 0.09 | 0.10 |
| | $[M{\bullet}H{\bullet}A^R]^+_{homo}$ | 1.34±0.05 | 1.26±0.09 | 1.07±0.03 | 0.91±0.06 | 1.25±0.09 | 0.11 | 0.09 |
| | $[M{\bullet}H{\bullet}(A^R)2{\bullet}HCl]^+_{hetero}$ | 3.05±0.11 | | 2.60±0.03 | | 1.17±0.06 | 0.26 | 0.22 |
| | $[M{\bullet}H{\bullet}(A^R)2{\bullet}HCl]^+_{homo}$ | 3.54±0.02 | 1.16±0.05 | 3.20±0.03 | 1.23±0.03 | 1.11±0.01 | 0.30 | 0.27 |
| | $[M{\bullet}H{\bullet}A^R{\bullet}B{\bullet}HCl]^+_{hetero}$ | 0.71 | | 0.68 | | 1.04 | 0.06 | 0.06 |
| | $[M{\bullet}H{\bullet}A^R{\bullet}B{\bullet}HCl]^+_{homo}$ | 0.82 | 1.15 | 0.64 | 0.94 | 1.28 | 0.07 | 0.05 |

[a] Reaction efficiency expressed by the ratio between the measured rate constants and the corresponding collision constant $k_C$, calculated using the trajectory evaluation method (Reference 16).

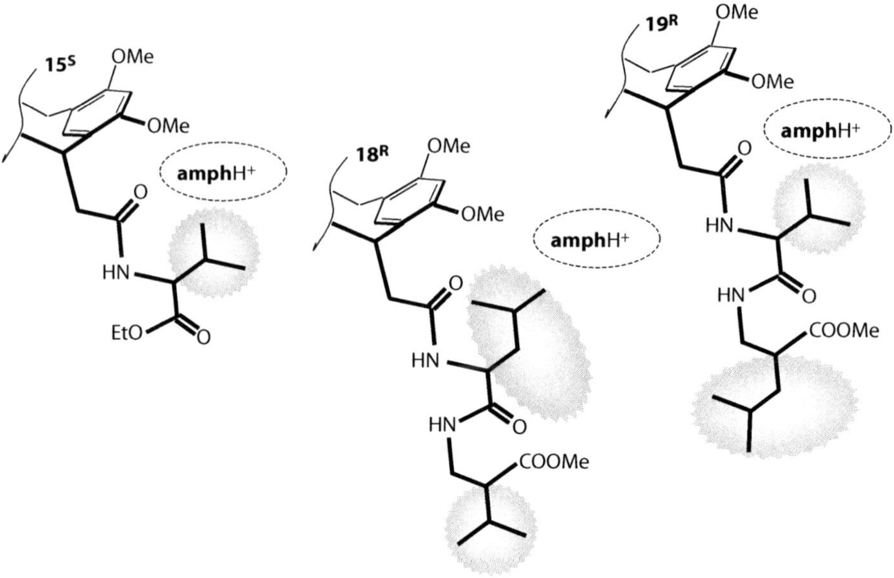


**FIGURE 6.11** Most favored proton-bond interactions between amphetamine (**amph**) and the pendants of M in the *ext* [M•H•**amph**]$^+$ complexes (M = **14**$^S$, **18**$^R$, and **19**$^R$).

valine methyl ester ones, as in [**18**$^R$•H•$(A)_2$•HCl]$^+$, the enantioselectivity of the first B-to-A displacement is strongly reduced and even inverted [$\rho = 1.05 \pm 0.02$ ($B^R$); 1.87 $\pm$ 0.18 ($B^S$); Table 6.8], whereas that of the further B-to-A displacement is essentially unchanged. Such a pronounced variability of the enantioselectivity factors as a function of the natures and isomeric structures of the chiral pendants of the flexible hosts **15**, **18**, and **19** does not find any correspondence in the family of the rigid hosts **20** and **21**. Here, in fact, both the measured $\rho$ and $\xi$ factors are close to unity (Table 6.8). A major reason for such different behavior can be found in the "open" cavity of the host, caused by the specific position of the chiral cyclohexane (M = **20**) and diphenylethane (M = **21**) moieties placed at the largest distance from each other. With this rigid arrangement, the host is not able to exert any significant chiral discrimination either toward the amphetamine moiety or toward the incoming 2-aminobutane reactant.

#### 6.3.2.4 Tetra-Amide Macrocycles as Chiral Hosts

Interest in macrocyclic amides as hosts comes from the amphiphilic properties of their amido groups in dipolar or H-bonding interactions, the carbonyl acting as a dipole donor and a H-bond acceptor and the N-H as a dipole acceptor and a H-bond donor (Chart 6.4). A comprehensive kinetic investigation was undertaken several years ago by Speranza and coworkers on the gas-phase chiral recognition of selected amino acid derivatives (A), i.e., phenylalanine (**phe**), its amide (**pheNH$_2$**), its ethyl ester (**pheOEt**), and the ethyl ester of naphthylalanine (**naphOEt**) by some representative tetra-amidic macrocycles (M = **22**$^R$, **23**$^{R/S}$ (and $d_6$-**23**$^S$), and **24**$^{R/S}$).[98,99] The aim of this study was to (1) verify the effects of the functional groups of guest A on the enantioselectivity of the gas-phase reaction (Equation 6.1), (2) elucidate the

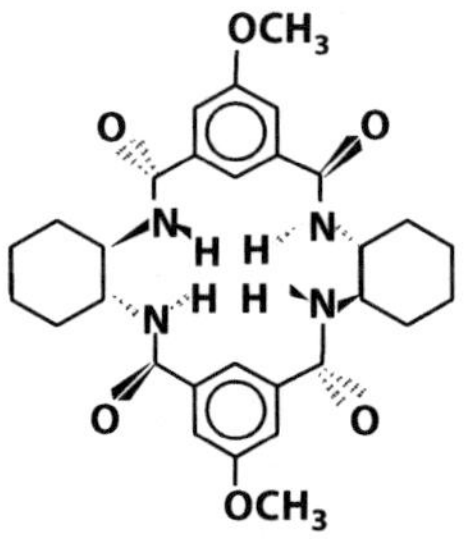

**22^R**

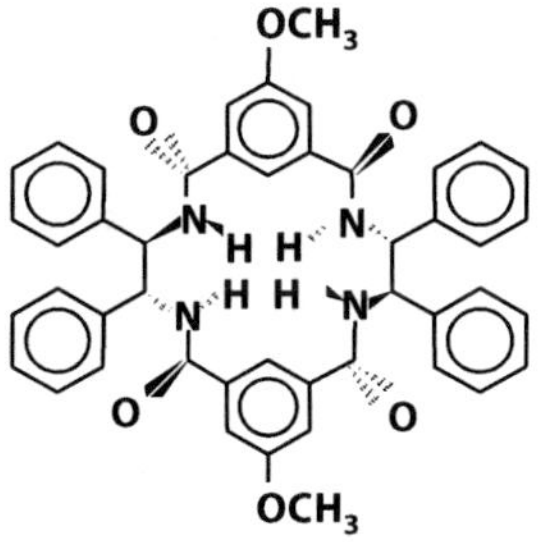

**23^R**

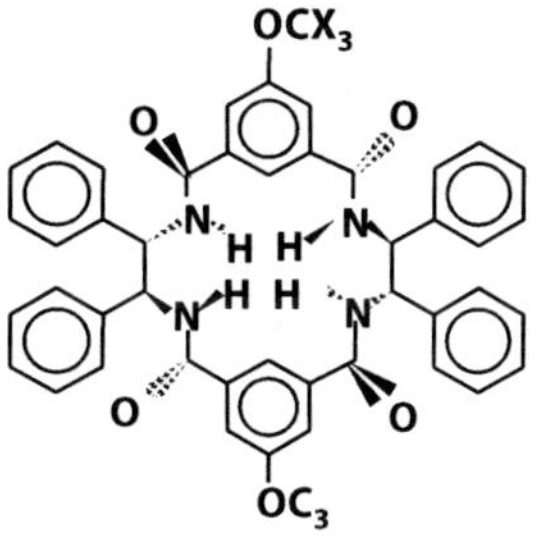

**23^S** (X = H)
d$_6$-**23^S** (X = D)

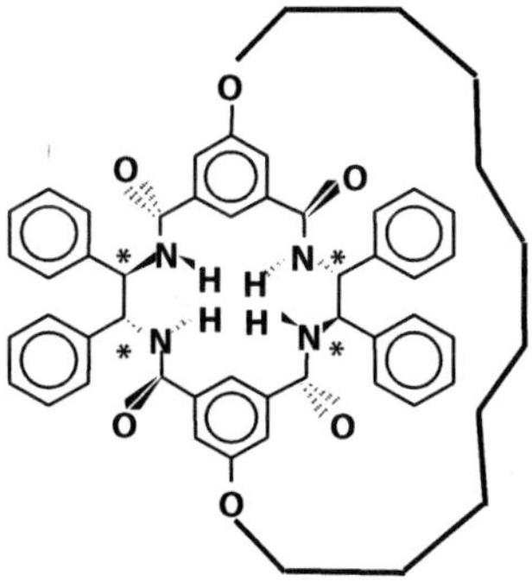

**24^R**

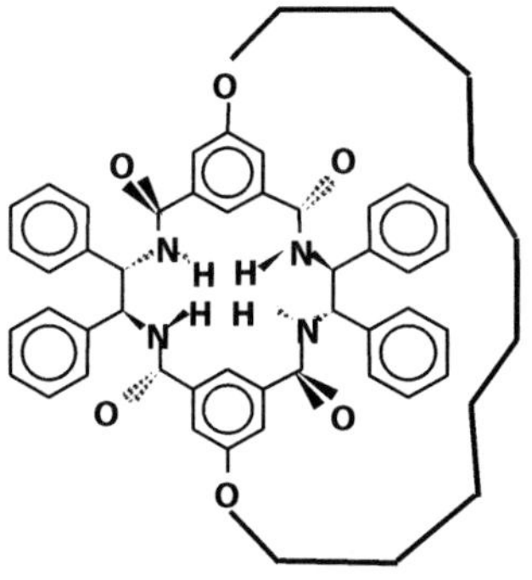

**24^S**

**CHART 6.4**

structural features of the diastereomeric $[M\bullet H\bullet A]^+$ complexes, and (3) shed some light on the dynamics and the mechanism of Equation 6.1 involving the selected compounds.

Table 6.9 reports the relevant ESI-FT-ICR-MS kinetic results.[98] Monoexponential kinetic curves were obtained for all the systems with M = **22**$^R$ and A = **phe**, **pheNH2**, and **pheOEt**, and with M = **23**$^R$ and A = **phe** and **pheNH**$_2$. The heterochiral [**23**$^R$•H•**pheOEt**$^L$]$^+$ complex was found to follow monoexponential kinetics, while its homochiral [**23**$^R$•H•**pheOEt**$^D$]$^+$ analogue follows biexponential kinetics under the same experimental conditions. This means that at least two stable isomeric structures of [**23**$^R$•H•**pheOEt**$^D$]$^+$ coexist in the gas phase at room temperature, one less reactive ([**23**$^R$•H•**pheOEt**$^D$]$^+_{slow}$) and the other more reactive ([**23**$^R$•H•**pheOEt**$^D$]$^+_{fast}$) toward B. In contrast, the monoexponential kinetics, exhibited by the heterochiral [**23**$^R$•H•**pheOEt**$^L$]$^+$ complex, is attributed to the occurrence of a single structure or, alternatively, of several stable isomers, but with comparable reactivity toward B. The same rationale applies to the monoexponential kinetics observed for the guest exchange reactions in the [**22**$^R$•H•A]$^+$ (A = **phe**, **pheNH**$_2$, and **pheOEt**) and [**23**$^R$•H•A]$^+$ (A = **phe** and **pheNH**$_2$) complexes. The exchange reaction (Equation 6.1) on the diastereomeric [**23**$^{R/S}$•H•**naphOEt**$^L$]$^+$ and [**24**$^{R/S}$•H•**naphOEt**$^L$]$^+$ complexes was found to follow biexponential kinetics, which suggests the presence of two persistent isomeric

**TABLE 6.9**
**Exchange Rate Constants ($\times10^{11}$ cm$^3$ mol$^{-1}$ s$^{-1}$)**

| Host (M) | Guest (A) | (%) | (R)-(-)- $C_4H_9NH_2$ | | | (S)-(+)- $C_4H_9NH_2$ | | | |
|---|---|---|---|---|---|---|---|---|---|
| | | | *k* | eff | *ρ* | *k* | eff | *ρ* | ξ |
| $22^R$ | phe$^D$ | | 89.6±1.9 | 0.78 | | 62.1±2.4 | 0.54 | | 1.44±0.08 |
| $22^R$ | phe$^L$ | | 64.6±1.8 | 0.56 | 1.39±0.07 | 62.9±1.1 | 0.55 | 0.99±0.05 | 1.03±0.03 |
| $22^R$ | pheNH$_2$$^D$ | | 43.8±1.4 | 0.38 | | 39.6±0.5 | 0.35 | | 1.11±0.07 |
| $22^R$ | pheNH$_2$$^L$ | | 41.9±1.3 | 0.37 | 1.05±0.06 | 42.5±1.3 | 0.37 | 0.93±0.06 | 0.99±0.06 |
| $22^R$ | pheOEt$^D$ | | 19.1±0.2 | 0.17 | | 19.3±0.6 | 0.17 | | 0.99±0.04c |
| $22^R$ | pheOEt$^L$ | | 29.2±0.3 | 0.26 | 0.65±0.02 | 31.0±1.2 | 0.27 | 0.62±0.05 | 0.94±0.05c |
| $23^R$ | phe$^D$ | | 77.5±1.0 | 0.68 | | 59.3±0.6 | 0.52 | | 1.31±0.03 |
| $23^R$ | phe$^L$ | | 89.7±1.5 | 0.79 | 0.86±0.03 | 61.9±0.8 | 0.55 | 0.96±0.02 | 1.45±0.04 |
| $23^R$ | pheNH$_2$$^D$ | | 33.6±0.7 | 0.30 | | 42.1±0.5 | 0.37 | | 0.80±0.02c |
| $23^R$ | pheNH$_2$$^L$ | | 40.4±0.7 | 0.36 | 0.83±0.03 | 48.1±1.1 | 0.42 | 0.87±0.03 | 0.84±0.03c |
| $23^R$ | (pheOEt$^D$)$_{fast}$ | 37±2 | 26.3±3.0 | 0.24 | | 25.2±0.6 | 0.22 | | 1.04±0.17 |
| $23^R$ | pheOEt$^L$ | 100 | 20.6±0.3 | 0.18 | 1.28±0.18 | 26.3±0.3 | 0.23 | 0.96±0.02 | 0.78±0.02 |
| $23^R$ | (pheOEt$^D$)$_{slow}$ | 63±2 | 4.0±0.2 | 0.04 | | 3.6±0.2 | 0.03 | | 1.11±0.12 |
| $23^R$ | pheOEt$^L$ | 100 | 20.6±0.3 | 0.18 | 0.19±0.02 | 26.3±0.3 | 0.23 | 0.14±0.01 | 0.78±0.02 |
| d$_6$-$23^S$ | (pheOEt$^L$)$_{fast}$ | 35±8 | 26.2±1.4 | 0.23 | | 23.0±1.9 | 0.20 | | 0.88±0.12d |
| d$_6$-$23^S$ | pheOEt$^D$ | 100 | 26.3±0.7 | 0.23 | 1.00±0.08 | 15.6±1.9 | 0.14 | 1.47±0.35 | 0.59±0.09d |
| d$_6$-$23^S$ | (pheOEt$^L$)$_{slow}$ | 65±8 | 4.44±0.27 | 0.039 | | 2.77±0.15 | 0.024 | | 0.62±0.06d |
| d$_6$-$23^S$ | pheOEt$^D$ | 100 | 26.3±0.7 | 0.23 | 0.17±0.01 | 15.6±1.9 | 0.14 | 0.18±0.03 | 0.59±0.09d |
| $23^S$ | (naphOEt$^L$)$_{fast}$ | 12±3 | 1.22±0.04 | 0.010 | | 1.25±0.07 | 0.011 | | 0.98±0.09c |
| $23^R$ | (naphOEt$^L$)$_{fast}$ | 29±4 | 26.3±1.3 | 0.22 | 0.046±0.004 | 25.0±2.5 | 0.21 | 0.050±0.009 | 1.05±0.18c |
| $23^S$ | (naphOEt$^L$)$_{slow}$ | 88±3 | 0.064±0.003 | 0.0005 | | 0.058±0.007 | 0.0005 | | 1.10±0.18c |
| $23^R$ | (naphOEt$^L$)$_{slow}$ | 71±4 | 1.23±0.09 | 0.010 | 0.052±0.07 | 1.20±0.06 | 0.010 | 0.048±0.09 | 1.02±0.04c |
| $24^S$ | (naphOEt$^L$)$_{fast}$ | 13±2 | 23.9±4.3 | 0.19 | | 24.9±0.6 | 0.21 | | 0.96±0.05 |
| $24^R$ | (naphOEt$^L$)$_{fast}$ | 34±2 | 20.8±1.1 | 0.18 | 1.15±0.28 | 34.7±2.0 | 0.29 | 0.72±0.06 | 0.60±0.07 |
| $24^S$ | (naphOEt$^L$)$_{slow}$ | 87±2 | 0.042±0.003 | 0.0004 | | 0.051±0.003 | 0.0004 | | 0.83±0.14 |
| $24^R$ | (naphOEt$^L$)$_{slow}$ | 66±2 | 1.32±0.12 | 0.012 | 0.032±0.005 | 1.15±0.09 | 0.010 | 0.044±0.005 | 1.15±0.19 |

structures with largely different reactivity toward B. Deconvolution of the biexponential kinetic curves provides an estimate of the relative abundance of the $[M\bullet H\bullet A]^+_{slow}$ and $[M\bullet H\bullet A]^+_{fast}$ isomeric structures of [**23**$^R$•H•**pheOEt**$^D$]$^+$, [**23**$^R$•H•**pheOEt**$^D$]$^+$, [$d_6$-**23**$^S$•H•**pheOEt**$^L$]$^+$, and [**24**$^{R/S}$•H•**naphOEt**$^L$]$^+$ (% in Table 6.9).

According to the kinetic results of Table 6.9, the [**22**$^R$•H•A]$^+$ (A = **phe**, **pheNH$_2$**, and **pheOEt**) and [**23**$^R$•H•A]$^+$ (A = **phe** and **pheNH$_2$**) complexes exhibit a limited enantioselectivity with regard to the A and B configurations. In contrast, more pronounced enantioselectivities are observed in the reaction with [**23**$^R$•H•**pheOEt**]$^+$ and [$d_6$-**23**$^S$•H•**pheOEt**$^L$]$^+$ ($0.14 < \rho < 1.28$; $0.59 < \xi < 1.11$). Both the more and less reactive isomers of the diastereomeric [**23**$^{R/S}$•H•**naphOEt**$^L$]$^+$ complexes display the same exceptional enantioselectivity, referred to as the A configuration ($0.046 < \rho < 0.052$), and a negligible selectivity toward the B enantiomers ($0.98 < \xi < 1.10$). Such a large enantioselectivity is not observed with the [**24**$^{R/S}$•H•**naphOEt**$^L$]$^+$fast structures ($0.72 < \rho < 1.15$). In contrast, the less reactive [**24**$^{R/S}$•H•**naphOEt**$^L$]$^+_{slow}$ structures showed a selectivity ($0.032 < \rho < 0.044$) even greater than that exhibited by the [**23**$^{R/S}$•H•**naphOEt**$^L$]$^+_{slow}$ analogue ($\rho = 0.050$).

The most stable MM2*-calculated geometries of **22**$^R$, **23**$^R$, and **24**$^R$ are summarized in Figure 6.12.[99] Three different geometries were discerned, equatorial-equatorial (*eq-eq*), axial-axial (*ax-ax*), and axial-equatorial (*ax-eq*), relative to the dispositions assumed in each conformer by the alkyl or phenyl fragments on two contiguous stereogenic carbons of the host. In **22**$^R$, only the *eq-eq* disposition is possible. In both **23**$^R$

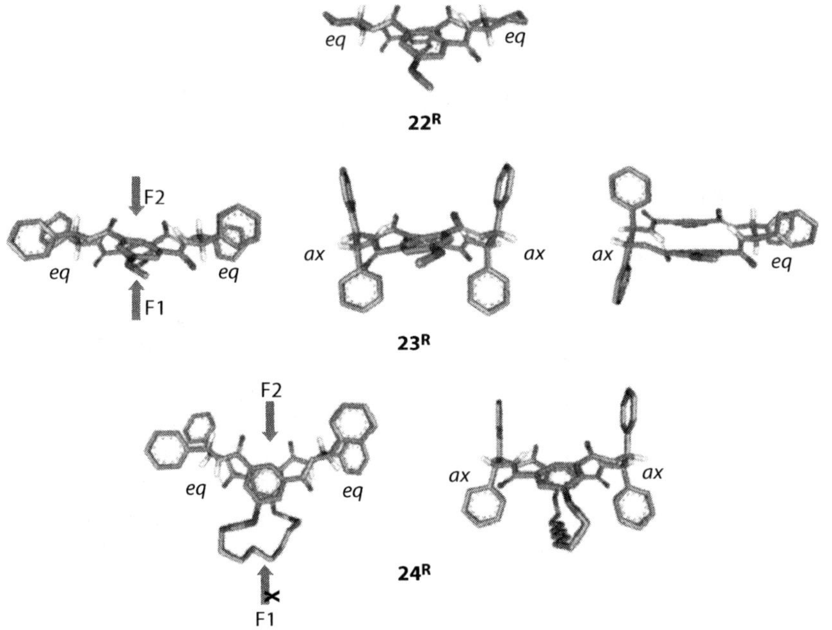


**FIGURE 6.12** Relevant minimum energy structures of the **22**$^R$, **23**$^R$, and **24**$^R$ hosts found by conformational search. The possible directions of approach by the guest are denoted by F1 (on the concave side of the host) and F2 (on the convex side of the host) (see text).

and **24$^R$**, the *eq-eq* dispositions are energetically most populated (98.40 and 99.95%, respectively). In all *eq-eq* geometries, the macrocyclic host presents a C2-symmetric folded structure with a concave side F1 and a convex one F2 (Figure 6.12). Alternate and diverging CO and NH groups are located on the outer margins of F1, whereas the same, but now converging functionalities are placed on the central folding of F2. Such molecular framework, reminding one of the structure of a saddle roof, is stabilized by two strong H-bonds between two facing amide moieties.

Multiconformational molecular docking procedures on the diastereomeric pairs of complexes [**22$^R$**•H•**pheOEt**]$^+$, [**23$^R$**•H•**pheOEt**]$^+$, [**23$^R$**•H•**pheNH$_2$**]$^+$, [**23$^R$**•H•**naphOEt**]$^+$, and [**24$^R$**•H•**naphOEt**]$^+$ show that, irrespective of the type and configuration of host and guest species, a pattern of strong H-bonds is established between the two molecular units at the center of the macrocycle convex side F2.[99] In particular, the protonated amino group of the guest is always H$^+$ bonded to the converging C=O groups present on the F2 surface. By contrast, all the adducts formed with the guest approaching the host from the concave surface were computed to be about 17 kcal mol$^{-1}$ higher in energy. Only the *eq-eq* and *ax-ax* conformations of the macrocycles were found in the ensembles of the more stable adducts and in proportions different from those of the uncomplexed hosts. In particular, host **23$^R$**, which in the free form is calculated to assume almost exclusively the *eq-eq* conformation (*eq-eq* > 98%), acquires a predominant *ax-ax* geometry by induced fit on complexation with **pheOEt$^L$**, **pheNH$_2$$^L$**, and both **naphOEt** enantiomers (68% < *ax-ax* < 97%; Table 6.10). Similarly, host **24$^R$**, which is almost exclusively *eq-eq* in the free state (*eq-eq* ~ 100%), is calculated to assume a predominant *ax-ax* geometry by induced fit on complexation with **naphOEt$^L$** (*eq-eq* ~ 8%) and, to a lesser extent, on complexation with the **naphOEt$^D$** (*eq-eq* ~ 48%).

For the homochiral [**23$^R$**•H•**pheOEt$^D$**]$^+$ and [d$_6$-**23$^S$**•H•**pheOEt$^L$**]$^+$ complexes, MM calculations point to their *eq-eq* structures as 0.9 kcal mol$^{-1}$ more stable than the corresponding *ax-ax* ones [ΔΔH°(*fast-slow*) in Table 6.10]. The order is reversed in the heterochiral [**23$^R$**•H•**pheOEt$^L$**]$^+$ and [d$_6$-**23$^S$**•H•**pheOEt$^D$**]$^+$ complexes, whose *eq-eq* forms are 2.0 kcal mol$^{-1}$ less stable than the relevant *ax-ax* structures. For the [**23$^R$**•H•**naphOEt**]$^+$ diastereomers, the *ax-ax* form is more stable than the *eq-eq* congener by 1.1 kcal mol$^{-1}$, in the case of the homochiral complex, and by 0.8 kcal mol$^{-1}$in the case of the heterochiral one. The same stability order is calculated for the heterochiral [**24$^R$**•H•**naphOEt$^L$**]$^+$ complex [ΔΔH°(*fast-slow*) = 1.5 kcal mol$^{-1}$], while the *eq-eq* and *ax-ax* structures of its homochiral congener are almost degenerate.

The relative abundances, estimated for the more and less stable couple of isomers of each complex, are compared in Table 6.10 with those derived from the experimental kinetic curves. The occurrence of two stable isomeric forms for the homochiral [**23$^R$**•H•**pheOEt$^D$**]$^+$ (and [d$_6$-**23$^S$**•H•**pheOEt$^L$**]$^+$) complex accounts for its biexponential decay with B. On the grounds of their relative stability, the most stable *eq-eq* structure is associated with the less reactive [**23$^R$**•H•**pheOEt$^D$**]$^+_{slow}$ (and [d$_6$-**23$^S$**•H•**pheOEt$^L$**]$^+_{slow}$) complex and the less stable *ax-ax* structure to the more reactive [**23$^R$**•H•**pheOEt$^D$**]$^+_{fast}$ (and [d$_6$-**23$^S$**•H•**pheOEt$^L$**]$^+_{fast}$) one. In contrast, the pronounced stability gap (2.0 kcal mol$^{-1}$) between the *ax-ax* and *eq-eq* forms of the heterochiral [**23$^R$**•H•**pheOEt$^L$**]$^+$ (and [d$_6$-**23$^S$**•H•**pheOEt$^D$**]$^+$) complex justifies the large predominance of a single stable *ax-ax* isomer, which is responsible for the

**TABLE 6.10**
**Percent Distribution of *eq-eq* and *ax-ax* Conformations for the [M•H•A]$^+$ Complexes**

| Host (M) | Guest (A) | [M•H•A]$^+$ | [M•H•A]$^+_{fast}$ experimental | [M•H•A]$^+_{slow}$ experimental | [M•H•A]$^+_{fast}$ MM-calculated (conformation) | [M•H•A]$^+_{slow}$ MM-calculated (conformation) | ΔΔH°(fast-slow)[a] (kcal mol$^{-1}$) |
|---|---|---|---|---|---|---|---|
| **22** | **pheOEt** | *homochiral* | 100 | | 100 (*eq-eq*) | | |
| | | *heterochiral* | 100 | | 100 (*eq-eq*) | | |
| **23** | **pheNH$_2$** | *homochiral* | 100 | | 27 (*ax-ax*) | 73 (*eq-eq*) | 0.6 |
| | | *heterochiral* | 100 | | 32 (*eq-eq*) | 68 (*ax-ax*) | 0.4 |
| **23** | **pheOEt** | *homochiral* | 37±2 | 63±2 | 17 (*ax-ax*) | 83 (*eq-eq*) | 0.9 |
| | | *heterochiral* | 100 | | 3 (*eq-eq*) | 97 (*ax-ax*) | 2.0 |
| d$_6$-**23** | **pheOEt** | *homochiral* | 35±8 | 65±8 | 17 (*ax-ax*) | 83 (*eq-eq*) | 0.9 |
| | | *heterochiral* | 100 | | 3 (*eq-eq*) | 97 (ax-ax) | 2.0 |
| **23** | **naphOEt** | *homochiral* | 12±3 | 88±3 | 13 (*eq-eq*) | 87 (ax-ax) | 1.1 |
| | | *heterochiral* | 29±4 | 71±4 | 21 (*eq-eq*) | 79 (ax-ax) | 0.8 |
| **24** | **naphOEt** | *homochiral* | 13±2 | 87±2 | 48 (*eq-eq*) | 52 (ax-ax) | 0.0 |
| | | *heterochiral* | 34±2 | 66±2 | 8 (*eq-eq*) | 92 (ax-ax) | 1.5 |

[a] Derived from the MM-calculated [M•H•A]$+_{fast}$ and [M•H•A]$+_{slow}$ relative abundances (columns 6 and 7). ΔΔH°(*fast-slow*) = $\Delta H°_{fast}$ - $\Delta H°_{slow}$ = -RTln([M•H•A]$+_{fast}$/[M•H•A]$+_{slow}$), T = 298K.

observed monoexponential decay with B. An analogous rationale can be advanced for the biexponential decay registered for the diastereomeric [**23**•H•**naphOEt**$^{L}$]$^{+}$ complexes. Indeed, both diastereomeric complexes show a stable *ax-ax* structure, associated with the less reactive [**23**•H•**naphOEt**$^{L}$]$^{+}_{slow}$ complexes, accompanied by a less stable *eq-eq* form, associated with the more reactive [**23**•H•**naphOEt**$^{L}$]$^{+}_{fast}$ isomers. The same stability order is observed with the heterochiral [**24**$^{R}$•H•**naphOEt**$^{L}$]$^{+}$ complex, whereas the *ax-ax* and *eq-eq* forms of the homochiral [**24**$^{S}$•H•**naphOEt**$^{L}$]$^{+}$ one are almost degenerate. At variance with the heterochiral [**23**$^{R}$•H•**pheOEt**$^{L}$]$^{+}$ and [d$_6$-**23**$^{S}$•H•**pheOEt**$^{D}$]$^{+}$ complexes, the apparent monoexponential decay of the diastereomeric [**23**$^{R}$•H•**pheNH**$_2$]$^{+}$ pair when reacting with B was not ascribed to the large predominance of a single isomer, but rather to the limited stability difference between their *ax-ax* and *eq-eq* forms (0.4 ÷ 0.6 kcal mol$^{-1}$; Table 6.10).

The relationships between the experimental and calculated enantioselectivity factors for hosts **22**, **23**, and **24** are illustrated in Figure 6.13. A linear correlation ($r^2$ = 0.961; slope = 0.921; Y intercept = –0.205) is found between the MM-calculated $\Delta\Delta H°_{th}$ = $(\Delta H°_{hetero} - \Delta H°_{homo})_{th}$ difference and the $\Delta\Delta G^{\#} = \Delta G^{\#}_{homo} - \Delta G^{\#}_{hetero}$ gap derived from the kinetic experiments (thick solid line in Figure 6.13). The linear relationship suggests that the kinetic enantioselectivity, measured in the FT-ICR experiments at 298 K, is essentially an expression of the stability gap between the corresponding diastereomeric [M•H•A]$^{+}$ reactants. This is not true for the [**24**$^{R}$•H•**naphOEt**$^{L}$]$^{+}$ diastereoisomers (Figure 6.13f), as demonstrated by their large deviation from linearity. This conclusion is further supported by the linear correlation between $\Delta\Delta G_{CID}$ = $(\Delta G_{hetero} - \Delta G_{homo})_{CID} = RT_{eff}\ln R$ (taken a $T_{eff}$ = 298 K) stability gap between diastereomeric [M•H•A]$^{+}$, measured by collision-induced decomposition (CID) of the corresponding [M$_2$•H•A]$^{+}$ precursors, and $\Delta\Delta G^{\#} = \Delta G^{\#}_{homo} - \Delta G^{\#}_{hetero}$ ($r^2$ = 0.971; slope = 0.601; Y intercept = +0.290; thin solid line in Figure 6.13). Indeed, if $T_{eff}$ is taken as equal to 457 K, the $\Delta\Delta G^{\#}$ vs. $\Delta\Delta G_{CID}$ linear correlation becomes parallel to that of the $\Delta\Delta G^{\#}$ vs. $\Delta\Delta H_{th}$ one, though shifted upward by ca. 0.65 kcal mol$^{-1}$ (broken line in Figure 6.13). The pronounced deviation of the [**24**•H•**naphOEt**$^{L}$]$^{+}$ systems from the relationships of Figure 6.13 implies that the enantioselectivity, measured in the FT-ICR experiments at 298 K, reflects not only their stability difference, but also the effects of the decamethylene chain of **24** on the relative stability of the diastereomeric transition structures of Equation 6.1. This conclusion is consistent with the observation of biexponential kinetics for the reaction of B with the homochiral [**24**$^{S}$•H•**naphOEt**$^{L}$]$^{+}$ complex, despite the degeneracy of its *ax-ax* and *eq-eq* structures (Table 6.10).

In conclusion, the formation of the selected [M•H•A]$^{+}$ complexes is mainly driven by the establishment of a pattern of three strong H-bonds between the protonated amino group of guest A and the couple of converging C=O functionalities placed on the F2 surface of the host M. The guest is placed above the center of the host convex side and establishes further weaker interactions with the peripheral zones of the host. An additional H-bond is established between the carbonyl oxygen of A and one-amide N-H of the M in the *eq-eq* [M•H•A]$^{+}$ structures. In the *ax-ax* structures with aromatic guests, additional attractive edge-to-face and face-to-face interactions operate between the aromatic moiety of the guest and one isophthalic ring of the host. The extent to which the side chain groups of the guest either favor or hinder

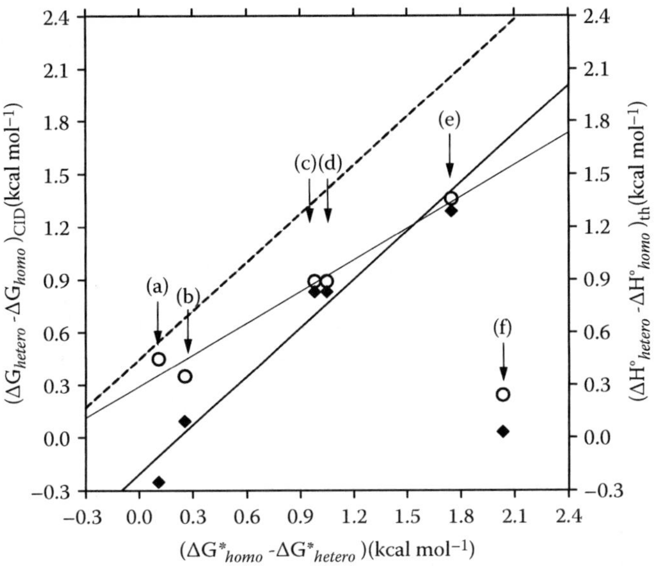


**FIGURE 6.13** Relationship between the gas-phase thermodynamic ($\Delta\Delta G_{CID}$ and $\Delta\Delta H°_{th}$) and kinetic ($\Delta\Delta G^{\#}$) enantioselectivities of $[M•H•A]^+$ (M/A = **23$^R$/pheNH$_2$** (a), **22$^R$/pheOEt** (b), d$_6$-**23$^S$/pheOEt** (c), **23$^R$/pheOEt** (d), **23/naphOEt$^L$** (e), and **24/naphOEt$^L$** (f) complexes. The thick solid line describes the relationship between the experimental activation barriers of the slow reaction of several two-body $[M•H•A]^+$ complexes with amine B (obtained from the relevant ρ values from Table 6.9; $\Delta\Delta G^{\#} = \Delta G^{\#}_{homo} - \Delta G^{\#}_{hetero} = -RT\ln\rho$) and the MM-computed relative enthalpies of the same $[M•H•A]^+$ complexes [$\Delta\Delta H°_{th} = (\Delta H°_{hetero} - \Delta H°_{homo})_{th}$]. The thin solid line describes the relationship between the experimental $\Delta\Delta G^{\#} = \Delta G^{\#}_{homo} - \Delta G^{\#}_{hetero} = -RT\ln\rho$ values and the relative stability of the same $[M•H•A]^+$ complexes, calculated at $T_{eff}$ = 298 K, from CID of the corresponding three-body $[M_2•H•A]^+$ complexes ($\Delta\Delta G_{CID} = (\Delta G_{hetero} - \Delta G_{homo})_{CID} = RT_{eff}\ln R$). If the effective temperature $T_{eff}$ is taken equal to 457 K (broken line), both the slopes of the $\Delta\Delta G^{\#}$ vs. $\Delta\Delta H°_{th}$ and the $\Delta\Delta G^{\#}$ vs. $\Delta\Delta G_{CID}$ straight lines coincide and are close to unit (slope = 0.921).

complexation with a given host depends on the stereochemistry of the guest, and on the size and structure of the side chain itself. The aptitude of the **23** and **24** hosts to assume the *ax-ax* conformation by complexation enhances their enantiodiscrimination abilities compared to *eq-eq*-locked host **22**. The reduced stability gap between the diastereomeric complexes with **24** does not seem to play any appreciable role on the large kinetic enantioselectivity, measured for the same complexes in Equation 6.1, because the presence of the decamethylene chain probably influences dramatically the relative stability of the corresponding transition structures.

## 6.4 CONCLUSIONS AND OUTLOOK

Understanding enantioselective processes in chemical and biochemical systems requires knowledge of the intrinsic noncovalent interactions involved. The advantages connected with gas-phase studies come from the possibility to make precise statements on the structure and stability of ion-molecule complexes as well as on the intrinsic factors governing their dynamics and reactivity in the lack of any perturbing environmental effects (solvation, ion pairing, cage viscosity, etc.). By virtue of the very high sensitivity and the wealth of information it generates, MS plays an increasingly important role in chiral recognition studies. Tandem mass spectrometry as well as ion trap and especially FT-ICR mass spectrometric techniques, equipped with ESI sources, play an important role in this field, since these methods can be applied to obtain quantitative information on the enantioselectivity of ion-molecule reactions that is difficult to obtain using other techniques. ESI-FT-ICR-MS is especially useful for the measurement of rate and equilibrium constants for chiral supramolecular systems, and supported by MM calculations and MD simulations, it may provide significant information on the behavior of noncovalent chiral guest–chiral host complexes mimicking enzyme recognition and catalysis.

This chapter provides several examples of ion-molecule enantioselectivity that parallels and, in some instances, even surpasses enantioselectivities measured in classical solution reactions. They distinctly indicate that enantioselectivity in ion-molecule reactions depends not only on the specific configuration of the species at play, but also on their conformation in the encounter complex, which often is different from that in the separated subunits. Reciprocal conformational distortions may take place in the subunits of the encounter complex to optimize their interaction energy. These situations are reminiscent of the induced-fit phenomenon and the allosteric effects in biological processes.

Given the growing interest in the factors underlying enantioselective processes in chemistry and biochemistry, it is easy to predict for the near future increasing research efforts in gas-phase chiral discrimination and enantioselective catalysis with special regard to (1) enantioselective ion-molecule reactions in axial and planar dissymmetric hosts, (2) enantioselective reactions on achiral molecules induced by chiral hosts, (3) remote asymmetric induction in prochiral systems, and (4) conformational analysis of diastereomeric ion-molecule complexes by laser spectroscopy.

## ACKNOWLEDGMENTS

The author thanks his coworkers and colleagues for their contributions, which are reported in the list of references. This work was supported by the Ministero dell'Istruzione dell'Università e della Ricerca (MiUR-COFIN; PRIN grant 2007H9S8SW_002) and ASI (contract grant NI/015070).

## REFERENCES

1. S. C. Stinson. 2001. *Chem. Eng. News* 79:79.
2. T. H. Webb, C. S. Wilcox. 1993. *Chem. Soc. Rev.* 22:383.
3. M. Tsukamoto, H. B. Hagan. 2002. *Adv. Synth. Catal.* 344:453.
4. See, for instance, (a) P. A. Fitzpatrick, A. M. Klibanov. 1991. *J. Am. Chem. Soc.* 113:3166. (b) L. Yang, J. S. Dordick, S. Garde. 2004. *Biophys. J.* 87:812. (c) G. Pescitelli, L. Di Bari, P. Salvadori. 2006. *J. Organomet. Chem.* 691:2311. (d) H. Bao, J. Zhou, Z. Wang, Y. Guo, T. You, K. Ding. 2008. *J. Am. Chem. Soc.* 130:10116.

5. F. Hollfelder, A. J. Kirby, D. S. Tawfik. 1996. *Nature* 383:60.
6. A. Filippi, A. Giardini, S. Piccirillo, M. Speranza. 2000. *Int. J. Mass Spectrom.* 198:137.
7. M. Speranza. 2004. *Int. J. Mass Spectrom.* 232:277.
8. M. Sawada. 1997. *Mass Spectrom. Rev.* 16:73.
9. W. A. Tao, R. G. Cooks. 2003. *Anal. Chem.* 75:25A.
10. G. Finn. 2002. *Chirality* 14:534.
11. J. S. Splitter, F. Tureček. 1994. *Applications of mass spectrometry to organic stereochemistry*. New York: VCH Publishers.
12. M. Sawada. 1994. In *Biological mass spectrometry: Present and future*, ed. T. Matsuo, R. M. Caprioli, M. L. Gross, T. Seyama. 639–46. New York: Wiley.
13. C. A. Schalley. 2000. *Int. J. Mass Spectrom.* 194:11.
14. C. B. Le brilla. 2001. *Acc. Chem. Res.* 34:653.
15. D. V. Dearden, Y. Liang, J. B. Nicoll, K. A. Kellersberger. 2001. *J. Mass Spectrom.* 36:989.
16. K. A. Schug, W. Lindner. 2005. *J. Sep. Sci.* 28, 1932.
17. M. Speranza, F. Gasparrini, B. Botta, C. Villani, D. Subissati, C. Fraschetti, F. Subrizi. 2009. *Chirality* 21:69.
18. P. Dwivedi, C. Wu, L. M. Matz, B. H. Clowers, W. F. Seims, H. H. Hill, Jr. 2006. *Anal. Chem.* 78:8200.
19. E. N. Nikolaev, E. V. Denisov, V. S. Rakov, J. H. Futrell. 1999. *Int. J. Mass Spectrom.* 182/183:357.
20. Y. Liang, J. S. Bradshaw, D. V. Dearden. 2002. *J. Phys. Chem. A* 106:9665.
21. K. Vékey, G. Czira. 1995. *Rapid Commun. Mass Spectrom.* 9:783.
22. W. A. Tao, L. Wu, R. G. Cooks. 2000. *Chem. Commun.* 20:2023.
23. W. A. Tao, D. Zhang, E. N. Nikolaev, R. G. Cooks. 2000. *J. Am. Chem. Soc.* 122:10598.
24. W. A. Tao, R. G. Cooks. 2002. *Eur. J. Mass Spectrom.* 8:107.
25. W. A. Tao, R. L. Clark, R. G. Cooks. 2002. *Anal. Chem.* 74:3783.
26. W. A. Tao, R. G. Cooks. 2001. *Angew. Chem. Int. Ed. Engl.* 40:757.
27. W. A. Tao, F. G. Gozzo, R. G. Cooks. 2001. *Anal. Chem.* 73:1692.
28. L. Wu, R. G. Cooks. 2003. *Anal. Chem.* 75:678.
29. J. Ramirez, F. He, C. B. Lebrilla. 1998. *J. Am. Chem. Soc.* 120:7387.
30. J. Ramirez, S. Ahn, G. Grigorean, C. B. Lebrilla. 2000. *J. Am. Chem. Soc.* 122:6884.
31. J. F. Gal, M. Stone, C. B. Lebrilla. 2003. *Int. J. Mass Spectrom.* 222:259.
32. S. Ahn, J. Ramirez, G. Grigorean, C. B. Lebrilla. 2001. *J. Am. Soc. Mass Spectrom.* 12:278.
33. G. Grigorean, C. B. Lebrilla. 2001. *Anal. Chem.* 73:1684.
34. A. G. Marshall, C. L. Hendrickson, M. R. Emmett, R. P. Rodgers, G. T. Blakney, C. L. Nilsson. 2007. *Eur. J. Mass Spectrom.* 13:57, and references therein.
35. T. T. Dang, S. F. Pedersen, J. A. Leary. 1994. *J. Am. Soc. Mass Spectrom.* 5:452.
36. P. Nemes, S. Goyal, A. Vertes. 2008. *Anal. Chem.* 80:387.
37. C. L. Sherman, J. S. Brodbelt. 2003. *Anal. Chem.* 75:1828.
38. J. E. Ham, B. Durham, J. R. Scott. 2005. *Rev. Sci. Instrum.* 76:014101/1.
39. R. L. Grimm, J. L. Beauchamp. 2005. *J. Phys. Chem. B* 109:8244.
40. Z. Tian, S. R. Kass. 2008. *J. Am. Chem. Soc.* 130:10842.
41. B stands for magnetic and B for electric sector. J. S. Cottrell, R. J. Greathead. 1986. *Mass Spectrom. Rev.* 5:215.
42. Q stands for quadrupole and H for hexapole. W. Paul, H. Steinwedel. 1953. *Z. Naturforsch.* 8a:448.
43. Q stands for quadrupole and TOF stands for time of flight. H. Wollnik. 1993. *Mass Spectrom. Rev.* 12:89.
44. R. E. March. 2000. *Int. J. Mass Spectrom.* 200:285.
45. J. C. Schwartz, M. W. Senko, J. E. P. Syka. 2002. *J. Am. Soc. Mass Spectrom.* 13:659.
46. M. B. Comisarow, A. G. Marshall. 1974. *Chem. Phys. Lett.* 25:282.
47. A. G. Marshall, C. L. Hendrickson, G. S. Jackson. 1998. *Mass Spectrom. Rev.* 17:1.

48. Q. Hu, R. J. Noll, H. Li, A. Makarov, M. Hardman, R. G. Cooks. 2005. *J. Mass Spectrom.* 40:430.
49. McLaffert, F. W., ed. 1983. *Tandem mass spectrometry*. New York: John Wiley & Sons.
50. K. L. Busch, G. L. Glish, S. A. McLuckey. 1988. *Mass spectrometry/mass spectrometry*. Deerfield, FL: VCH Publishers.
51. H. M. Fales, G. J. Wright. 1977. *J. Am. Chem. Soc.* 99:2339.
52. F. J. Winkler, D. Stahl, F. Maquin. 1986. *Tetrahedron Lett.* 27:335.
53. F. J. Winkler, J. S. Splitter. 1994. In *Applications of mass spectrometry to organic stereochemistry*, ed. J. S. Splitter, F. Tupecek. New York: VCH Publishers.
54. M. A. Baldwin, S. A. Howell, K. J. Welham, F. J. Winkler. 1988. *Biomed. Environ. Mass Spectrom.* 16:357.
55. F. J. Winkler, R. Medina, J. Winkler, H. Krause. 1994. *J. Chromatog. A* 666:549.
56. R. Sussmuth, G. Jung, F. J. Winkler, R. Medina. 1999. *Eur. Mass Spectrom.* 5:298.
57. E. N. Nikolaev, E. V. Denisov, M. I. Nikolaeva, J. H. Futrell, V. S. Rakov, F. J. Winkler. 1998. *Adv. Mass Spectrom.* 14:279.
58. F. J. Winkler, R. Medina, J. Winkler, H. Krause. 1997. *J. Mass Spectrom.* 32:1072.
59. E. V. Denisov, V. Shustryakov, E. N. Nikolaev, F. J. Winkler, R. Medina. 1997. *Int. J. Mass Spectrom. Ion Proc.* 167/168:259.
60. E. N. Nikolaev, G. T. Goginashvili, V. L. Tal'rose, R. G. Kostyanovsky. 1988. *Int. J. Mass Spectrom. Ion Proc.* 86:249.
61. E. N. Nikolaev, T. B. McMahon. 1995. *Proceedings of the 43rd Annual Conference on Mass Spectrometry and Allied Topics*, Atlanta, Georgia, p. 973.
62. E. N. Nikolaev, E. V. Denisov. 1996. *Proceedings of the 44th Annual Conference on Mass Spectrometry and Allied Topics*, Portland, Oregon, p. 415.
63. E. N. Nikolaev, E. V. Denisov, V. S. Rakov, J. H. Futrell. 1999. *Int. J. Mass Spectrom.* 182/183:357.
64. H. Suming, C. Yaozu, J. Longfei, X. Shuman. 1986. *Org. Mass Spectrom.* 21:7.
65. Y. Z. Chen, H. Li, H. J. Yang, S. M. Hua, H. Q. Li, F. Z. Zhao, N. Y. Chen. 1988. *Org. Mass Spectrom.* 23:821.
66. J. Martens, S. Lübben, W. Schwarting. 1991. *Z. Naturforsch.* 46b:320.
67. H. J. Yang, Y. Z. Chen. 1992. *Org. Mass Spectrom.* 27:736.
68. K. Hashimoto, Y. Sumida, S. Terada, K. Okamura. 1993. *J. Mass Spectrom. Soc. Jpn.* 41:87.
69. K. Hashimoto, Y. Sumida, S. Terada, K. Okamura. 1993. *J. Mass Spectrom. Soc. Jpn.* 41:95.
70. K. Okamura, Y. Sumida, Y. Fujiwara, S. Terada, H. Kim, K. Hashimoto. 1995. *J. Mass Spectrom. Soc. Jpn.* 43:97.
71. K. Hashimoto, K. Okamura, Y. Fujiwara, Y. Sumida, S. Terada. 1998. *Adv. Mass Spectrom.* 14:1.
72. R. G. Cooks, T. L. Kruger. 1977. *J. Am. Chem. Soc.* 99:1279.
73. D. Schröder, H. Schwarz. 2004. *Int. J. Mass Spectrom.* 231:139.
74. F. R. Novara, H. Schwarz, D. Schröder. 2007. *Helv. Chim. Acta* 90:2274.
75. F. R. Novara, P. Gruene, D. Schröder, H. Schwarz. 2008. *Chem. Eur. J.* 14:5957.
76. V. Znamenskiy, I. Marginean, A. Vertes. 2003. *J. Phys. Chem. A* 107:7406.
77. P. Nemes, S. Goyal, A. Vertes. 2008. *Anal. Chem.* 80:387.
78. C. L. Sherman, J. S. Brodbelt. 2003. *Anal. Chem.* 75:1828.
79. J. E. Ham, B. Durham, J. R. Scott. 2005. *Rev. Sci. Instrum.* 76:014101.
80. R. L. Grimm, J. L. Beauchamp. 2005. *J. Phys. Chem. B* 109:8244.
81. C. Fraschetti, M. Aschi, A. Filippi, A. Giardini, M. Speranza. 2008. *Chem. Commun.* 2544.
82. T. T. Dang, S. F. Pedersen, J. A. Leary. 1994. *J. Am. Soc. Mass Spectrom.* 5:452.
83. G. Grigorean, X. Cong, C. B. Lebrilla. 2004. *Int. J. Mass Spectrom.* 234:71.
84. S. Ahn, X. Cong, C. B. Lebrilla, S. Gronert. 2005. *J. Am. Soc. Mass Spectrom.* 16:166.

85. X. Cong, G. Czerwieniec, E. McJimpsey, E. Ahn, F. A. Troy, C. B. Lebrilla. 2006. *J. Am. Soc. Mass Spectrom.* 17:442.
86. E. Camara, M. K. Green, S. Penn, C. B. Lebrilla. 1996. *J. Am. Chem. Soc.* 118:8751.
87. D. E. Clemmer, R. R. Hudgins, M. F. Jarrold. 1995. *J. Am. Chem. Soc.* 117:10141.
88. B. Botta, M. Botta, A. Filippi, A. Tafi, G. Delle Monache, M. Speranza. 2002. *J. Am. Chem. Soc.* 124:7658.
89. A. Tafi, B. Botta, M. Botta, G. Delle Monache, A. Filippi, M. Speranza. 2004. *Chem. Eur. J.* 10:4126.
90. B. Botta, F. Caporuscio, I. D'acquarica, G. Delle Monache, D. Subissati, A. Tafi, M. Botta, A. Filippi, M. Speranza. 2006. *Chem. Eur. J.* 12:8096.
91. T. Su. 1988. *J. Chem. Phys.* 88: 4102, 5355.
92. B. Botta, F. Caporuscio, D. Subissati, A. Tafi, M. Botta, A. Filippi, M. Speranza. 2006. *Angew. Chem. Int. Ed.* 45:2717.
93. B. Botta, G. Delle Monache, P. Salvatore, F. Gasparrini, C. Villani, M. Botta, F. Corelli, A. Tafi, E. Gacs-Baitz, S. Santini, C. Carvalho, D. Misiti. 1997. *J. Org. Chem.* 62:932.
94. B. Botta, D. Subissati, A. Tafi, G. Delle Monache, A. Filippi, M. Speranza. 2004. *Angew. Chem. Int. Ed.* 43:4767.
95. B. Botta, G. Delle Monache, C. Fraschetti, L. Nevola, D. Subissati, M. Speranza. 2007. *Int. J. Mass Spectrom.* 267:24.
96. B. Botta, I. D'Acquarica, G. Delle Monache, D. Subissati, G. Uccello-Barretta, M. Mastrini, S. Nazzi, M. Speranza. 2007. *J. Org. Chem.* 72:9283.
97. B. Botta, A. Tafi, F. Caporuscio, M. Botta, L. Nevola, I. D'Acquarica, C. Fraschetti, M. Speranza. 2008. *Chem. Eur. J.* 14:3585.
98. A. Filippi, F. Gasparrini, M. Pierini, M. Speranza, C. Villani. 2005. *J. Am. Chem. Soc.* 127:11912.
99. F. Gasparrini, M. Pierini, C. Villani, A. Filippi, M. Speranza. 2008. *J. Am. Chem. Soc.* 130:522.

# 7 Equilibrium Methods for Characterizing Gas Phase Chiral Recognition

*David V. Dearden and Nannan Fang*

**CONTENTS**

7.1 Introduction .......... 133
7.2 Experimental Techniques .......... 135
7.3 Computational Methods .......... 136
7.4 Results .......... 136
Acknowledgment .......... 140
References .......... 141

## 7.1 INTRODUCTION

Chirality is built into all living things. Many important biomolecules, including all but one of the common amino acids and most sugars, are chiral. Frequently, these materials are present only in small quantities. Likewise, many synthetic drugs are chiral (for example, see Figure 7.1) and an increasing number are being marketed as single enantiomers.[1] Often, only one enantiomer is therapeutically effective; the other may even be harmful, as in the case of thalidomide.

For both regulatory and fundamental reasons, a critical need exists for analytical methods that can measure small enantiomeric impurities and do so rapidly, with minuscule samples. All analytical methods that distinguish between enantiomers work by presenting the analyte with a chiral environment, because only in a chiral environment do the differences between enantiomers become apparent.

This chapter deals with the application of equilibrium methods to the characterization of chiral compounds in the gas phase. The immediate goal of this work is to use the isolated molecule environment of the gas phase to probe the interactions responsible for enantiodiscrimination, using systems of intrinsic interest. Gas phase studies are ideal for studying fundamental interactions because no solvent or counter-ion species is present, and the intrinsic interactions between the species of interest are directly probed. In addition, the results of gas phase experiments can be directly compared with those of high-level computational studies, which almost always deal with isolated molecules.

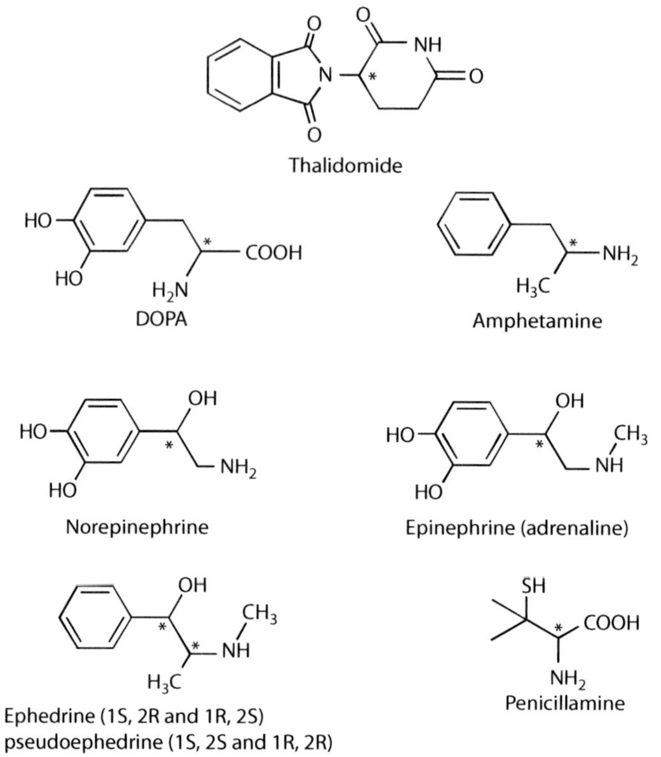


**FIGURE 7.1** Some chiral drugs containing amine functional groups. Stereocenters are marked with asterisks.

Although mass spectrometry is inherently a sensitive, rapid analytical tool, determination of enantiomeric excess using mass spectrometry is particularly challenging because the vacuum environment experienced by ions during mass spectrometric analysis is intrinsically achiral. Several methods have been developed to overcome this problem. All of them involve presenting the analyte with a chiral environment, usually by exposing it to a chiral reagent either prior to ionization or by using ion-molecule reactions with chiral reagents after the analyte ions are produced.

The products of enantiomeric ions binding to an enantiomerically pure neutral are diastereomers, which have different chemical and physical properties and therefore can be distinguished. The product diastereomers may appear in the mass spectrum with different intensities, depending on the abundances of the analyte enantiomers; react with different rates; dissociate differently; or have different equilibrium abundances in exchange reactions. The aim of this chapter is to describe the practice and application of equilibrium methods.

Our group was among the first to demonstrate enantiodiscrimination in an ion-molecule reaction.[2] We have focused on ion-molecule equilibrium measurements (low-pressure equilibrium techniques have recently been reviewed[3]) to probe the thermochemistry of chiral host-guest systems, emphasizing comparison of the behavior of the systems in condensed media with that in the gas phase. Because we are using equilibrium techniques, our results directly yield free energy, enthalpy, and entropy differences that measure the extent of chiral recognition in the system under study. Although we have shown that equilibrium methods can be successfully applied to analytical determination of enantiomeric excess,[4] equilibrium techniques are much better suited to probing the underlying factors responsible for chiral recognition than they are to analytical work. The necessity of achieving and characterizing equilibrium, and the requirement that the neutral species involved in the equilibrium have significant vapor pressures, make equilibrium methods relatively slow and much less general than some other approaches. Therefore, our focus is on elucidating enantiodiscrimination mechanisms rather than on the direct development of analytical protocols; we hope that our work will lay the foundation necessary for improved analytical methods.

## 7.2 EXPERIMENTAL TECHNIQUES

Virtually all of our experimental work uses Fourier-transform ion cyclotron resonance (FTICR) mass spectrometry.[5–10] To briefly summarize, this involves using a high-field superconducting solenoidal magnet to trap ions in two dimensions inside an ultra-high-vacuum (base pressures of $10^{-9}$ mbar) chamber. An electrostatic field provides trapping in the third dimension. Ions can be formed and introduced into the trap using any of a wide variety of methods.[11]

We primarily make ions outside the high field region of the magnet (mainly using electrospray ionization[12–14]) and inject them into the trap. Ions can be held in the trap for periods of hours if necessary. Ions in the trap are easily manipulated using RF electric fields; this allows the contents of the trap to be precisely controlled, because ions of any mass/charge ratio can be selectively ejected or retained. The trapped ions are nondestructively detected via the image currents they generate in the walls of the trap. The ions collide (and sometimes react) with background neutral gases, which can be introduced through controlled variable leak valves, pulsed valves, or direct-exposure probes. Reactions involving a change in mass/charge ratio can be followed as a function of time, facilitating kinetic measurements, and reactions are often observed to proceed to equilibrium in the trap, facilitating equilibrium constant measurements.[15] These two types of measurement form the basis for most of our work. Chiral systems are particularly well suited to these measurements because many possible errors cancel when dealing with enantiomers.

More specific procedures for the measurement of equilibria to characterize enantiodiscrimination are discussed below. The general approach involves exchange of chiral and achiral guest molecules on a chiral host molecule in such a way that all the neutral pressures are eliminated as variables in the experiment. Although it is less ideal, we can also perform exchange experiments where a chiral and achiral host exchange a chiral analyte guest.[2] These experiments typically require measurement of the pressure ratio of the two hosts and so are more prone to error, but sometimes are required because the guest vapor pressure is too low to use guest exchange.

## 7.3 COMPUTATIONAL METHODS

We use an array of powerful computational methods to study enantiodiscrimination and to compare with the experimental results. Typically, we begin with Monte Carlo conformational searches with a molecular mechanics force field, followed by *ab initio* geometry optimization and energy computation, as detailed below. We conduct conformational searches using the Macromodel package (Schrödinger, Inc., Portland, Oregon). We use both the AMBER*[16] and MMFF94[17–21] force fields supplied with Macromodel.

The lowest-energy conformations found in the conformational search are typically used as starting points for *ab initio* geometry optimization; alternatively, where X-ray structures are available, they can be used as starting geometries. We set up and manage all of the higher-level calculations using the ECCE package (Pacific Northwest National Laboratory, Richland, Washington) and use either the Gaussian 98 (Gaussian, Inc., Pittsburgh, Pennsylvania) or NWChem (Pacific Northwest National Laboratory) computational engines to carry out the calculations.

## 7.4 RESULTS

Our work began with a chiral crown ether host, dimethyldiketopyridino-18-crown-6, hereafter referred to as **1** (Figure 7.2). This ligand has two stereocenters, which must have the same absolute configuration for the molecule to be chiral. Thus, (*R,R*)-**1** and (*S,S*)-$_1$ are chiral enantiomers, whereas (*R,S*)-**1** has a mirror plane and is therefore a *meso* compound. In condensed media, $_1$ preferentially binds ammonium ions that have absolute configurations *opposite* those of the stereocenters of **1** (heterochiral complexes are favored over homochiral ones).[22,23]

The reactions involved in probing enantiodiscrimination are listed below. Experiments are carried out as follows. First, one of the crown ether host enantiomers (for example, (*R,R*)-**1**) is electrosprayed to produce the protonated chiral crown, which is then trapped in a FTICR mass spectrometer cell containing a constant

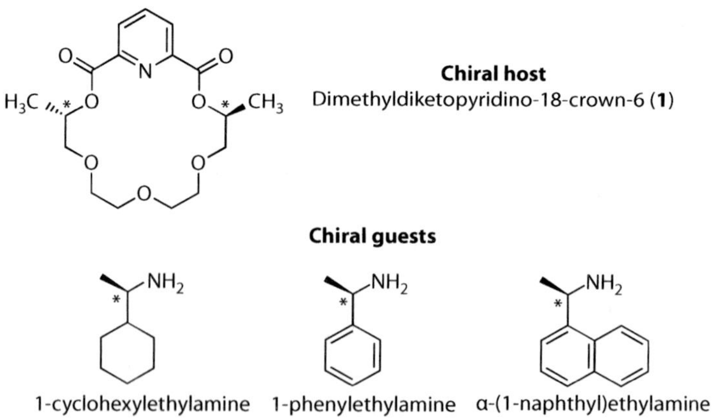


**FIGURE 7.2** Chiral host and guest molecules used in gas phase equilibrium studies. Stereocenters are marked with asterisks.

pressure of one enantiomer of the chiral amine of interest (in the reactions below, (*R*)-amine), plus a constant pressure of an achiral reference amine (Ref), which typically is cyclohexylamine (chosen because its affinity for the host is similar to that of the chiral species of interest). Exchange of the two amines on the protonated crown (Equation 7.1) is allowed to proceed to equilibrium, which is verified both by establishment of a constant reactant/product ratio and by attainment of the same reactant/product ratio following ejection of either the reactant or the product ion (Figure 7.3).

$$(R)-\text{amine} \bullet (R,R)-\mathbf{1}H^{+} + \text{Ref} \rightleftharpoons \text{Ref} \bullet (R,R)\text{-}\mathbf{1}\text{H}^{+} + (R)-\text{amine} \quad (7.1)$$

$$\text{Ref} \bullet (S,S)\text{-}\mathbf{1}\text{H}^{+} + (R)-\text{amine} \rightleftharpoons (R)-\text{amine} \bullet (S,S)\text{-}\mathbf{1}\text{H}^{+} + \text{Ref} \quad (7.2)$$

$$\begin{aligned}&\text{Ref} \bullet (S,S)\text{-}\mathbf{1}\text{H}^{+} + (R)-\text{amine} \bullet (R,R)\text{-}\mathbf{1}\text{H}^{+} \rightleftharpoons \\ &(R)-\text{amine} \bullet (S,S)\text{-}\mathbf{1}\text{H}^{+} + \text{Ref} \bullet (R,R)\text{-}\mathbf{1}\text{H}^{+}\end{aligned} \quad (7.3)$$

Following this measurement, the ion source is flushed to remove (*R,R*)-**1**, and then the other enantiomer, (*S,S*)-**1**, is electrosprayed and measurement of the equilibrium constant is again performed (Equation 7.2). Addition of Equations 7.1 and 7.2 yields

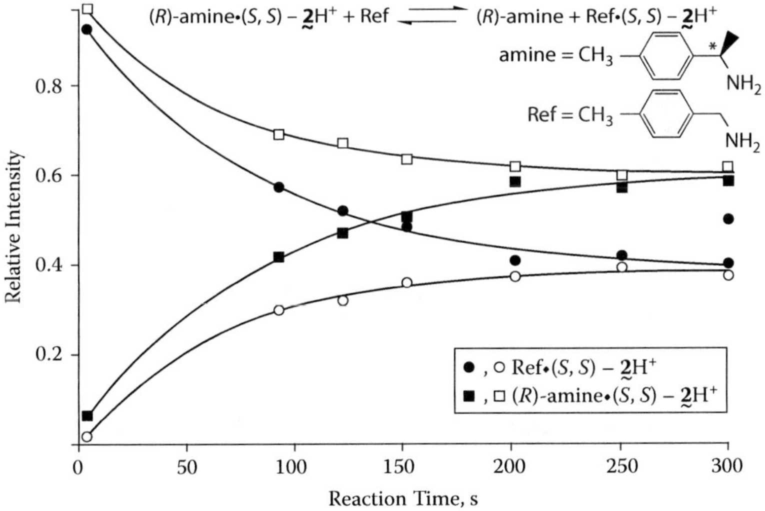


**FIGURE 7.3** This kinetic plot shows attainment of equilibrium in the forward and reverse directions for exchange of a chiral amine and an achiral reference amine on a chiral crown ether host. The host, (*S,S*)-**2**, is dimethylphenazino-18-crown-6. At long times, the same reactant/product ratio is reached, regardless of the direction of approach; the system reaches equilibrium.

Equation 7.3, which has an equilibrium constant equal to the product of the equilibrium constants for Equations 7.1 and 7.2.

Equation 7.3 is an ion-ion reaction that would have a negligibly slow rate at room temperature, but the thermochemistry of this reaction is nevertheless accessible. Equation 7.3 has the advantage that all the reactants and products are ions, whose abundances are all measured using the same technique (FTICR). The pressure of the neutral reference amine is not a factor, eliminating one of the primary sources of error in the equilibrium measurement. Further, mass discrimination problems are minimal because the pair of ions Ref•(*S*,*S*)-**1**H$^+$ and Ref•(*R*,*R*)-**1**H$^+$ have identical masses, as do (*R*)-amine•(*R*,*R*)-**1**H$^+$ and (*R*)-amine•(*S*,*S*)-**1**H$^+$. In Equation 7.3, Ref•(*S*,*S*)-**1**H$^+$ and Ref•(*R*,*R*)-**1**H$^+$ are enantiomers, so their thermochemical values cancel, whereas (*R*)-amine•(*R*,*R*)-**1**H$^+$ and (*R*)-amine•(*S*,*S*)-**1**H$^+$ are diastereomers. Equation 7.3 therefore measures the thermochemical difference between diastereomers (*R*)-amine•(*R*,*R*)-**1**H$^+$ and (*R*)-amine•(*S*,*S*)-**1**H$^+$, which is a measure of the enantiodiscrimination of the host for the amine guest. Note that Equation 7.3 is written so that the favored, heterochiral complex is the product.

We have shown[24] that **1** distinguishes between the enantiomers of 1-cyclohexylethylammonium, 1-phenylethylammonium, and α-(1-naphthyl)ethylammonium, but does not discriminate between the enantiomers of *sec*-butylammonium. Further, the degree of recognition (as measured by the free energy change for Equation 7.3) is much greater for 1-phenylethylammonium than for 1-cyclohexylethylammonium (Table 7.1); the degree of recognition increases further for the guest with the largest π system, α-(1-naphthyl)ethylammonium. Recognition is greater in the gas phase than in solutions of achiral solvents,[2,24,25] where solvation competes with the enantiospecific host-guest interaction.

The equilibrium constant, *K*, is related to standard entropy, $\Delta S°$, and enthalpy, $\Delta H°$, through Equation 7.4, where *R* is the ideal gas constant and *T* is the absolute temperature. This equation shows that if R ln*K* is plotted as a function of reciprocal temperature (a van't Hoff plot), the slope of the resulting line will be $\Delta H°$ and the intercept will be $\Delta S°$. A van't Hoff study of Equation 7.3 for two different ammonium ions[25] found that in both cases formation of the favored heterochiral complex is enthalpically driven and that entropy works against the favored complex (Table 7.1). Further, the enthalpy is more favorable for the guest with the larger π system, suggesting π-π stacking is important to enantiodiscrimination in this system.

**TABLE 7.1**
**Experimental Degree of Recognition of Dimethyldiketopyridino-18-Crown-6 (1) for Chiral Ammonium Cations**

| Ammonium Cation | $\Delta G°_{R}$,[24] kJ mol$^{-1}$ | $\Delta H°_{R}$,[25] kJ mol$^{-1}$ | $\Delta S°_{R}$,[25] J mol$^{-1}$ K$^{-1}$ |
|---|---|---|---|
| *sec*-butyl | −0.3 ± 0.4 | — | — |
| 1-cyclohexylethyl | −0.9 ± 0.2 | — | — |
| 1-phenylethyl | −2.4 ± 0.5 | −6.7 ± 0.7 | −14.8 ± 2.2 |
| α-(1-naphthyl)ethyl | −3.5 ± 0.6 | −10.0 ± 1.2 | −20.0 ± 3.9 |

$$R \ln K = \Delta S^\circ - \frac{\Delta H^\circ}{T} \tag{7.4}$$

We have also studied the complexes of **1** with chiral ammonium ions using computational methods,[25] ranging from molecular mechanics calculations with the AMBER[16] and MMFF94[17–21] force fields to quantum calculations with a moderate-sized basis set (6–31 + G* on all atoms except C, which lacked the diffuse functions) at Hartree–Fock, B3LYP, and MP2 levels of theory. The calculations complement the experimental work, in that they give structural information that is not available from the experiments and thermochemical information for comparison with the experimental values. The computed and experimental thermochemical data are in general agreement: the calculations correctly find that the heterochiral complexes are more stable than the homochiral ones, and the magnitudes of the computed and experimental enthalpies for Equation 7.3 are similar. The calculations are unsuccessful in showing the experimentally observed influence of π-π stacking: the angle between host and guest aromatic rings is calculated to be larger for the favored, heterochiral complexes than for the disfavored, homochiral ones, and at most levels of theory (and especially at the higher levels) the calculations do not find greater recognition for the α-(1-naphthyl)ethylammonium enantiomers than for the 1-phenylethylammonium enantiomers.

More recent work has investigated the role of π stacking in enantiodiscrimination, both by varying the π electron density of the chiral guest and by varying the extent of the π system of the host (see Figure 7.4). Varying the substituent in the *para* position of 1-phenylethylammonium was used to vary the π electron density of the guest. We have examined 1-(*p*-methylphenyl)ethylammonium, which has greater π electron density than the unsubstituted compound, and 1-(*p*-nitrophenyl)ethylammonium, which has lower π electron density than the unsubstituted compound. Results with the *p*-methyl-substituted ammonium ion (Table 7.2) indicate enantiodiscrimination by **1** is nearly twice as large for this ion than for the unsubstituted one. Experiments with 1-(*p*-nitrophenyl)ethylammonium were unsuccessful, likely because we could not achieve sufficient vapor pressure of the neutral amine to generate significant

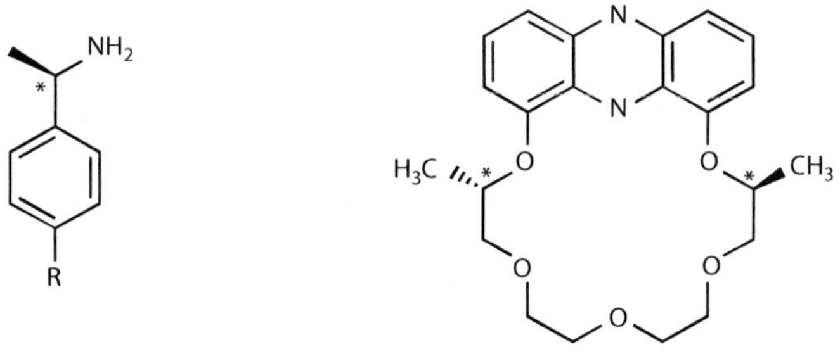


**Guests**
R = H, 1-phenylethylamine
R = $CH_3$, 1-(*p*-methylphenyl)ethylamine
R = $NO_2$, 1-(*p*-nitrophenyl)ethylamine

**Host**
Dimethylphenazino-18-crown-6 (**2**)

**FIGURE 7.4** Guests and host for π stacking experiments. Hosts with bulkier substituents than methyl are also available.

**TABLE 7.2**
**Experimental $\Delta G^\circ_R$ Values (kJ $mol^{-1}$) for Hosts 1 and 2**

| Ammonium Cation | Host 1 | Host 2 |
|---|---|---|
| 1-phenylethyl | –2.4 ± 0.5 | –1.2 ± 0.4 |
| α-(1-naphthyl)ethyl | –3.5 ± 0.6 | –2.4 ± 0.5 |
| 1-(*p*-methylphenyl)ethyl | –4.5 ± 0.5 | –1.4 ± 0.5 |

amounts of the complex. This highlights one of the main limitations of the technique, the requirement for volatile-neutral species.

The dimethylphenazino-18-crown-6 host **2** (Figure 7.4) has a more extensive π system than host **1**. X-ray crystallographic studies[26] of the complexes of **2** with the enantiomers of α-(1-naphthyl)ethylammonium suggest that π-π host-guest interactions in these complexes are stronger than those in the corresponding complexes of host **1**. In particular, the angle between host and guest aromatic rings is smaller (essentially 0°) for the heterochiral complex of **2** than for the corresponding complex of **1**, and circular dichroism studies[27] also indicate the π-π interactions in complexes of **2** are strong. As a result, enantiodiscrimination in condensed media for α-(1-naphthyl)ethylammonium by **2** is greater than that by **1**.[27] Molecular mechanics calculations[26] suggest **2** is also better at distinguishing enantiomers of this guest in the gas phase; the computed value of $\Delta E$ comparing the diastereomers of the α-(1-naphthyl)ethylammonium complexes of **2** is about –26 kJ $mol^{-1}$ in favor of the heterochiral complex, whereas the corresponding (experimentally measured) free energy difference for complexes of **1** is only –3.5 ± 0.6 kJ $mol^{-1}$ and the enthalpy difference is –10.0 ± 1.2 kJ $mol^{-1}$ (Table 7.1).

We have obtained samples of both enantiomers of ligand **2** through a collaboration with Professor Peter Huszthy, and have carried out gas phase studies of enantiodiscrimination by this ligand (Table 7.2). Interestingly, in the gas phase the more extensive π system of host **2** does not lead to increased enantiodiscrimination relative to host **1** for any of the guests examined. The reason for this is not entirely clear, but it is possible that the large planar area of the three-ring π system of host **2** flattens the potential energy surface sufficiently that there is a great deal of flexibility in guest binding. If this is the case, close proximity of the host and guest chiral centers might not be required, decreasing the possible amount of enantiodiscrimination even if the overall binding affinity increases (as suggested by the condensed phase data). Alternatively, it is possible that as the overall affinity increases, the ability of the system to discriminate between guests decreases. This latter explanation is consistent with the condensed phase and computational results.

## ACKNOWLEDGMENT

We are grateful for support of this work by the U.S. National Science Foundation (CHE-0615964) and for computer time from the Ira and Marylou Fulton Supercomputing Center at Brigham Young University.

## REFERENCES


1. Stinson, S. C. 2000. Chiral drugs. *Chemical and Engineering News*, 55–78.
2. Chu, I.-H., Dearden, D. V., Bradshaw, J. S., Huszthy, P., Izatt, R. M. 1993. Chiral host-guest recognition in an ion-molecule reaction *J. Am. Chem. Soc.* 115:4318–20.
3. Kellersberger, K. A., Dearden, D. V. 2003. Low-pressure ion-molecule equilibrium. In *The Encyclopedia of mass spectrometry*, ed. P. B. Armentrout, 338–45. Vol. 1. San Diego: Elsevier.
4. Liang, Y., Bradshaw, J. S., Izatt, R. M., Pope, R. M., Dearden, D. V. 1999. Analysis of enantiomeric excess using mass spectrometry: Fast atom bombardment/sector and electrospray ionization/Fourier transform mass spectrometric approaches. *Int. J. Mass Spectrom.* 185/186/187:977–88.
5. Marshall, A. G. 1985. Fourier transform ion cyclotron resonance mass spectrometry. *Acc. Chem. Res.* 18:316–22.
6. Marshall, A. G., Verdun, F. R. 1990. *Fourier transforms in NMR, optical, and mass spectrometry: A user's handbook.* Amsterdam: Elsevier.
7. Marshall, A. G., Schweikhard, L. 1992. Fourier transform ion cyclotron resonance mass spectrometry: Technique development. *Int. J. Mass Spectrom. Ion Proc.* 118/119:37–70.
8. Dunbar, R. C. 1991. History and general introduction. In *FT-ICR/MS: Analytical applications of Fourier transform ion cyclotron resonance mass spectrometry*, ed. B. Asamoto, 1–28. New York: VCH.
9. Dunbar, R. C., Asamoto, B. 1991. Instrumentation. In FT *FT-ICR/MS: Analytical applications of Fourier transform ion cyclotron resonance mass spectrometry*, ed. B. Asamoto, 29–82. New York: VCH.
10. Amster, I. J. J. 1996. Fourier transform mass spectrometry. *Mass Spectrom.* 31:1325–37.
11. McLafferty, F. W., Turecek, F. 1993. *Interpretation of mass spectra.* 4th ed. Mill Valley, CA: University Science Books.
12. Fenn, J. B., Mann, M., Meng, C. K., Wong, S. F., Whitehouse, C. M. 1990. Electrospray ionization-principles and practice. *Mass Spectrom. Rev.* 9:37–70.
13. Fenn, J. B. 1993. Ion formation from charged droplets: Roles of geometry, energy, and time. *J. Am. Soc. Mass Spectrom.* 4:524–35.
14. Kebarle, P., Tang, L. 1993. From ions in solution to ions in the gas phase: The mechanism of electrospray mass spectrometry. *Anal. Chem.* 65:972A–86A.
15. Operti, L., Tews, E. C., Freiser, B. S. 1988. Determination of gas-phase ligand binding energies to $Mg^+$ by FTMS techniques. *J. Am. Chem. Soc.* 110:3847–53.
16. Weiner, S. J., Kollman, P. A., Nguyen, D. T., Case, D. A. 1986. An all atom force field for simulations of proteins and nucleic acids. *J. Comput. Chem.* 7:230–52.
17. Halgren, T. A. J. 1996. Merck molecular force field. I. Basis, form, scope, parameterization, and performance of MMFF94. *Comput. Chem.* 17:490–519.
18. Halgren, T. A. J. 1996. Merck molecular force field. II. MMFF94 van der Waals and electrostatic parameters for intermolecular interactions. *Comput. Chem.* 17:520–52.
19. Halgren, T. A. J. 1996. Merck molecular force field. III. Molecular geometries and vibrational frequencies for MMFF94. *Comput. Chem.* 17:553–86.
20. Halgren, T. A. J. 1996. Merck molecular force field. IV. Conformational energies and geometries for MMFF94. *Comput. Chem.* 17:587–615.
21. Halgren, T. A. J. 1996. Merck molecular force field. V. Extension of MMFF94 using experimental data, additional computational data, and empirical rules. *Comput. Chem.* 17:616–41.

22. Bradshaw, J. S., Maas, G. E., Lamb, J. D., Izatt, R. M., Christensen, J. J. 1980. Cation complexing properties of synthetic macrocyclic polyether-diester ligands containing the pyridine subcyclic unit. *J. Am. Chem. Soc.* 102:467–74.
23. Davidson, R. B., Bradshaw, J. S., Jones, B. A., Dalley, N. K., Christensen, J. J., Izatt, R. M., Morin, F. G., Grant, D. M. 1984. Enantiomeric recognition of organic ammonium salts by chiral crown ethers based on the pyridino-18-crown-6 structure. *J. Org. Chem.* 49:353–57.
24. Dearden, D. V., Dejsupa, C., Liang, Y., Bradshaw, J. S., Izatt, R. M. 1997. Intrinsic contributions to chiral recognition: Discrimination between enantiomeric amines by dimethyldiketopyridino-18-crown-6 in the gas phase. *J. Am. Chem. Soc.* 119:353–59.
25. Liang, Y., Bradshaw, J. S., Dearden, D. V. 2002. The thermodynamic basis for enantiodiscrimination: Gas phase measurement of the enthalpy and entropy of chiral amine recognition by dimethyldiketopyridino-18-crown-6. *J. Phys. Chem. A* 106:9665–71.
26. Gerczei, T., Bocskei, Z., Keseru, G. M., Samu, E., Huszthy, P. 1999. Enantiomeric recognition of a-(1-naphthyl)ethylammonium perchlorate by enantiomerically pure dimethylphenazino-18-crown-6 ligand in solid and gas phases. *Tetrahedron Asymmetry* 10:1995–2005.
27. Samu, E., Huszthy, P., Somogyi, L., Hollosi, M. 1999. Enantiomerically pure chiral phenazino-crown ethers: Synthesis, preliminary circular dichroism spectroscopic studies and complexes with the enantiomers of 1-arethyl ammonium salts. *Tetrahedron Asymmetry* 10:2775–95.

# 8 Deoxy Oligonucleotides as Chiral References for the Discrimination of Enantiomeric Amino Acids under Mass Spectrometry

*M. Vairamani and Sangeeta Kumari*

## CONTENTS

8.1 Introduction ........ 143
8.2 Mass Spectrometry and Chiral Discrimination ........ 144
8.3 Enantioselectivity Using the Chiral Recognition Ratio (CR) Method ........ 147
8.4 Deoxy Oligonucleides as Chiral Auxiliaries ........ 148
8.5 Amino Acid as a Co-Selector ........ 154
8.6 Relative GCA Binding Order ........ 158
8.7 DNA Quartets as Chiral Auxiliaries ........ 159
8.8 Conclusion ........ 164
Acknowledgments ........ 164
References ........ 164

## 8.1 INTRODUCTION

Until the mid-1980s, nucleic acids were seen merely as informative molecules in opposition to proteins that carry the conformational information. This conception changed upon discovering the catalytic ability of nucleic acids (ribozymes), and especially the high affinity and specificity of short single-stranded DNA and RNA sequences (aptamers) toward a great variety of molecules, ranging from ions to whole cells.[1,2] Single-stranded oligonucleotide sequences have an exceptional propensity to assume an array of secondary and tertiary structural motifs with different shapes. The number of possible thermodynamically stable structural variants of an oligonucleotide sequence is much higher than the number of variants available for a peptide sequence of the same length. This is simply based on the ability of nucleotide bases to interact with each other through canonical Watson–Crick as well as unusual base pairing. The existence of oligonucleotide sequences that could assume myriad shapes within a random sequence library is the basis for the remarkable success of generating aptamers to a wide variety of target molecules. Most of the aptamers reported in the literature are related to RNA sequences, but it was found that such an immobilized ligand was very quickly degraded by RNases under conventional chromatographic condition and storage. In order to overcome this severe limitation, it was important to develop an RNA molecule resistant to the classical cleaving RNases. A very interesting strategy involving the mirror-image approach was successfully developed to design biostable L-RNA ligands, also known as spiegelmers, for potential therapeutic or diagnostic applications.[3] As the structure of nucleases is inherently chiral, the RNases accept only a substrate in the correct chiral configuration, i.e., the "natural" D-oligonucleotide. So, L-oligonucleotides are expected to be unsusceptible to the naturally occurring enzymes. This concept has been successfully applied to create a biostable RNA chiral stationary phase (CSP). It was demonstrated that a CSP based on L-RNA, that is, the mirror image of the natural D-RNA aptamer, was stable for an extended period of time under usual chromatographic conditions of storage and experiments.[4,5] Aptamers have been used in flow cytometry,[6] biosensors,[7,8] affinity probe capillary electrophoresis, capillary electrochromatography, affinity chromatography, and liquid chromatography.[9–15] For chiral compounds, the efficient monitoring of the selection procedure has allowed in most cases a very high specificity, exemplified by the capability of the aptamer to bind enantioselectively the target.[16–19] Eric Peyrin, for the first time, used an immobilized DNA aptamer as a new target-specific CSP for high-performance liquid chromatography.[20] The enantiomers of arginine-vasopressin were separated using an immobilized 55-base DNA aptamer known to bind stereospecifically the (all-D)-isomer of the oligopeptide. In a further work, such an approach has been extended to the chiral resolution of small molecules of biological interest.[21] The DNA aptamers used have been selected against D-adenosine and L-tyrosinamide enantiomers. An apparent enantio separation factor of around 3.5 was observed for the anti-D-adenosine aptamer CSP, while a very high enantioselectivity was obtained with the immobilized anti-L-tyrosinamide aptamer. Structures of aptamer complexes reveal the key molecular interactions conferring specificity to the aptamer-ligand association, including the precise stacking of flat moieties, specific hydrogen bonding, and molecular shape complementarities. These basic principles of discriminatory molecular interactions in aptamer complexes parallel recognition events central to many cellular processes involving nucleic acids.

## 8.2 MASS SPECTROMETRY AND CHIRAL DISCRIMINATION

Mass spectrometry has been used for differentiating enantiomers by probing the interactions with chiral references. Many research groups have reported chiral discrimination and enantiomeric excess (ee) determination of amino acids using cyclodextrins, chiral crown ethers, modified sugars, modified amino acids, and peptides as selectors.[22–27] The mass-spectrometry-based chiral recognition experiments are mainly

1. The measurement of the relative abundance of diastereomeric adduct ions formed between the analyte of interest and a chiral reference: Vairamani et al.[27] studied the chiral recognition properties of monosaccharide hosts (D-glucose, D-mannose, and D-galactose) as co-matrices with glycerol for the chiral discrimination of amino acid ester salts (alanine, leucine, and phenylglycine) under liquid secondary ion mass spectral conditions. In the studied amino acid esters, the L-isomers form more abundant adducts with sugars than the D-isomers. The same technique was also found to work for the determination of the enantiomeric purity of chiral phenyl ethylamines.[28]
2. Based on ion-molecule reactions: Diastereomeric adducts (host-guest complexes) are allowed to exchange the chiral analyte with a chiral or an achiral molecule, wherein chiral discrimination is possible as the rates of exchange depend on the chirality of the guest.[29–32]
3. Dissociation of diastereomeric adducts formed between the analyte and a chiral reference that results in distinct spectra:[33–46] While studying the decomposition of diastereomeric adduct ions for the purpose of chiral discrimination two methods are widely used: the kinetic method developed by Cooks et al.[33–41] and the chiral recognition ratio (CR) method introduced by Che et al.[42–45]
4. Chiral recognition by the kinetic method: One of the most versatile mass spectrometric methods for chiral recognition is certainly the kinetic method, developed by Graham Cooks and coworkers, and is widely used for determining the thermochemical properties such as proton affinity,[46–49] gas phase acidity,[50,51] metal ion affinity,[52,53] electron affinity,[54,55] etc., and is capable of differentiating processes with energy differences of even <1 kJ/mol. It does not involve wet chemistry, isotopic labeling, or chromatography, and therefore is particularly useful for fast analysis and under circumstances in which chromatography and derivatization are ineffective or inconvenient.

To achieve chiral recognition, multiple-point interactions are required;[56] this means that the chiral analyte and chiral reference need to be bound together in a polydentate complex, even if only transiently. Transition metal ions, for this reason, are much better choices to form such complexes than are proton or alkali metal cations, a fact that is well known for the condensed phase. It has become increasingly evident that most inter- and intramolecular interactions in biological systems are metal-cation mediated, and metal-ligand interactions in the gas phase have also become a major area of research. The role of transition metal ion is more significant than that of the alkali and alkaline earth metal ion toward chiral discrimination due to stereochemical effects of the d-orbital.[34,56]

Generally, the chiral analyte (enantiomers $A_R$ or $A_S$) and the chiral reference compound (ref*) are complexed with a transition metal ion ($M^{II}$) to generate diastereomeric cluster ions: in particular, trimeric cluster ions $[M^{II}(A_R)(ref^*)_2 - H]^+$ and $[M^{II}(A_S)(ref^*)_2 - H]^+$. The cluster ion is defined as "homo" when the analyte and

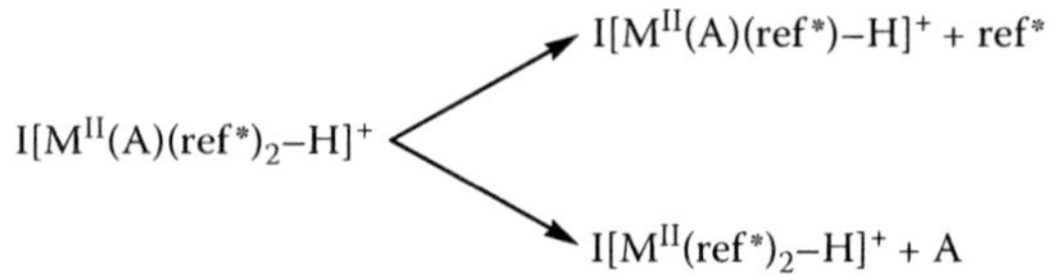


**SCHEME 8.1**

reference possess the same chiral configuration and as "hetero" in the other case. If $M^{II}$ is a divalent transition metal ion (e.g., $Cu^{II}$, $Zn^{II}$, $Ni^{II}$), the singly deprotonated ions $[M^{II}(A_R)(ref^*)_2 - H]^+$ and $[M^{II}(A_S)(ref^*)_2 - H]^+$ are detected in mass spectrometry experiments under positive ionization conditions. The collision-induced dissociation (CID) of these ions produces, respectively, the dimeric cluster ions: $[M^{II}(A_R)(ref^*) - H]^+$ and $[M^{II}(ref^*)_2 - H]^+$ or $[M^{II}(A_S)(ref^*) - H]^+$ and $[M^{II}(ref^*)_2 - H]^+$ (Scheme 8.1). This concept is equally applicable to the proton-bound trimeric cluster ions. In case of proton-bound trimeric cluster ion there is no deprotonation from either the reference or analyte ligand. The metal-mediated complexation to form a trimeric cluster ion has been observed in most of the reported studies. The kinetic method is based on the competitive dissociation of trimeric cluster ions. The abundances (I) of the product ions in the CID experiments directly reflect the difference in the stabilities of the diastereomeric complexes $[M^{II}(A_R)(ref^*) - H]^+$ and $[M^{II}(A_S)(ref^*) - H]^+$ relative to the reference complex $[M^{II}(ref^*)_2 - H]^+$. The chiral selectivity, in fact, depends on the difference in the free energy of activation ($\Delta(\Delta G_{chiral})$; Figure 8.1) for the dissociation reaction, which generates the dimeric complexes $[M^{II}(A_R)(ref^*) - H]^+$ and $[M^{II}(A_S)(ref^*) - H]^+$. The chiral selectivity, $R_{chiral}$, is determined as a measurement of the ratio of the individual ratios $(I[M^{II}(A_{R\ or\ S})(ref^*) - H]^+/I[M^{II}(ref^*)_2 - H]^+)$ of homo vs. hetero fragment ion abundances.

$$R_{chiral} = \frac{R_{homo}}{R_{hetero}}$$

If the chiral reference is an (R)-enantiomer, then the following equation holds:

$$R_{chiral} = \frac{I[M^{II}(A_R)(ref^*) - H]^+ / I[M^{II}(ref^*)_2 - H]^+}{I[M^{II}(A_S)(ref^*) - H]^+ / I[M^{II}(ref^*)_2 - H]^+}$$

The $R_{chiral}$ factor indicates the degree of chiral distinction achieved; the more different the $R_{chiral}$ value is from unity, the higher the degree of chiral recognition. Cooks and coworkers have been the major contributors to the development of chiral recognition techniques by mass spectrometry using the kinetic method. In a series of elegant studies,[56–64] they have shown the discrimination of enantiomeric amino acids, peptides, pharmaceuticals, and drugs.

Structural studies of the dimeric clusters $[M^{II}(A_R)(ref^*) - H]^+$ and $[M^{II}(A_S)(ref^*) - H]^+$, arising from CID of the corresponding $[M^{II}(A_R)(ref^*)_2 - H]^+$ and $[M^{II}(A_S)(ref^*)_2 - H]^+$ precursors, reveal that two ligands are noncovalently bound to the metal ion through

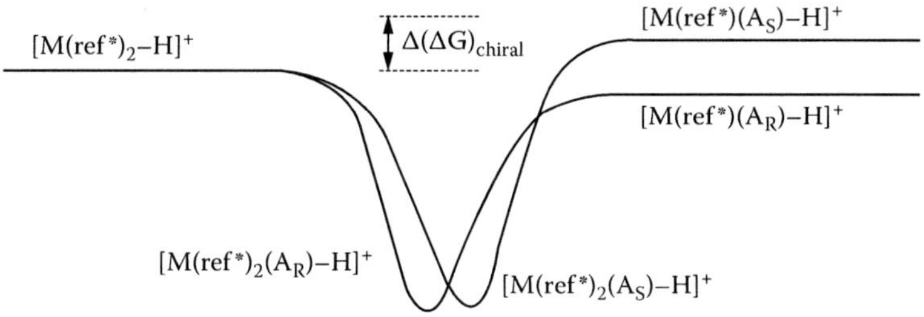


**FIGURE 8.1** Energy diagram for competitive dissociations of metal-ion-bound cluster ions of two deprotonated trimeric cluster ions that differ in the chirality of one ligand. (Taken from W.A. Tao, D. Zhang, E.N. Nikolaev, R.G. Cooks, *J. Am. Chem. Soc.* 122 (2000): 10598.)

multiple binding sites, which provide the basis for efficient chiral discrimination.[54] In the case of amino acids, two of the interactions between the two ligands are $M^{II}$ mediated, resulting from the coordination of the amino and carboxylate groups to the central metal ion, whereas the third interaction involves the substituents at or near the asymmetric α-carbon of each of the two ligands. The superior chiral recognition achieved when reference has an aromatic side chain suggests that π-d-orbital interactions play an important role in the stereoselectivity. In this connection, the nature of the metal ion is expected to play an important role, and the experimental data indeed show intriguing effects.[62] In the case of amino acids, for example, $Cu^{II}$ offers much larger chiral selectivity than $Zn^{II}$ or $Ni^{II}$, which is due to the formation of a square planar complex.[34]

It is also possible to rapidly determine ee of an amino acid sample by a single measurement of ln $R_{chiral}$ in a tandem mass spectrometer.[56] This behavior is typified by Clevudine (2-fluoro-5-methyl-β,L-arabinofuranosyluracil), a potent antiviral nucleoside against hepatitis B.[57] To optimize its chiral discrimination, several metal ions have been checked together with a variety of amino acids as chiral references. Different from amino acid analytes that clearly prefer $Cu^{II}$ as metal ion center, Clevudine prefers transition metals $Co^{II}$ and $Zn^{II}$, as these ions have a high affinity for its heteroaromatic ring, which is distant from the stereocenters. Using N-acetyl-L-proline as chiral selector and $Co^{II}$ as the metal ion, the data for various enantiomeric mixtures of Clevudine display a linear relationship between ln $R_{chiral}$ and ee with a correlation coefficient of 0.9995. This calibration curve is then used to measure the percent ee of various unknown samples. Calibration curves can be established for the simultaneous chiral analysis of different amino acids in mixtures.[36] The same procedure has been applied to the chiral analysis of some peptides,[37,38,62] neurotransmitters,[63] thalidomide,[64] and antibiotics.[41]

## 8.3 ENANTIOSELECTIVITY USING THE CHIRAL RECOGNITION RATIO (CR) METHOD

The measurement of CR is based on the dissociation of a diastereomeric complex. Instead of the ratio of product ions resulting from dissociation, calculation of CR relies on the ratio of the intensity of one fragment ion to that of the parent ion. Similar to the kinetic method, this technique relies on the formation, isolation, and fragmentation of a diastereomeric cluster ion composed of a chiral analyte of interest and an enantiomerically pure reference. The dissociation of the specified diastereomeric cluster ion leads to the loss of one reference stereoisomer from the cluster. This method would be more useful in cases where the diastereomeric trimeric cluster ion leads to only one fragment. The CR value is given by the following equation:

$$CR = \frac{I[M''(A_S)(ref_R^*) - H]^+ / I[M''(A_S)(ref_R^*)_2 - H]^+}{I[M''(A_R)(ref_R^*) - H]^+ / I[M''(A_R)(ref_R^*)_2 - H]^+}$$

Thus, a CR value of unity would mean there is no observable chiral discrimination. A CR value larger than unity indicates that dissociation is more favorable in the heterochiral case, whereas a CR value smaller than unity indicates that dissociation is more favorable in the homochiral case. The more deviation the CR value is from unity, the more significant is the chiral discrimination observed.

## 8.4 DEOXY OLIGONUCLEIDES AS CHIRAL AUXILIARIES

With a view to testing the ability of deoxy oligonucleotides for the discrimination of enantiomers of amino acids under electrospray ionization (ESI) mass spectral conditions, the author's group selected a series of DNA triplets as auxiliaries. The selected trinucleotides include AAA, AAG, AGA, AGG, ACA, ACG, CCA, CGA, CCG, GCA, GCC, CGC, CCC, CGG, and GCG. Since the oligonucleotides are more amenable to negative ion mass spectral analysis due to the presence of phosphate groups, all the mass spectral experiments were confined to negative mode only. The ESI mass spectra of all the trinucleotides comprise dominant $[M - H]^-$ and $[M - 2H]^{2-}$ ions of the trinucleotide. Like the spectra of trinucleotides, the spectra of a mixture of amino acid and trinucleotide also predominantly show $[M - H]^-$ and $[M - 2H]^{2-}$ ions of trinucleotide; however, they also include the $[X + Y - H]^-$ ion in the high mass region in considerable abundance. Based on the expected isotopic pattern for singly and doubly charged ions of the same *m/z* value, the $[X + Y - H]^-$ ion is found to be a combination of two ions in which the $[2X + 2Y - 2H]^{2-}$ ion is predominant (70–80%) over the $[X + Y - H]^-$ ion. A typical ESI spectrum obtained from the mixture of GCA and D-Trp is shown in Figure 8.2, along with simulated spectra.

The source spectra obtained for two enantiomeric amino acids do not show significant differences among the D- and L-isomers. Nevertheless, it is worth mentioning that the oligonucleotide forms dominant monomeric $[Y - H]^-$ ions, whereas in the presence of an amino acid it dimerizes, resulting in the formation of $[2X + 2Y - 2H]^{2-}$ adduct ion. It implies that the trinucleotides prefer to form a stable dimeric complex ion with amino acid rather than a monomeric complex $[X + Y - H]^-$, possibly due to strong bridging between amino acids and trinucleotides.

One cannot expect substantial differences in the abundance of the diastereomeric adduct ions formed between the chiral analyte of interest and a chiral reference unless the adduct formation is highly specific. Also, the source spectra are very much dependent on the stoichiometry ratio of the analytes, which is very difficult to maintain under the mass spectrometry experimental conditions, especially at the

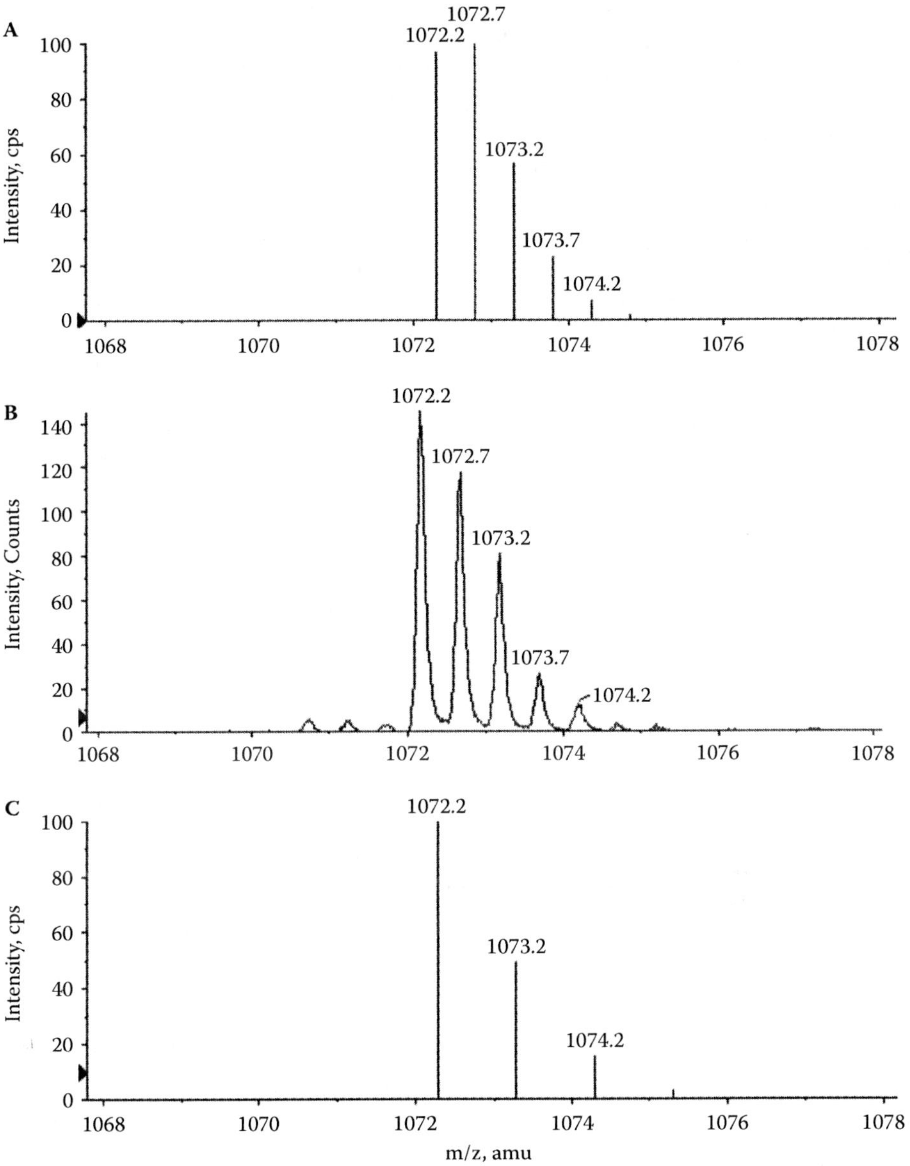


**FIGURE 8.2** (A) Simulated spectrum of $[2Trp + 2GCA - 2H]^{2-}$ ion. (B) Expanded ESI spectrum of the mixture of GCA and D-Trp showing contribution of $[Trp + GCA - H]^{-}$ ion in $[2Trp + 2GCA - 2H]^{2-}$ ion. (C) Simulated spectrum of $[Trp + GCA - H]^{-}$ ion.

micromolar level. But the small energy difference between the two diastereomeric adducts can get amplified in the product ion spectra of these adducts obtained under CID conditions. It was found that the CID spectra of $[2X + 2Y - 2H]^{2-}$ ions are distinct for D- and L-isomers of tryptophan, phenylalanine, glutamic acid, aspartic acid, asparagine, and tyrosine when GCA is used as a selector. The CID spectra of $[2X + 2Y - 2H]^{2-}$ of D- and L-Glu enantiomers with GCA triplet as the chiral selector are given in Figure 8.3 as an example. No such differences are observed with any of the other DNA triplets studied. This first observation suggested that deoxy oligonucleotides can be selective and specific for chiral amino acids. This led to the detailed investigation of the use of short single-strand deoxy oligonucleotides as selectors toward the gas phase chiral discrimination of the amino acids.

In the preliminary studies with DNA triplets, the $[2X + 2Y - 2H]^{2-}$ ions were used and results were encouraging, though this ion was found to be a mixture of $[2X + 2Y - 2H]^{2-}$ and $[X + Y - H]^{-}$. However, there is no control over the extent of contribution of $[X + Y - H]^{-}$ ion to the $[2X + 2Y - 2H]^{2-}$ ion in the source; hence, the

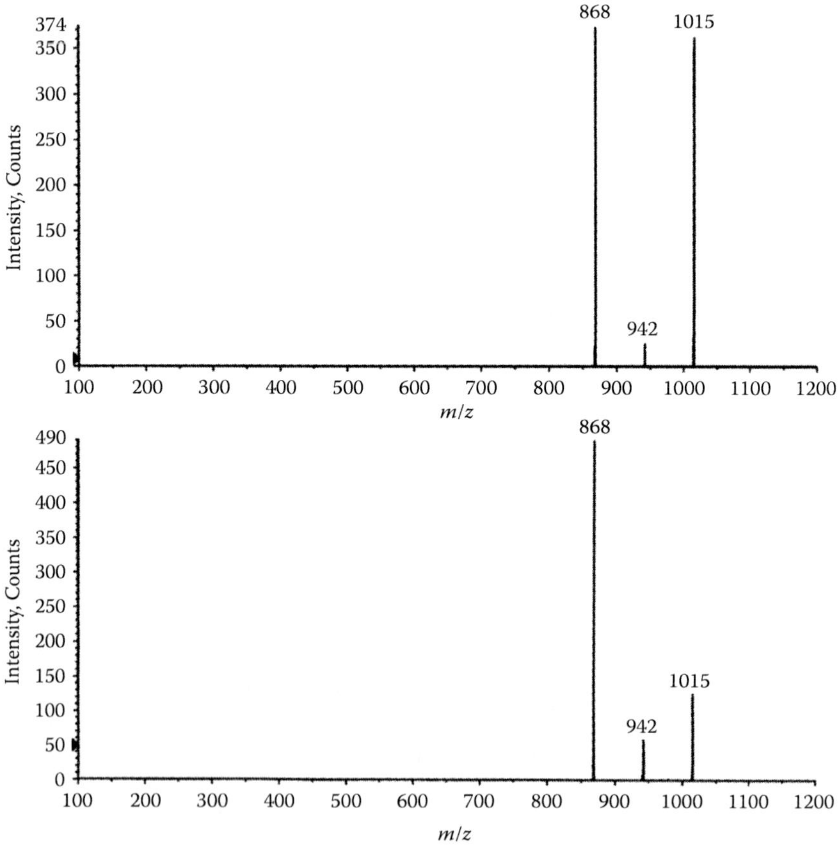


**FIGURE 8.3** CID spectra of $[2X + 2Y - 2H]^{2-}$ ion, X = D-glutamic acid (top) or L-glutamic acid (bottom) and Y = GCA.

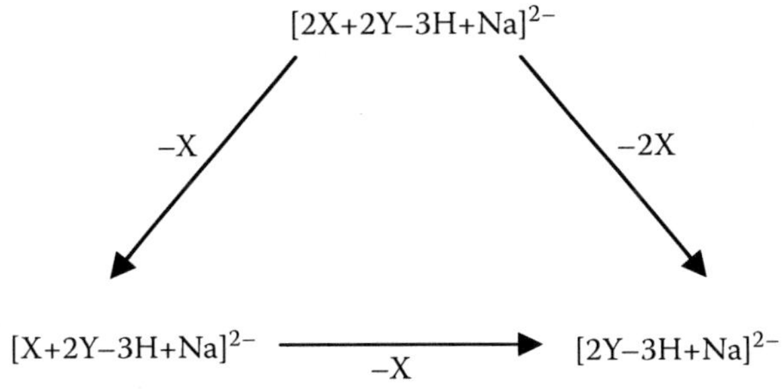


**SCHEME 8.2**

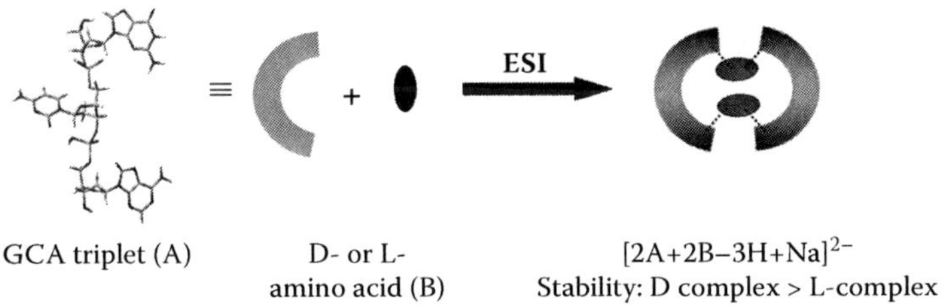


**FIGURE 8.4** The graphical tentative picture of the $[2X + 2Y - 3H + Na]^{2-}$ ions.

contribution of fragment ions from $[X + Y - H]^-$ may affect the extent of chiral discrimination. At the same time, it is not possible to resolve the $[2X + 2Y - 2H]^{2-}$ and $[X + Y - H]^-$ ions by any mass spectrometric methods, as both the ions essentially appear at the same *m/z* value. Therefore, it was necessary to produce purely a doubly charged ion that is similar to the $[2X + 2Y - 2H]^{2-}$ ion. One such ion, $[2X + 2Y - 3H + Na]^{2-}$, could be generated by the replacement of a hydrogen in the trinucleotide of the $[2X + 2Y - 2H]^{2-}$ ion by a metal ion, M (M = Na, K, etc.). Indeed, the ion $[2X + 2Y - 3H + Na]^{2-}$ appeared in the pure form in the spectrum recorded for the mixture of trinucleotide and amino acid in the presence of NaCl. It is not possible to pinpoint the location of $Na^+$ in the adduct, though it is expected to replace one of the acidic protons from the phosphate group. The decomposition pathway of $[2X + 2Y - 3H + Na]^{2-}$ ions (Scheme 8.2) clearly suggests the localization of two negative charges on phosphate groups of trinucleotides, and the amino acids are linked between trinucleotides through hydrogen bonding. The tentative graphical picture of the $[2X + 2Y - 3H + Na]^{2-}$ ions is shown in Figure 8.4. The importance of H-bonding in the noncovalent interaction involving nucleic acids and peptides in the gas phase has been well studied. Vertes et al.[65] first demonstrated ionic interaction in DNA-peptide complexes. Woods et al.[66] highlighted the role of electrostatic interactions between peptides containing several basic residues or acidic residues with phosphate groups in the DNA strands.

$$CR = \frac{[2X_D + 2Y - 3H + Na]^{2-}/[2Y - 3H + Na]^{2-}}{[2X_L + 2Y - 3H + Na]^{2-}/[2Y - 3H + Na]^{2-}} \quad (8.1)$$

$$CR = \frac{[2X_D + 2Y - 3H + Na]^{2-}/[X_D + 2Y - 3H + Na]^{2-}}{[2X_L + 2Y - 3H + Na]^{2-}/[X_L + 2Y - 3H + Na]^{2-}} \quad (8.2)$$

The results obtained from the CID experiments on $[2X + 2Y - 3H + Na]^{2-}$ ions are similar to those obtained for the $[2X + 2Y - 2H]^{2-}$ ions, wherein chiral discrimination for all the used amino acids is achieved only when GCA is used as a chiral selector, and the CID spectra of the $[2X + 2Y - 3H + Na]^{2-}$ ion of GCA and D/L-glutamic acid are given in Figure 8.5. The CR method was used for studying the

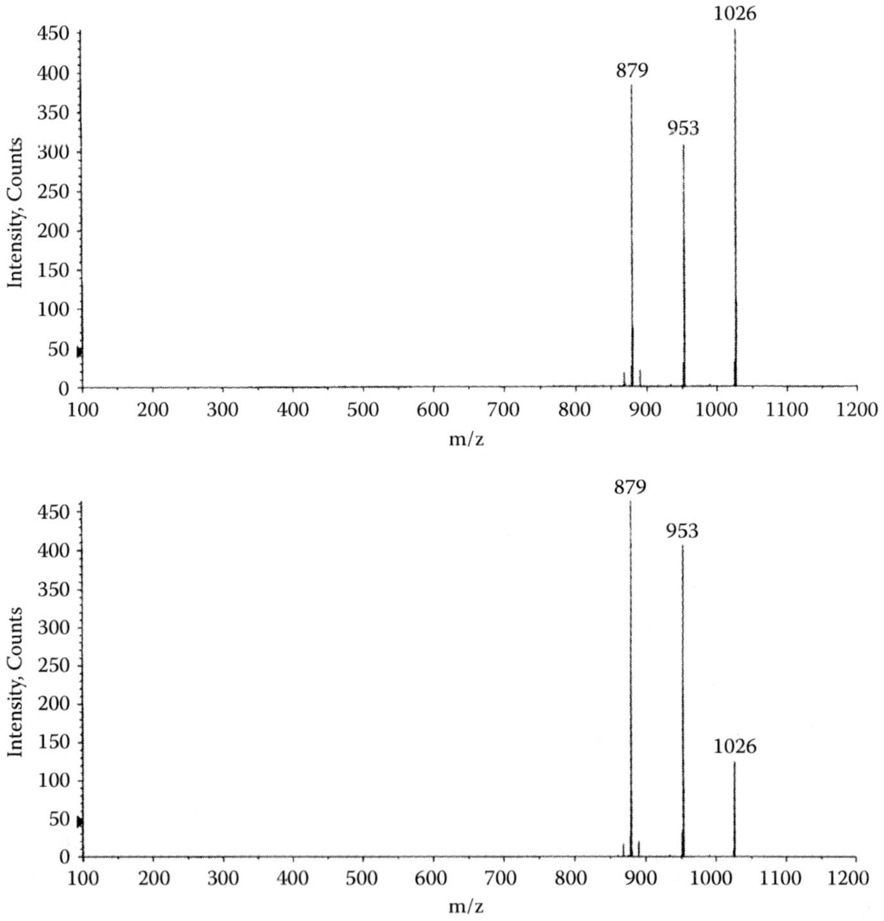


**FIGURE 8.5** CID spectra of $[2X + 2Y - 3H + Na]^{2-}$ion, X = D-glutamic acid (top) or L-glutamic acid (bottom) and Y = GCA.

chiral discrimination process. In the CR method, the ratio of precursor ion to either of the product ions is used for the calculation of CR values (Equations 8.1 and 8.2):

The calculated CR values using Equations 8.1 and 8.2 showed remarkable chiral recognition; moreover, the CR values obtained from both the equations gave the same results. The calculated CR values for the studied amino acids are presented in Table 8.1. The observed results for all the studied amino acids demonstrate that the GCA is preferentially binding with the D-isomer rather than the L-isomer, irrespective of the amino acid used. Among the amino acids used, Phe, Trp, Tyr, Asn, Asp, and Glu resulted in abundant $[2X + 2Y - 3H + Na]^{2-}$ ions, and their CID spectra are distinct for the two enantiomers. The D-selectivity of GCA can be clearly visualized when experiments are performed by taking a mixture of Glu enantiomers, out of which one enantiomer is deuterium labeled. The ESI spectrum recorded for the mixture of GCA, D-Glu, and L-Glu-$d_5$ showed a $[(\text{D-Glu}) + (\text{L-Glu-}d_5) + 2\text{GCA} - 2H]^{2-}$ ion at *m/z* 1017.7. The CID spectrum of the ion at *m/z* 1017.7 shows the fragment ions corresponding to the loss of D-Glu and L-Glu-$d_5$ at *m/z* 944.2 and 941.7, respectively (Figure 8.6). As anticipated, the relative abundance of the ion at *m/z* 941.7 is higher than that at *m/z* 944.2, which demonstrates that the D-Glu is binding stronger than L-Glu. The data obtained for all the amino acids reveal that the chiral discrimination by GCA is dramatic for acidic and aromatic amino acids (Table 8.1), whereas it is poor for basic and aliphatic amino acids. Better selectivity observed for the aromatic and acidic amino acids may be due to the effective secondary interactions of the aromatic ring/acidic group of amino acids with the trinucleotide. Among the many trinucleotides selected for chiral discrimination, only the GCA is showing the discrimination of D- and L-amino acids. It is reasonable to anticipate that the ACG, which has the same type of bases but in the 3'-5' direction, might show the same selectivity as GCA, but such selectivity is not observed with ACG experimentally. This implies that it is not just the presence of the bases that is important, but that the particular sequence of the bases in one direction (5' → 3') is responsible for the remarkable chiral discrimination of GCA among many other trinucleotides that were attempted.

**TABLE 8.1**
**Calculated CR Values from CID Spectra of $[2X + 2Y - 3H + Na]^{2-}$ Ions (X = D- or L-Amino Acid, Y = GCA)**

| Amino Acid | a/b | a/c |
|---|---|---|
| Tryptophan | 6.4 | 5.6 |
| Phenylalanine | 3.6 | 5.4 |
| Glutamic acid | 4.9 | 4.2 |
| Aspartic acid | 2.8 | 2.5 |
| Asparagine | 3.7 | 5 |
| Tyrosine | 2.5 | 2.1 |
| Proline | 1.2 | 1.5 |
| Alanine | 1.1 | 1.5 |

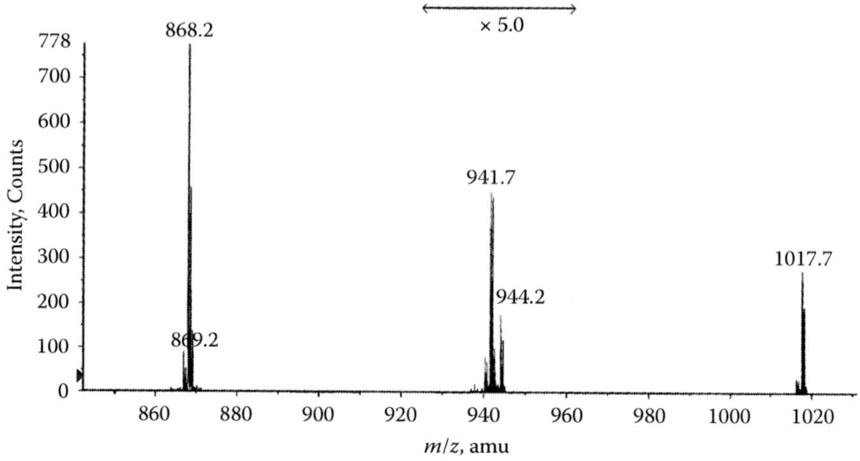


**FIGURE 8.6** CID spectrum of [(D-Glu) + (L-Glu-d5) + 2GCA – 2H]$^{2-}$ ion (*m/z* 1017.7).

Zhu et al.[67] reported the formation of an unusual hairpin structure for the CAATGCAATG sequence, and based on NMR, X-ray, and computational studies, they proved that the unusual behavior of CAATGCAATG is due to the presence of a unique GCA sequence in the middle of the chain. Existence of CAATGCAATG as a hairpin structure was explained by the formation of a GCA cap, involving a key G-A pairing between G and A of the GCA sequence. Such hairpin structures were not found for CAG or AGC triplets containing nucleotides. Similarly, in the present case, the cap-like structure of GCA involving sheared G-A pairing between G and A may be playing a crucial role in differential stability for the dimeric complexes with D- and L-isomers of amino acids. The higher stability for the dimeric complexes of D-amino acids might be due to suitable orientation of D-isomers in effectively binding with nucleotide. In order to prove the importance of the position of C in the middle of sequence GCA in forming a cap-like structure, experiments were performed with three other trinucleotides, by replacing the C with other nucleo-bases, i.e., GGA, GTA, and GAA, using Glu and Trp enantiomers. With this set of trinucleotides also abundant [2X + 2Y – 3H + Na]$^{2-}$ ions were formed, but no chiral discrimination was obtained with GTA and poor discrimination was seen with GGA and GAA. The poor discrimination by GGA, GTA, and GAA could be attributed to failing of the formation of a cap-like structure involving a G and A base pair. Based on these experiments it is clear that all three amino groups of the trinucleotide and their orientations are responsible for the discrimination. This method can be used not only for the chiral discrimination of individual enantiomers, but also for the determination of enantiomeric excess of amino acids when they are present in an enantiomeric mixture.

## 8.5 AMINO ACID AS A CO-SELECTOR

In the above studies a DNA triplet, GCA, was successfully used as a chiral selector for the chiral discrimination of amino acids, and the formation of a dimeric doubly charged ion made up of two DNA molecules and two amino acids was useful to get the desired information. This suggested that there is a scope for using another chiral compound as a reference, and the obvious choice was another amino acid so that the interactions that are responsible for the formation of adducts can be maintained to the maximum extent. Hence, in the next set of experiments in the author's laboratory, amino acids themselves were used as co-selectors along with the GCA triplet. The negative ESI mass spectra obtained for a mixture with two different amino acids, i.e., a reference (D-amino acid, $X_R$) and an analyte (D- or L-amino acids, $X_A$), and GCA triplet (Y) show abundant ions corresponding to $[Y - H]^-$ and $[Y - 2H]^{2-}$, and the adduct ions $[2X_R + 2Y - 2H]^{2-}$, $[2X_A + 2Y - 2H]^{2-}$, and $[X_A + X_R + 2Y - 2H]^{2-}$. The spectra also include low abundant $[2X_A + Y - 2H]^{2-}$, $[X_A + 2Y - 2H]^{2-}$, $[2X_R + Y - 2H]^{2-}$, and $[X_R + 2Y - 2H]^{2-}$ adduct ions. The ion of interest, $[X_A + X_R + 2Y - 2H]^{2-}$, generated with the combination of two different amino acids, is absolutely pure, as the molecular weights of the two amino acids are different. The relative abundances of the fragment ions that resulted in the CID spectra of $[X_A + X_R + 2Y - 2H]^{2-}$ ions from D- and L-analytes are considered for measuring the degree of chiral discrimination by applying the kinetic method. Based on the observations in the previous study on the generation of adduct ions of GCA and amino acids,[68] D-Trp, which binds efficiently with GCA, was selected as a reference in experiments for the discrimination of a set of D- and L-amino acids. The preliminary results showed that Gln, Glu, and Tyr enantiomers form abundant $[X_A + X_R + 2Y - 2H]^{2-}$ ions (**a**). The CID spectra of ion **a** show three fragment ions corresponding to $[X_A + 2Y - 2H]^{2-}$ (**b**), $[X_R + 2Y - 2H]^{2-}$ (**c**), and $[2Y - 2H]^{2-}$ (**d**), which are formed by the loss of the reference amino acid ($X_R$), analyte amino acid ($X_A$), and both amino acids ($X_R$ and $X_A$), respectively. The relative abundances of ions **b** and **c** reflect the competitive loss between the reference and the analyte, and the abundance ratio of these two ions could be used for the calculation of $R_{chiral}$ values. Ions **b** and **c** can be seen only when the GCA binding efficiency of both reference and analyte amino acids are closer. The appearance of both **b** and **c** ions for Gln, Glu, and Tyr enantiomers using D-Trp as a reference gives a hint that all these amino acids must have closer GCA binding efficiency. This observation is comparable to the metal-mediated chiral discrimination experiments where analyte and reference should have closer metal ion affinity to give two daughter ions. Indeed, when experiments were performed among Trp, Glu, Gln, and Tyr using one of the D-amino acids as a reference for the discrimination of other enantiomeric pairs, all the combinations formed abundant $[X_A + X_R + 2Y - 2H]^{2-}$ ions and their CID spectra showed both ions **b** and **c**. Typical CID spectra of ion **a** for D-Glu and L-Glu using D-Trp as the reference are shown in Figure 8.7. The comparative loss of reference amino acid to that of the analyte amino acid is different for different enantiomers of analyte amino acid. The calculated $R_{chiral}$ values with D-amino acid as reference using the relative abundance ratios of ions **b** and **c**, as expressed by Equation 8.3, are summarized in Table 8.2.

$$R_{chiral} = R_D/R_L = (\mathbf{b}/\mathbf{c})_D/(\mathbf{b}/\mathbf{c})_L \quad (8.3)$$

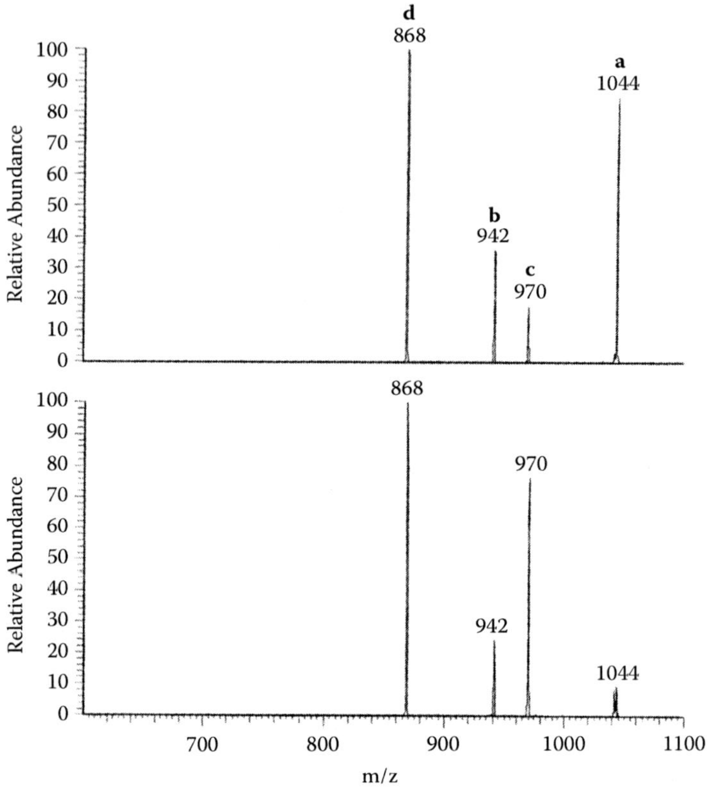


**FIGURE 8.7** CID spectra of $[X_A + X_R + 2Y - 2H]^{2-}$ ion, $X_A$ = D-Glu (top) or L-Glu (bottom), $X_R$ = D-Trp, and Y = GCA.

**TABLE 8.2**
**$R_{chiral}$ Values for the Amino Acids Using Different D-Amino Acids as References**

| | $R_{chiral}$ Values | | | | | |
|---|---|---|---|---|---|---|
| **Reference** | **Trp** | **Glu** | **Tyr** | **Gln** | **Lys** | **Asp** |
| D-Trp | — | 5.5 | 0.15 | 10.1 | 0.62 | 2.5 |
| D-Glu | 8.3 | — | 0.17 | — | — | — |
| D-Tyr | 5.9 | 5.3 | — | 10.2 | — | |
| D-Gln | 7.8 | — | 0.17 | — | — | 2.5 |
| D-Asp | — | — | — | — | 0.66 | — |

A high degree of $R_{chiral}$ values could be obtained for the enantiomers of Trp, Glu, Gln, and Tyr. The $R_{chiral}$ values for Trp, Glu, and Gln in this study are better than the earlier reported literature values.[34,57]

The experiments were extended to discriminate other amino acids using Trp, Glu, Gln, and Tyr as references. The enantiomers of Lys could be discriminated using only D-Trp as the reference. The Gln or Glu could not be used as reference for the enantiomers of Lys because of the same mass of analyte and reference, or 1 amu difference between the two, where isotopic overlapping problem is encountered. The enantiomers of Asp could be discriminated using only D-Trp/D-Gln as the reference. In the case of other combinations for the enantiomers of Asp and Lys, the experiments failed as the yields of the required adduct ion are poor, or the CID of the adduct ion did not result in both **b** and **c** ions. The GCA binding efficiency of Asp and Lys must be closer to that of D-Trp when compared to Gln, Glu, and Tyr; hence, these combinations are successful in chiral discrimination by the present method. Further, the GCA binding efficiency of Asp and Lys is found to be closer because the enantiomers of Lys are successfully discriminated by using D-Asp as the reference.

Arg, Asn, Phe, Ser, and Thr form the adduct ion of interest in good yields when Trp, Glu, Gln, or Tyr is used as the reference; however, the present method cannot be applied to discriminate these enatiomeric pairs because the CID spectra show absence of either **b** or **c**. Ion **b** is absent for enantiomers of Asn, Phe, Ser, and Thr, and this indicates strong binding of the reference, rather than these amino acids. Ion **c** is absent for Arg enantiomers, and absence of ion **c** indicates strong binding of the analyte with GCA. These results clearly demonstrate that there is a large difference in the binding efficiency of GCA with the analyte (Arg, Asn, Phe, Ser, and Thr) and the reference (Trp, Glu, Gln, and Tyr). The Arg isomers fail to show ion **c** by using any other amino acid as the reference, which proves the highest GCA binding efficiency of Arg among all the amino acids. The Asn and Phe enantiomers could be discriminated by using D-Ser as the reference; the calculated $R_{chiral}$ values for Asn and Phe are 9.8 and 9.0, respectively. The Ser enantiomers ($R_{chiral}$ = 6.3) and Thr enantiomers ($R_{chiral}$ = 9.7) could be discriminated by using D-Asp and D-Asn as the reference, respectively. Therefore, selection of a suitable reference having closer GCA binding efficiency to that of the analyte is crucial for the success of this method. The aliphatic amino acids Ala, Val, Leu, Pro, Cys, and Met form very low abundant $[X_A + X_R + 2Y - 2H]^{2-}$ ions when Trp, Glu, Gln, and Tyr are used as the reference, and hence cannot be discriminated by applying this method of using amino acids themselves as the reference. Formation of low abundant $[X_A + X_R + 2Y - 2H]^{2-}$ ion for the aliphatic amino acids reveals their poor binding efficiency with GCA, which could be due to lack of extra functional groups in the side chain.

With a view to check the effect of the chirality of the reference in chiral discrimination of GCA, the experiments were performed with Trp, Glu, Gln, and Tyr using L-isomer as the reference to discriminate the enantiomers of the other three amino acids, saving the combinations of Glu and Gln. As in the case of D-isomers, all the used combinations showed abundant $[X_A + X_R + 2Y - 2H]^{2-}$ ion, where $X_R$ = L-amino acid, and its CID spectra show both the expected fragment ions. The $R_{chiral}$ values calculated for all the successful combinations using L-amino acids are summarized in Table 8.3. If the $R_{chiral}$ value is >1, the D-enantiomer of analyte amino

**TABLE 8.3**
**$R_{chiral}$ Values for the Amino Acids Using Different L-Amino Acids as References**

| | $R_{chiral}$ Value | | | | |
|---|---|---|---|---|---|
| **Reference** | **Trp** | **Glu** | **Tyr** | **Gln** | **Lys** |
| L-Trp | — | 6.1 | 0.14 | 10.7 | 0.7 |
| L-Glu | 7.6 | — | 0.16 | — | — |
| L-Tyr | 6.7 | 6.1 | — | 10.8 | — |
| L-Gln | 7.3 | — | 0.16 | — | — |

acid is binding relatively more strongly to GCA than the L-enantiomer of analyte amino acid, and if the value is <1.0, it is the L-amino acid that binds strongly with GCA. From Table 8.2, it can be noticed that D-isomers of all the amino acids, except Tyr and Lys, are binding strongly with GCA when compared to the corresponding L-isomers. The same D-selectivity of GCA is reported in the previous method, also. In the present method, by the use of a reference, the chiral selectivity of GCA is reversed in the case of Tyr and Lys, where the L-isomer of reference amino acid binds preferentially with GCA. Structurally both Phe and Tyr are closely related, and the difference between the two is the presence of a hydroxyl group on the phenyl ring in Tyr. However, Phe isomers are behaving similar to those of other amino acids (D-selectivity). Hence, interactions of the hydroxyl group in Tyr must be playing a crucial role in showing unusual L-selectivity for Tyr. Similarly, the selectivity of GCA is also reversed in the case of Lys having a free $-NH_2$ group in the side chain. However, Gln and Asn, which consist of a $-CONH_2$ group in the side chain, retain the D-selectivity of GCA. Therefore, the free amino group of Lys might be responsible for the L-selectivity. Nevertheless, it is difficult to propose a concrete explanation with the present experimental data without the support of theoretical calculations. The interesting point to note is that the chiral selectivity of GCA did not change with the chirality of the reference used. For example, the calculated $R_{chiral}$ values are similar when D- and L-Trp are used as the references. Though the selectivity of GCA is reversed for Tyr enantiomers, the selectivity of GCA is not changed for other amino acids when D-Tyr or L-Tyr is used as the reference. Therefore, the discrimination ability of GCA improves by the use of a reference, saving the preference of GCA.

## 8.6 RELATIVE GCA BINDING ORDER

The relative binding order of GCA for the D- and L-amino acids was studied based on the abundances of fragment ions produced in the CID experiments. Arg occupies first place in the binding order, because the CID spectra of these enantiomers resulted in only ion **c**, irrespective of the reference used. Trp, Gln, Glu, and Tyr result in both ions **b** and **c** using any of the amino acids (D- or L-isomer) in this group as a reference, and use of these amino acids as references for other amino acids (Asp, Phe, Asn, and Ser) failed to show ion **b**. Therefore, the D- and L-isomers of Trp, Gln, Glu, and Tyr come next to Arg, and the set of Asp, Phe, Asn, and Ser follows next.

Therefore, the GCA binding order for all the studied amino acids can be given as: D/L-Arg >> L-Tyr > D-Glu > D-Trp ≈ D-Gln > D-Tyr > L-Glu > L-Trp > L-Gln > D-Asp, L-Lys, D-Phe, D-Asn, D-Ser > D-Lys, L-Asp, L-Phe, L-Asn, L-Ser.

## 8.7 DNA QUARTETS AS CHIRAL AUXILIARIES

The authors' group extended the above studies to DNA tetranucleotides in place of trinucleotides. For doing this, the GCA motif was retained in DNA tetranucleotides, in which an extra base is added at either the 3'- or 5'-end of GCA. Thus, GCAA, GCAG, GCAC, GCAT, AGCA, GGCA, CGCA, and TGCA were selected to test whether these tetranucleotides act as chiral selectors for the discrimination of amino acids. Moody et al.[69] studied the stabilities of expanded versions of DNA hairpin loops in the condensed phase and reported that the known stable d(cGNAg) tri-loop motif can be embedded into a tetra-loop, with the extra nucleotide inserted either into the middle of the loop, d(cGNNAg), or at the 3'-end of the loop, d(cGNABg) (where N= A, C, G, or T; and B = C, G, or T).

The negative ion ESI mass spectra of tetranucleotides comprise dominant $[M-2H]^{2-}$ and $[M-3H]^{3-}$ ions; the $[M-H]^{-}$ ion is negligible in the spectra. In general, the negative ESI mass spectra obtained for a mixture of individual tetranucleotide (X) and D- or L-amino acid (Y) (1:1 to 1:4 ratio of tetranucleotide and amino acid) show abundant ions corresponding to $[X-2H]^{2-}$ and $[X+Y-2H]^{2-}$. In addition to the aforementioned ions, $[X+(Y)_n-2H]^{2-}$, where n = 2 and 3, are also observed at variable abundances depending on the amino acid used and the ratio of amino acid with respect to the oligonucleotide, but these ions are always low in abundance when compared to the $[X+Y-2H]^{2-}$ ion. On the basis of the spectra recorded at high resolution, all the above-mentioned ions are found to be pure, and there is no contribution from other ions. It is interesting to note that all the tetranucleotides preferably form 1:1 adducts with amino acids, whereas earlier experiments with GCA showed a formation of 2:2 adducts preferably.[68]

The experiments in the present study were restricted to the CID of $[X+Y-2H]^{2-}$ ions to probe the utility of tetranucleotides as chiral selectors because of the low abundance of the other ions. In the previous report aromatic amino acids showed better selectivity when GCA was used as a selector.[70] Hence, two aromatic amino acids, Trp and Tyr, were selected for the experiments to check the chiral selectivity with the selected tetranucleotides. CID spectra were recorded for $[X+Y-2H]^{2-}$ ions formed from all the combinations with the tetranucleotides and the two studied amino acids. All the CID spectra of $[X+Y-2H]^{2-}$ ions (**a**) from Trp and Tyr, irrespective of the tetranucleotide used, exclusively yield $[X-2H]^{2-}$ ion (**b**) as the product ion. The $[X-2H]^{2-}$ ion corresponds to the loss of bound amino acid from the $[X+Y-2H]^{2-}$ ion. The CR method was used to measure the extent of chiral discrimination because the CID spectra of the $[X+Y-2H]^{2-}$ ion yielded only one product ion. The CR values calculated by measuring the abundance ratio of precursor ion (**a**) to that of the product ion (**b**), as shown in Equation 8.4, are presented in Table 8.4.

$$CR = R_D/R_L = [\mathbf{a}/\mathbf{b}]_D/[\mathbf{a}/\mathbf{b}]_L \quad (8.4)$$

**TABLE 8.4**
**CR Values[a] Obtained for Trp and Tyr Isomers**

| | CR value $(R_D/R_L)^b$ | |
|---|---|---|
| **Tetranucleotide** | **Trp ± SD** | **Tyr ± SD** |
| GCAA | 2.54 ± 0.3 | 2.60 ± 0.3 |
| GCAG | 1.44 ± 0.03 | 1.27 ± 0.05 |
| GCAC | 1.16 ± 0.1 | 1.07 ± 0.2 |
| GCAT | 2.03 ± 0.2 | 1.94 ± 0.3 |
| AGCA | 1.14 ± 0.1 | 1.30 ± 0.1 |
| GGCA | 0.34 ± 0.02 | 0.35 ± 0.02 |
| CGCA | 0.68 ± 0.1 | 0.77 ± 0.02 |
| TGCA | 1.73 ± 0.04 | 1.82 ± 0.1 |

[a] Values shown are the average of three separate measurements done at different occasions.
[b] $R_D = [a/b]_D$, $R_L = [a/b]_L$.

It can be noted from Table 8.4 that among the tetranucleotides, GCAA, GCAG, GCAC, and GCAT, which are extended versions of GCA at the 3′-end, the CR values are relatively better (far from unity) for GCAA and GCAT than for GCAG and GCAC, where the discrimination is poor (close to unity). Both GCAA and GCAT show D-selectivity, and of the two, GCAA shows better selectivity than GCAT. If the observed solution phase phenomenon[69] is true in the gas phase, then all four tetranucleotides extended at the 3′-end are also expected to show chiral discrimination similar to that observed with GCA. In the present experiments only the GCAA and GCAT are able to shows discrimination with the same selectivity as found with GCA. This clearly indicates that the loop-like structure of the GCA sequence involving G-A pairing is retained in the gas phase in the case of GCAA and GCAT. Among the tetranucleotides with a 5′-end extension (AGCA, GGCA, CGCA, and TGCA) TGCA shows discrimination with D-selectivity, whereas AGCA does not discriminate. GGCA and CGCA show the discrimination with L-selectivity; however, the discrimination is poor in the case of CGCA. This demonstrates that the ability of chiral discrimination is retained when G or C is added at the 5′ end of GCA, but they show reverse selectivity when compared to the parent trinucleotide GCA. These results, using eight tetranucleotides, reveal that among the studied nucleotides GCAA and GGCA are showing better discrimination with D- and L-selectivity, respectively, for the two aromatic amino acids.

Among the other amino acids tested, Ala, Asp, Cys, and Pro with both the tetranucleotides and Val with GCAA did not form sufficient $[X + Y - 2H]^{2-}$ ion to perform CID experiments; hence, these combinations were excluded from the study. The other amino acids (Asn, Arg, Gln, Glu, Ile, Leu, Lys, Met, Phe, Ser, Thr, Trp, Tyr, and Val) form abundant $[X + Y - 2H]^{2-}$ ion, and their CID spectra also exclusively yielded the expected product ion $[X - 2H]^{2-}$ (X = GCAA, *m/z* 590; GGCA, *m/z* 598) that corresponds to the loss of bound amino acid. The CR values calculated

**TABLE 8.5**
**CR Values[a] Obtained for Trp and Tyr Isomers**

| | CR values $(R_D/R_L)^b$ | |
|---|---|---|
| **Amino acid** | **GCAA ± SD** | **GGCA ± SD** |
| Phe | 3.1 ± 0.2 | 0.37 ± 0.03 |
| Tyr | 2.6 ± 0.3 | 0.35 ± 0.02 |
| Trp | 2.5 ± 0.3 | 0.34 ± 0.02 |
| Val | | 0.32 ± 0.02 |
| Thr | 2.3 ± 0.3 | 0.80 ± 0.02 |
| Met | 2.1 ± 0.07 | 0.50 ± 0.03 |
| Leu | 2.1 ± 0.2 | 0.25 ± 0.02 |
| Ile | 2.0 ± 0.2 | 0.23 ± 0.02 |
| Ser | 2.0 ± 0.2 | 0.71 ± 0.2 |
| Asn | 1.51 ± 0.07 | 1.13 ± 0.03 |
| Lys | 1.33 ± 0.2 | 0.71 ± 0.2 |
| Arg | 1.08 ± 0.03 | 0.74 ± 0.04 |
| Gln | 0.98 ± 0.05 | 1.10 ± 0.02 |
| Glu | 1.02 ± 0.1 | 0.81 ± 0.02 |

[a] Values shown are the average of three separate measurements done on different occasions.
[b] $R_D = [a/b]_D$, $R_L = [a/b]_L$.

for all the successful combinations of amino acids with GCAA and GGCA are presented in Table 8.5.

Interestingly, the CID spectrum of $[X + Y - 2H]^{2-}$ ion (X = GCAA, *m/z* 664; GGCA, *m/z* 672) from Glu shows exclusively the $[X - H]^-$ ion (X = GCAA, *m/z* 1181; GGCA, *m/z* 1197) as the product. Formation of $[X - H]^-$ ion from $[X + Y - 2H]^{2-}$ ion is not observed for all the other amino acids. CID spectra of $[X + Y - 2H]^{2-}$ ion (X = GGCA and Y = Glu) is shown in Figure 8.8. It reveals that the $[X - H]^-$ ion is specifically formed when acidic amino acid is bound to the tetranucleotide. Because of the acidic nature of Glu, a proton might be transferred from amino acid to any basic site of GCAA or GGCA during the dissociation of $[X + Y - 2H]^{2-}$ ion. The CID spectra of $[X + Y - 2H]^{2-}$ ions are very much distinct for D- and L-isomers of Ile, Leu, Met, Phe, Ser, Trp, Tyr, and Val (CR values are far from unity, i.e., >1.5 for GCAA and <0.8 for GGCA). The typical CID spectra of $[X + Y - 2H]^{2-}$ ion of D- and L-Phe with GCAA is shown in Figure 8.9. The CID spectra recorded for the D- and L-isomers of Arg, Glu, and Gln with GCAA are similar, and the same is true for the D- and L-isomers of Asn, and Gln with GGCA; hence, no discrimination could be obtained for these combinations.

From Table 8.5, it can be noted that with GCAA as the selector, the CR values for aromatic amino acids are relatively higher than for the other amino acids. Multipoint

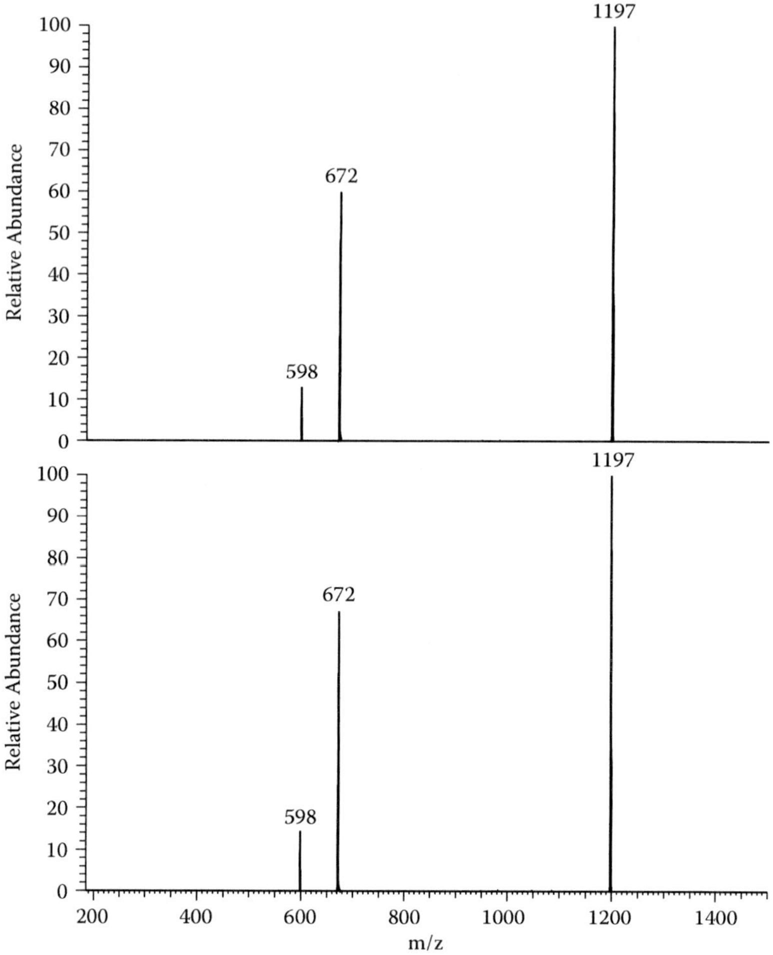


**FIGURE 8.8** CID spectra of $[X + Y - 2H]^{2-}$ ion (*m/z* 672), X = GGCA, Y = D-Glu (top), and L-Glu (bottom).

interactions are always expected to give better chiral recognition. Therefore, the higher chiral discrimination for aromatic amino acids could be due to π-π interactions of the aromatic ring of amino acid with the aromatic ring of nucleobases in the tetranucleotide. In fact, similar behavior was observed in the earlier study with GCA, also.[68] The hydroxyl- and sulfur-containing amino acids (Met, Ser, and Thr) show higher CR values (>2) that could be due to secondary interactions involving hydroxyl and sulfur groups. Leu and Ile also show discrimination comparable to that of the hydroxyl- and sulfur-containing amino acids. However, the chiral discrimination is least for Asn and Lys. There is no discrimination in the case of Arg isomers. Failure of discrimination for Arg isomers and the marginal discrimination for the Asn and Lys isomers suggests that GCAA is not a suitable selector for basic amino acids. In the case of Arg, $[X + Arg - 2H]^{2-}$ ions are found to dissociate at relatively higher collision

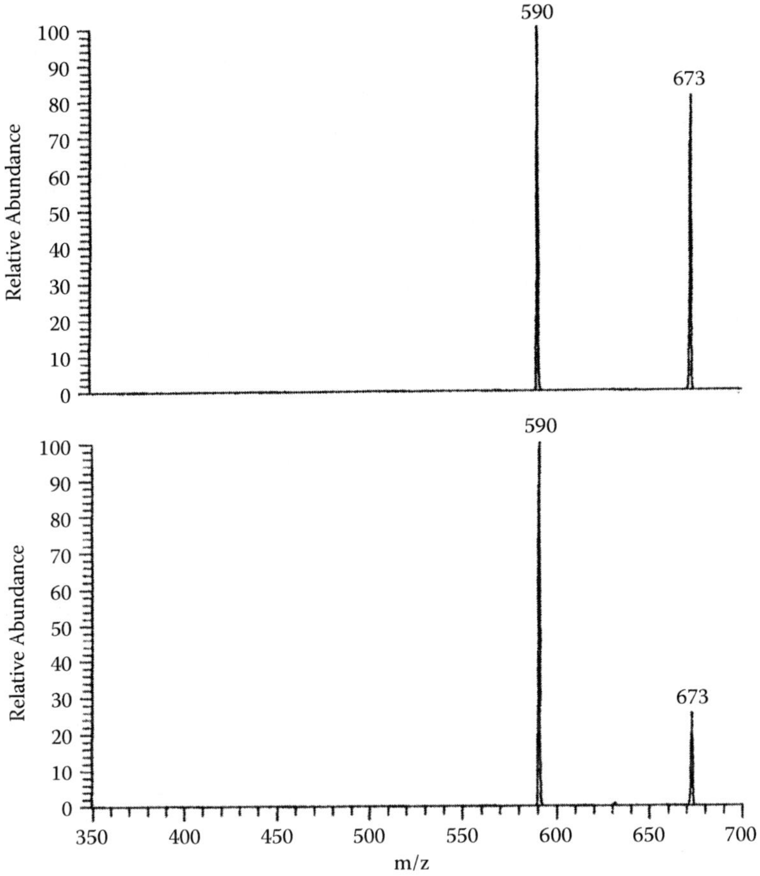


**FIGURE 8.9** CID spectra of $[X + Y - 2H]^{2-}$ ion (*m/z* 673), X = GCAA, Y = D-Phe (top), and L-Phe (bottom).

energy values (>16%) than at the values used for other adducts (<12%), which suggests relatively higher stability of the Arg adduct ions. It is reported in the literature that the interactions between the basic amino acids with DNA phosphate groups are ionic, and this might be the reason for the higher stability of adduct ion with Arg.[71] Similar behavior of Arg was also observed with GGCA and in the previous study during the dissociation of Arg adduct ions with GCA. In the case of GGCA, the aliphatic amino acids (Leu, Ile, and Val) show better discrimination than aromatic amino acids, and less discrimination is obtained for the hydroxy group containing amino acids (Ser and Thr) than for GCAA; it is difficult to propose a clear-cut explanation for this behavior of GGCA with available data. In the present study, the CR values obtained for all the successful amino acids (Table 8.5) are >1 with GCAA, and hence it can be concluded that D-amino acids are binding more strongly to GCAA than the L-amino acids. Similarly, the CR values are <1 with GGCA, and thus there is relative higher stability for GGCA and L-amino acids than for GGCA and D-amino acids.

## 8.8 CONCLUSION

The advantages connected with studying enantioselectivity in the gas phase come from the possibility to make precise measurements, using mass spectral techniques in the absence of any interfering effects such as solvation, and ion pairing. The study discussed in this chapter shows the successful use of DNA triplet GCA as a chiral selector for the discrimination of amino acids under negative ESI conditions. Chiral distinction of a few α-amino acids is achieved by examination of the dissociation of their doubly charged anionic ternary complexes formed with a GCA triplet. The chiral recognition ratio method is employed to measure the extent of discrimination, and the CR values show remarkable chiral discrimination for the studied amino acids, and the results demonstrate that the complexes formed with D-isomers are more stable than those formed with the L-isomers. The chiral selectivity of GCA is further amplified by the use of a reference amino acid and by applying the kinetic method. Selection of a suitable reference amino acid having closer GCA binding efficiency to that of analyte is crucial in the success of this method. The relative GCA binding order of the studied amino acids can help in choosing the correct reference. Further, tetranucleotides retaining the GCA sequence have also been tested as chiral auxiliaries for the discrimination of enantiomeric amino acids. Among the studied tetranucleotides, GCAA and GGCA show better discrimination for thirteen amino acids. The GCAA shows D-selectivity for all the successful amino acids, whereas the GGCA shows L-selectivity. The selectivity obtained with GCAA is similar to that observed with GCA in the earlier study, which suggests retention of a loop-like structure of GCA in GCAA, also. The work presented in this chapter clearly shows deoxy oligonucleotides may be developed as suitable chiral auxiliaries for carrying out chiral discrimination studies under mass spectrometry not only for amino acids but also for other types of chiral compounds.

## ACKNOWLEDGMENTS

The authors thank Dr. J. S. Yadav, director, IICT, for his encouragement, and a senior research fellowship from CSIR, New Delhi (to SK), is gratefully acknowledged.

## REFERENCES

1. J. Ciesiolka, J. Gorski, M. Yarus. 1995. *RNA* 1:538.
2. D. Shangguan, Y. Li, Z.W. Tang, Z.H.C. Cao, H.W. Chen, P. Mallikaratchy, K. Sefah, C.Y.J. Yang, W.H. Tan. 2006. *Proc. Natl. Acad. Sci. USA* 103:11838.
3. S. Klussmann, A. Nolte, R. Bald, A. Erdmann, J.P. Furste. 1996. *Nat. Biotechnol.* 14:1112–15.
4. A. Brumbt, C. Ravelet, C. Grosset, A. Ravel, et al. 2005. *Anal. Chem.* 77:1993–98.
5. S. Klussmann, A. Nolte, R. Bald, A. Erdmann, J.P. Furste. 1996. *Nat. Biotechnol.* 14:1112–15.
6. K.A. Davis, B. Abrams, Y. Lin, S.D. Jayasena. 1996. *Nucleic Acids Res.* 24:702.
7. F. Kleinjung, S. Klussmann, V.A. Erdmann, F.W. Scheller, J.P. Fuiste, F.F. Bier. 1998. *Anal. Chem.* 70:328–31.

8. R.A. Potyrailo, R.C. Conrad, A.D. Ellington, G.M. Hieftje. 1998. *Anal. Chem.* 70:3407–12.
9. M. Michaud, E. Jourdan, A. Villet, A. Ravel, C. Grosset, E. Peyrin. 2003. *J. Am. Chem. Soc.* 125:8672–79.
10. M. Michaud, E. Jourdan, C. Ravelet, A. Villet, A. Ravel, C. Grosset, E. Peyrin. 2004. *Anal. Chem.* 76:1015–20.
11. A. Brumbt, C. Ravelet, C. Grosset, A. Ravel, A. Villet, E. Peyrin. 2005. *Anal. Chem.* 77:1993–98.
12. C. Ravelet, R. Boulkedid, A. Ravel, C. Grosset, A. Villet, J. Fize, E. Peyrin. 2005. *J Chromatogr. A* 1076:62–70.
13. J. Ruta, C. Grosset, C. Ravelet, J. Fize, A. Villet, A. Ravel, E. Peyrin. 2007. *J Chromatogr. B* 845:186–90.
14. J. Ruta, C. Ravelet, C. Grosset, J. Fize, A. Ravel, A. Villet, E. Peyrin. 2006. Anal. Chem. 78:3032–39.
15. J. Ruta, C. Ravelet, I. Baussanne, J.L. Decout, E. Peyrin. 2007. *Anal. Chem.* 79:4716–19.
16. M. Michaud, E. Jourdan, A. Villet, A. Ravel, C. Grosset, E. Peyrin. 2003. *J. Am. Chem. Soc.* 125:8672.
17. M. Michaud, E. Jourdan, C. Ravelet, A. Villet, A. Ravel, C. Grosset, E. Peyrin. 2004. *Anal. Chem.* 76:1015.
18. A. Brumbt, C. Ravelet, C. Grosset, A. Ravel, A. Villet, E. Peyrin. 2005. *Anal. Chem.* 77:1993.
19. C. Ravelet, R. Boulkedid, A. Ravel, C. Grosset, A. Villet, J. Fize, E. Peyrin. 2005. *J. Chromatogr. A* 1076:62.
20. M. Michaud, E. Jourdan, A. Villet, A. Ravel, C. Grosset, E. Peyrin. 2003. *J. Am. Chem. Soc.* 125:8672.
21. M. Michaud, E. Jourdan, C. Ravelet, A. Villet, A. Ravel, C. Grosset, E. Peyrin. 2004. *Anal. Chem.* 76:1015.
22. R. Ramirez, F. He, C.B. Lebrilla. 1998. *J. Am. Chem. Soc.* 120:7387.
23. J.L. Seymour, F. Turecek, A.V. Malkov, P. Kocovsky. 2004. *J. Mass Spectrom.* 39:1044.
24. C. B. Lebrilla. 2001. *Acc. Chem. Res.* 34:653.
25. X. Cong, G. Czerwieniec, E. McJimpsey, S. Ahn, F.A. Troy, C.B. Lebrilla. 2006. *J. Am. Soc. Mass Spectrom.* 17:442.
26. M. Sawada, H. Yamaoka, Y. Takai, Y. Kawai, H. Yamada, T. Azuma, T. Fujioka, T. Tanaka. 1998. *Chem. Commun.* 1569.
27. P. Krishna, S. Prabhakar, M. Manoharan, E. D. Jemmis, M. Vairamani. 1999. *Chem. Commun.* 1215.
28. P. Krishna, S. Prabhakar, M. Manoharan, E.D. Jemmis, and M. Vairamani. 1999. *Eur. Mass Spectom.* 5:485–88.
29. C.B. Lebrilla. 2001. *Acc. Chem. Res.* 34:653.
30. G. Grigorean, J. Ramirez, S.H. Ahn, C.B. Lebrilla. 2000. *Anal. Chem.* 72:4275.
31. B. Botta, M. Botta, A. Filippi, A. Tafi, G.D. Monache, M. Speranza. 2002. *J. Am. Chem. Soc.* 124:7658.
32. A. Tafi, B. Botta, M. Botta, G.D. Monache, A. Filippi, M. Speranza. 2004. *Chem. Eur. J.* 10:4126.
33. R.G. Cooks, T.L. Kruger. 1977. *J. Am. Chem. Soc.* 99:1279.
34. D. Zhang, W.A. Tao, R.G. Cooks. 2001. *Int. J. Mass Spectrom.* 204:159.
35. W.A. Tao, R.L. Clark, R.G. Cooks. 2002. *Anal. Chem.* 74:3783.
36. L. Wu, W.A. Tao, R.G. Cooks. 2003. *J. Mass Spectrom.* 38:386.
37. W.A. Tao, L. Wu, R.G. Cooks. 2001. *J. Am. Soc. Mass Spectrom.* 12:490.
38. W.A. Tao, R.G. Cooks. 2001. *Angew Chem. Int. Ed. Engl.* 40:757.
39. W.A. Tao, L. Wu, R.G. Cooks. 2000. *Chem. Commun.* 2023.

40. W.A. Tao, R.L. Clark, R.G. Cooks. 2003. *Chem. Commun.* 136.
41. W.A. Tao, R.G. Cooks. 2003. *Anal. Chem.* 75:678.
42. Z.P. Yao, T.S.M. Wan, K.P. Kwong, C.T. Che. 1999. *Chem. Commun.* 2119.
43. Z.P. Yao, T.S.M. Wan, K.P. Kwong, C.T. Che. 2000. *Anal. Chem.* 72:5383.
44. Z.P. Yao, T.S.M. Wan, K.P. Kwong, C.T. Che. 2000. *Anal. Chem.* 72:5394.
45. K.A. Schug, W. Lindner. 2005. *J. Am. Soc. Mass Spectrom.* 16:825.
46. S.P. Mirza, N.P. Raju, M. Vairamani. 2004. *J. Am. Soc. Mass Spectrom.* 15:431.
47. V. Nagaveni, S. Prabhakar, M. Vairamani. 2004. *Anal. Chem.* 76:3505.
48. H. In-Su, C. Wesdemiotis. 2003. *Int. J. Mass Spectrom.* 222:465.
49. S.P. Mirza, P. Krishna, S. Prabhakar, M. Vairamani, P. Giblin, M.L. Gross. 2003. *Int. J. Mass Spectrom.* 230:175.
50. M. Shugang, W. Feng, R.G. Cooks. 1998. *J. Mass Spectrom.* 33:943.
51. L.G. Wright, S.A. McLuckey, R.G. Cooks, K.V. Wood. 1982. *Int. J. Mass Spectrom. Ion Processes* 42:115.
52. S. Rochut, C. Pepe, J.P. Paumard, J.C. Tabet. 2004. *Rapid Commun. Mass Spectrom.* 18:1686.
53. M.M. Kish, G. Ohanessian, C. Wesdemiotis. 2003. *Int. J. Mass Spectrom.* 227:509.
54. H.A. Lardin, R.R. Squires, P.G. Wenthold. 2001. *J. Mass Spectrom.* 36:607.
55. J.W. Denault, G. Chen, R.G. Cooks. 1998. *J. Am. Soc. Mass Spectrom.* 9:1141.
56. W.A. Tao, D. Zhang, E.N. Nikolaev, R.G. Cooks. 2000. *J. Am. Chem. Soc.* 122:10598.
57. W.A. Tao, R.G. Cooks. 2002. *Eur. J. Mass Spectrom.* 8:107.
58. W.A. Tao, L. Wu, R.G. Cooks, F. Wang, J.A. Begley, B. Lampert. 2001. *J. Med. Chem.* 44:3541.
59. L. Wu, K. Lemr, T. Aggerholm, R.G. Cooks. 2003. *J. Am. Soc. Mass Spectrom.* 14:152.
60. W.A. Tao, F.G. Gozzo, R.G. Cooks. 2001. *Anal. Chem.* 73:1692.
61. D.V. Augusti, R. Augusti, F. Carazza, R.G. Cooks. 2002. *Chem. Commun.* 2242.
62. L. Wu, K. Lemr, T. Aggerholm, R.G. Cooks. 2003. *J. Am. Soc. Mass Spectrom.* 14:152.
63. W.A. Tao, F.G. Gozzo, R.G. Cooks. 2001. *Anal. Chem.* 73:1692.
64. D.V. Augusti, R. Augusti, F. Carazza, R.G. Cooks. 2002. *Chem. Commun.* 2242.
65. X. Tang, J.H. Callahan, P. Zhou, A. Vertes. 1995. *Anal. Chem.* 67:4542.
66. S. Lin, R.J. Cotter, A.S. Woods. 1998. *Proteins Struct. Funct. Genet.* 2:12.
67. L. Zhu, S.-H. Chou, J. Xu, B.R. Reid. 1995. *Nat. Struct. Biol.* 2:1012.
68. M. Ravi Kumar, S. Prabhakar, M. Vairamani. 2007. *Chem. Commun.* 392.
69. E.M. Moody, P.C. Bevilacqua. 2004. *J. Am. Chem. Soc.* 126:9570.
70. M. Ravi Kumar, S. Prabhakar, T. Sivaleela, M. Vairamani. 2007. *J. Mass Spectrom.* 42:1218.
71. P. Terrier, J. Tortajada, W. Buchmann. 2007. *J. Am. Soc. Mass Spectrom.* 18:346.

# 9 Evaluating the Enantioselectivity of Asymmetric Catalytic Reactions and Screening Chiral Catalysts by ESI-MS

*Hao-Yang Wang and Yin-Long Guo*

**CONTENTS**

9.1 Introduction .......... 167
9.2 ESI-MS Studies on the Enantioselectivity of Asymmetric Catalytic Reactions .......... 168
9.2.1 MSEED for Products of Asymmetric Catalytic Reactions and Evaluating Chiral Catalysts .......... 168
9.2.2 MSICD for Evaluating and Screening Chiral Catalysts of Asymmetric Catalytic Reactions .......... 171
9.3 Summary and Outlook .......... 176
Acknowledgments .......... 176
References .......... 176

## 9.1 INTRODUCTION

Asymmetric reaction or synthesis is the organic reaction or synthesis that introduces one or more new and desired elements of chirality.[1,2] Today, the increasing number of industrial applications of asymmetric synthesis clearly demonstrates its importance,[3,4] especially in the pharmaceutical field, due to the fact that different enantiomers of a molecule often correspond to different biological activities.[5] High-throughput parallel screening of chiral catalysts is the important focus of asymmetric catalytic reactions. Normally combinatorial methods are widely used for this aim, including the techniques combining mass spectrometry and labeling/coding strategy.[6] Mass spectrometry is a highly selective and high-throughput analytical technique for analyzing reactants/products and even reactive intermediates/complexes in the reaction solution based on providing the information of molecular weight and even structural features from fragmentation patterns in $MS^n$.[7,8]

Considering the importance and general interest of asymmetric catalytic reactions and chiral catalysts screening, and its close connections with chiral analysis by electrospray ionization (ESI)–MS, we would like to provide an overview of the most recent methods applied in the studies on enantioselectivity of asymmetric catalytic reactions and high-throughput chiral catalysts screening by quasi-enantiomeric labeling strategy and mass spectrometry, especially ESI-MS, which transferred enantioselectivity information from the solution phase of reactions to the gas phase of mass spectrometry by measuring the enantiomeric excess (*ee*) value of products and detecting ratios of reactive intermediates/complexes.

This chapter can basically be divided into two parts: the first section deals with chiral catalysts screening methods based on mass spectrometry enantiomeric excess determination (MSEED) of the products of asymmetric reactions; the second section focuses on chiral catalysts screening methods developed recently by mass spectrometry intermediates/complexes detection (MSICD). This field is interdisciplinary, incorporating ESI mass spectrometry, coordination chemistry, homogeneous catalysis, reaction kinetics and mechanisms, and stereochemical concepts. We hope it will benefit and inspire a wide range of newcomers, specialists, and chemists contemplating the possibility of applying the ESI-MS technologies to solving the problems in asymmetric catalytic reactions for their own research.

## 9.2 ESI-MS STUDIES ON THE ENANTIOSELECTIVITY OF ASYMMETRIC CATALYTIC REACTIONS

### 9.2.1 MSEED FOR PRODUCTS OF ASYMMETRIC CATALYTIC REACTIONS AND EVALUATING CHIRAL CATALYSTS

The methods for MSEED developed together with soft ionization methods,[9,10] such as chemical ionization (CI),[11,12] fast atom bombardment (FAB),[13] matrix-assisted laser desorption/ionization (MALDI),[14–16] atmospheric pressure chemical ionization (APCI),[17] ESI,[18] and desorption electrospray ionization (DESI),[19] due to their high sensitivity, speed, general application, and the ability to provide qualitative, quantitative, and stoichiometric information. Mass spectrometric chiral recognition/analysis rests mainly on four methods: (1) solution-phase methods based on solution-phase kinetic resolution of the mass-coded racemic reactants followed by MS product analysis[20–22]; (2) solution-phase methods for studying chiral discrimination by measuring the diastereomeric adducts formed in solution with FAB,[23,24] MALDI,[25,26] and ESI-MS[27,28] or even DESI[29]; (3) gas-phase methods based on isolating diastereomeric ionic complexes in the gas phase by performing CID,[30–32] especially with data treatment of kinetic method[33–36]; and (4) gas-phase methods based on the ion-molecule reaction.[37–40]

Here we will focus on presenting the first kind of MSEED methods applied to products of asymmetric catalytic reactions and further applications to the high-throughput chiral catalysts screening; the remaining methods are discussed in other chapters. Such methods are mainly based on two concepts: quasi-enantiomeric labeling and kinetic resolution. Quasi-enantiomers[41] are defined as any pair of compounds that can be turned into true enantiomers by slightly changing the chemical composition of one or more substituents. Here some specific examples are shown in Scheme 9.1, such as tetrahedral quasi-enantiomers—(*S*)-1-benzoylpyrrolidine-2-carboxylic acid (**1a**) and (*R*)-1-(4-methylbenzoyl) pyrrolidine-2-carboxylic acid (**1b**),[42] and (*S*)-glycidyl phenyl ether (**2a**) and (*R*)-$D_5$-glycidyl phenyl ether (**2b**)[43]—as well as the axial chiral ligand quasi-enantiomer, (*S*)-BINOL (**3a**) and (*R*)-$F_8$BINOL(**3b**).[44] Kinetic resolution[45,46] describes that two enantiomeric isomers show different reaction rates in an enantioselective reaction, thereby creating an excess of the less reactive enantiomer. Finn[20,21] and Reetz[22] have reviewed such a strategy and its applications in high-throughput screening of chiral catalysts.

In general, the reactions of the enantiomeric isomers with any pair of chiral reagents will proceed with nonequal rate constants, when the quasi-enantiomeric reagents used in such derivatization reaction, the *ee* value of the enantiomer, could be deduced by measuring the ratios of quasi-enantiomeric derivation products.[47] Based on this strategy, the enantioselectivity of asymmetric catalytic reactions can be evaluated by determining the *ee* value of the enantiomeric products from asymmetric reactions, when suitable quasi-enantiomeric derivatization reagents are applied. Horeau and Nouaille[48] developed a method measuring the absolute configuration of optically active secondary alcohol by derivation of these compounds with an equimolecular mixture of (+)-2-phenylbutyric and (–)-2-phenyl-4-D-butyric anhydrides; thus, the configuration of the carbon atom bearing the OH group was deduced from the relative intensity peaks observed in the MS of the resulting diastereoisomeric esters due to the different reaction rate of chiral alcohol with the quasi-enantiomeric anhydrides.[48] Finn et al. reported an approach for the determination of *ee* value of chiral alcohols and chiral amines on the nanomole scale by diastereoselective derivatization and automated quantitative ESI-MS analysis[42] by performing the reaction of chiral alcohols **4** and chiral amines with quasi-enantiomeric acids (**1a/1b**) in the presence of DCC (1,3-dicyclohexylcarbodiimide) and a catalytic amount of DMAP (4-dimethylaminopyridine) (Scheme 9.2, $k_f > k_s$; f:

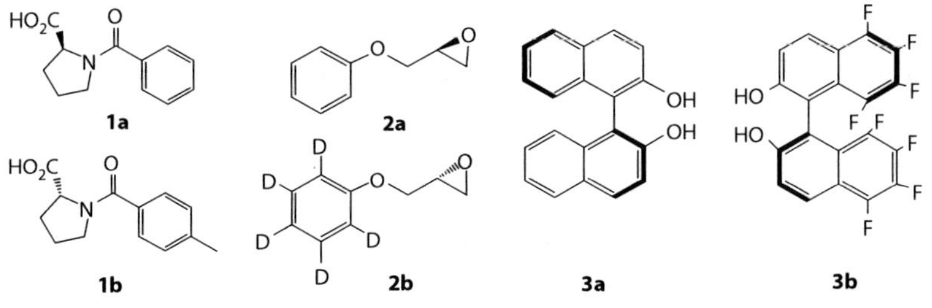


**SCHEME 9.1** Specific examples of quasi-enantiomers. Tetrahedral quasi-enantiomer—(S)-1-benzoyl pyrrolidine-2-carboxylic acid (**1a**) and (R)-1-(4-methylbenzoyl)pyrrolidine-2-carboxylic acid (1b), and (S)-glycidyl phenyl ether (**2a**) and (R)-D5-glycidyl phenyl ether (**2b**)—as well as the axial chiral quasi-enantiomer, (S)-BINOL (**3a**) and (R)-F8 BINOL(**3b**).

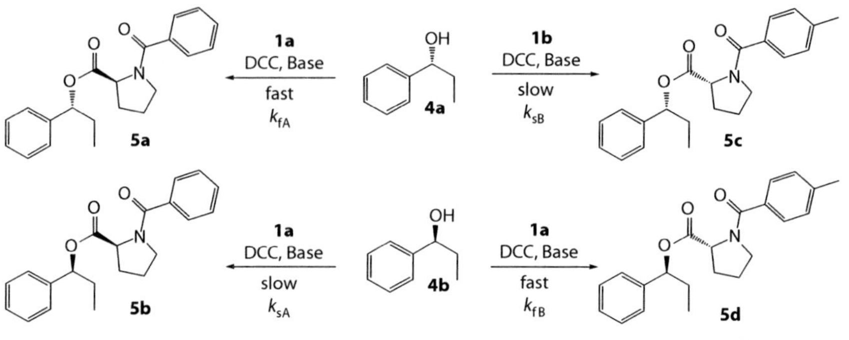


$$\% \text{ ee} = \left[\frac{(y-1)(s+1)}{(y+1)(s-1)}\right] \cdot 100 \quad \begin{cases} \dfrac{I\text{ mass 1}}{I\text{ mass 1}} = y \cdot q \\ s = \dfrac{k_f}{k_s} \end{cases}$$

**SCHEME 9.2** Determining *ee* value of chiral alcohols **4** by reacting with mass-tagged quasi-enantiomer chiral acids (**1a/1b**) in the presence of DCC and base. The selectivity factor *s*, ionization correction factor *q*, and corrected intensity ratio *y* are known or could be derived; then the *ee* value of starting sample **4** could be calculated by measuring the respective ratio mass spectrum peak intensity *I* of mass 1 and mass 2. This method could be used for high-throughput screening of the enantioselectivity of the chiral catalysts by measuring the *ee* value of the products from asymmetric catalytic reactions.

fast, s: slow). The correction factors *q* and *s* could be obtained through calibration using enantiomerically pure compounds. In general, the greater the enantiomeric discrimination in the acylation process (*s* or 1/*s*), the better is the *ee* value measurement, which falls within 10% of the true *ee* value. A similar method for determining *ee* value with ESI-MS was employed for measurement of the *ee* value of amines by ESI-MS following kinetic resolution with solid-phase chiral acylating agents[49] and screening a family of chiral phosphite P,N ligands for the Rh-catalyzed asymmetric hydrosilylation of ketones.[50]

Reetz et al. launched the application of mass spectrometry in high-throughput detection of asymmetric catalysis and biotransformations by isotopically labeling one of the starting enantiomers of a racemic pair. In the example, a quasiracemic mixture of **2a/2b** was used to mimic racemic glycidyl phenyl ether in a hydrolytic kinetic resolution by the enzyme epoxide hydrolase (Scheme 9.3).[51] This strategy can also be applied to the dissymmetrization of quasi-meso compound **7**. As demonstrated in Scheme 9.3, lipase-catalyzed hydrolysis of **4** provides a mixture of quasi-enantiomeric products **8a/8b** without separation being needed, and the difference in the mass of the quasi-enantiomers reflects the selectivity of the dissymmetrization process. A high-throughput method, based on ESI-MS using an eight-channel multiplexed (MUX) sprayer system connected to a time of flight (TOF)–MS, could screen the enantioselectivity of approximately ten thousand catalysts or biocatalysts per day,[43] and this method is useful for combinatorial enantioselective transition metal catalysis and in the directed evolution of enantioselective enzymes.[52]

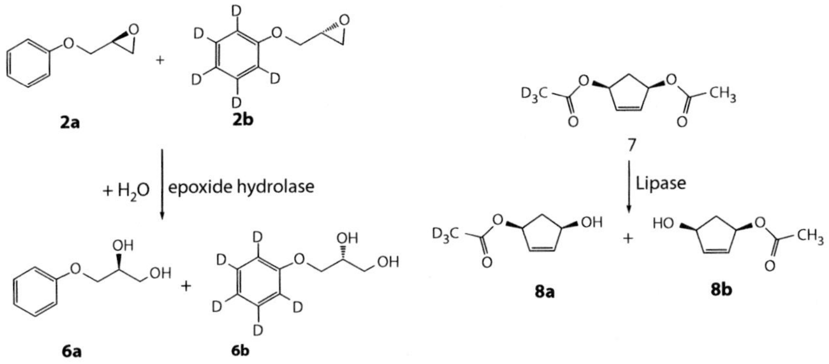

**SCHEME 9.3** Mass-coded analysis in enzymatic hydrolysis of quasi-racemate **2a/2b** to give **6a/6b** and lipase-catalyzed hydrolysis of quasi-meso compound $_7$ to give quasi-enantiomers **8a/8b**. The *ee* value of chiral compounds **6** and **8** could be obtained by measuring the respective ratio of **6a** and **6b** or **8a** and **8b** by ESI-MS; therefore, such a method could be used for high-throughput screening of the enantioselectivity of biological chiral catalysts such as enzymes.

### 9.2.2 MSICD for Evaluating and Screening Chiral Catalysts of Asymmetric Catalytic Reactions

The ESI-MS technique has also been proven to be an important and valuable tool for the high-throughput parallel screening of catalysts by Chen[53,54] and Irth[55] via the interception of the intermediates of catalytic cycles. Metzger,[56] Eberlin,[57] and Santos[58] have published several comprehensive reviews about studies of the reactive intermediates by ESI-MS. Recently, some reactive intermediates in the asymmetric catalytic reactions have been studied by ESI-MS to provide the mechanistic information. Finn et al. used ESI-MS to characterize the Ti complex catalyst in the study of asymmetric sulfoxidation of alkyl aryl sulfides.[59] Metzger et al. employed ESI-MS with the microreactor to identify the intermediates in the aldol reaction catalyzed by L-proline[60] and the transient catalytic intermediates in the L-prolinamide-catalyzed α-halogenation (Cl, Br, I) of aldehydes.[61] Schaus and coworkers intercepted two boronate intermediates with chiral diols in the enantioselective asymmetric allylboration of acyl imines catalyzed by the chiral binol-derived diols by ESI-MS.[62] Zhou detected the reactive Ti complexes in the enantioselective ring-opening aminolysis reaction by ESI-MS.[63] Berkessel studied Ti salalen catalysts and the degradation pathways of the asymmetric epoxidation by ESI-MS.[64] These studies for the asymmetric catalytic systems provided strong support to the mechanistic proposals; however, no direct enantioselectivity information of these asymmetric reactions could be obtained.

Due to the complicated aspects involved in the asymmetric catalysis reactions, such as the unselective background reaction, catalytically active impurities, or the dissociation of chiral complexes, the parallel screening based on product analysis could not completely reflect the enantioselectivity of an asymmetric reaction and the intrinsic selectivity of the catalyst, which could be avoided if the ability of the catalyst for enantiodiscrimination could be determined directly by examining catalyst-reactant complexes rather than from product analysis.

As the first example, Pfaltz et al. established a new method for screening asymmetric catalysts based on the detection of the catalytic intermediates rather than the precursors or products, and realized this concept for a palladium-catalyzed kinetic resolution of allylic esters by using ESI-MS (Scheme 9.4).[65] In the catalytic cycle, the formation of Pd-allyl intermediates **10A/10B** from reactants (**9a/9b**) is fast, the formation of products (**11a/11b**) by nucleophilic addition to the allyl complex is slower, and turnover is limiting. Thus, the cationic catalyst-reactant complexes **10A/10B** should exist in sufficient concentration to be detected by ESI-MS, and selectivity factor $s = k_A/k_B$ of a catalyst can be deduced from measuring the intensity ratio of the corresponding allyl intermediates **10A/10B** by ESI-MS. For example, after a certain reaction time the intensity ratio of quasi-enantiomer Pd intermediates **13A/13B** from **12a/12b** (Scheme 9.4, screening route 1) in ESI-MS represents the catalyst's ability to discriminate between the two enantiomeric substrates and the enantioselectivity imparted by the chiral ligand in the reaction. By using this technique, several ligands from a library of sixty members were able to be identified with selectivity factors of >20 and with lowering the reaction temperature to minimize ligand exchange. The allyl-bridged dinuclear palladium (I) complexes were observed and investigated in the studies for Pd-catalyzed allylic substitutions by ESI-MS.[66]

After screening of a large number of catalysts, Pfaltz et al. found that the selectivity of a catalyst in the kinetic resolution step for racemic reactants did not correlate with the enantioselectivity of the nucleophilic addition step, which determines the enantioselectivity of the overall reaction leading from the racemic allyl esters **9a/9b** to the optically active substitution products **11a/11b** (Scheme 9.4). Therefore, catalysts that gave high enantioselectivity in the overall reaction might be inefficient in the kinetic resolution of allyl esters.

In order to solve such problems, Pfaltz et al. recently introduced a new concept for screening chiral catalysts by ESI-MS monitoring of catalytic reactions in the reverse direction by using quasi-enantiomeric products instead of reactants, which makes it possible to distinguish catalyst-bound intermediates with an opposite sense of chirality. According to the principle of microscopic reversibility, the transition states of the forward and back reactions are identical, and therefore, the enantioselectivity determined from back-reaction screening is identical to that of the forward reaction. The corresponding allyl intermediates (**15A/15B**) were detected in the ESI-MS by elimination of the nucleophile from **14a/14b** (Scheme 9.4, screening route 2). It is important to study the reactions at the very beginning of the reaction because after longer reaction time, the slowly decreased signal ratios were observed due to the racemization of the quasi-enantiomers **14a** and **14b** by reversible elimination/readdition of acetyl acetone.[67]

The evaluation of several chiral ligands (Scheme 9.5) in one reaction was further achieved based on ESI-MS screening for the corresponding catalyst mixture and a double mass-labeling strategy[68] by preparing a library of three double mass-labeling chiral ligands (**18aa/18ab/18bb**) in two steps from two chiral diamines (**16a/16b**, with two different sulfonamide substitutes, $R^1$ and $R^2$) and a chiral diol. By varying different sulfonamide groups, a mixture of all possible ligands could be prepared and tested simultaneously.

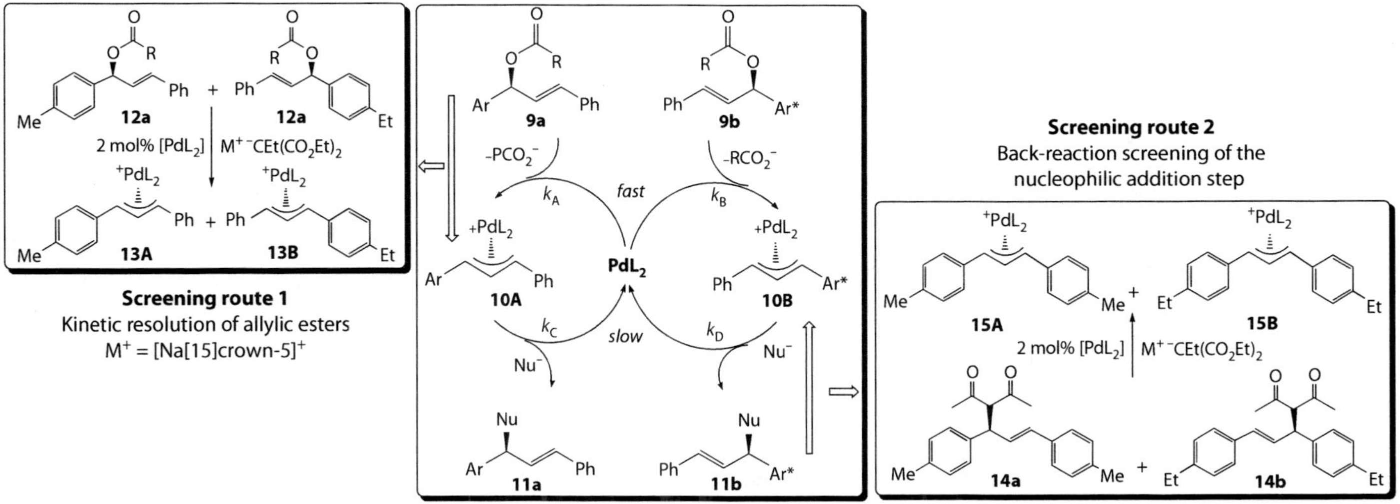


**SCHEME 9.4** Study of the enantioselectivity of Pd-catalyzed allylic ester substitution reactions by quasi-enantiomeric labeling and ESI-MS. Quasi-enantiomers of allylic ester (**9a/9b**, Ar and Ar* bearing two different mass codes) reacted with $Nu^-$ to give the quasi-enantiomeric products (**11a/11b**) in the presence of Pd catalyst ($PdL_2$, L represents chiral ligand); the ratio of allylic Pd intermediates **10A** and **10B** could reflect the enantioselectivity imparted by the chiral ligand L in reaction. Route 1 in the left frame of Scheme 9.4 is the ESI screening method based on the kinetic resolution of allylic ester reactants (**12a/12b**) for evaluating the chiral ligand L; route 2 in the right frame of Scheme 9.4 applies the back-reaction screening concept by studying the retro-nucleophilic addition step of acetyl acetonate products (**14a/14b**) for evaluating the chiral ligand.

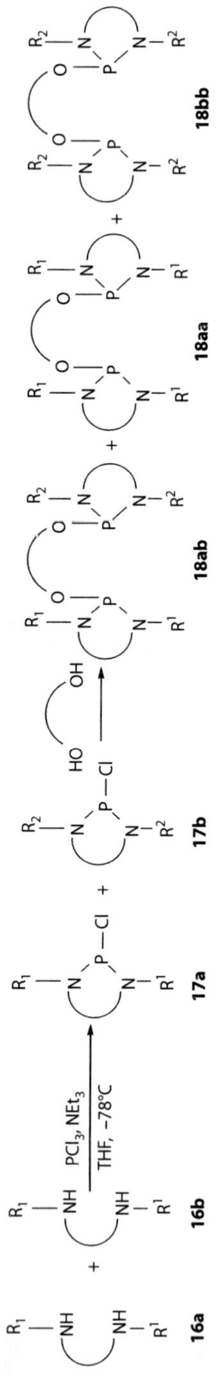


**SCHEME 9.5** Preparation of modular ligands combinational library (**18aa/18ab/18bb**, $R^1$ and $R^2$ represent two sulfonamide substitutes with different mass codes) from two chiral diamines (**16a/16b**) and a chiral diol. These chiral ligands in the ligands combinational library could be evaluated in one reaction by the ESI-MS screening method.

After having demonstrated the feasibility of the asymmetric catalyst screening methods for palladium-catalyzed allylic substitution reactions, Pfaltz et al. reported an extension to metal-catalyzed and organocatalytic Diels-Alder (DA) reactions based on reverse screening strategy.[69] The catalysts used in the DA reaction also catalyze the retro-DA reaction.[69] If formation of the complexes **20a/20b** from DA products **19a/19b** in the presence of catalysts is fast and reversible, and the following retro-DA reaction is slow and irreversible, the ratio of **21a/21b** directly reflects the enantioselectivity of the chiral $Cu^{II}$-bisoxazoline catalysts (Scheme 9.6). When the retro-DA reactions are performed at elevated temperature under dilute conditions, especially at the initial stage of the retro-DA reaction, the concentrations of cyclopentadiene are comparably low and the DA reaction of **20a/20b** with cyclopentadiene would be slow. In such conditions, the retro-DA reaction could be regarded as irreversible. Based on a similar principle, the retro-DA reactions of DA products (**22a/22b**) in the presence of a MacMillan catalyst were also studied, and the ratio of the iminium ion intermediates **23a/23b** generated by loss of cyclopentadiene in ESI-MS could show the enantioselectivity of the chiral organocatalysts (Scheme 9.6). Thus, the mixtures

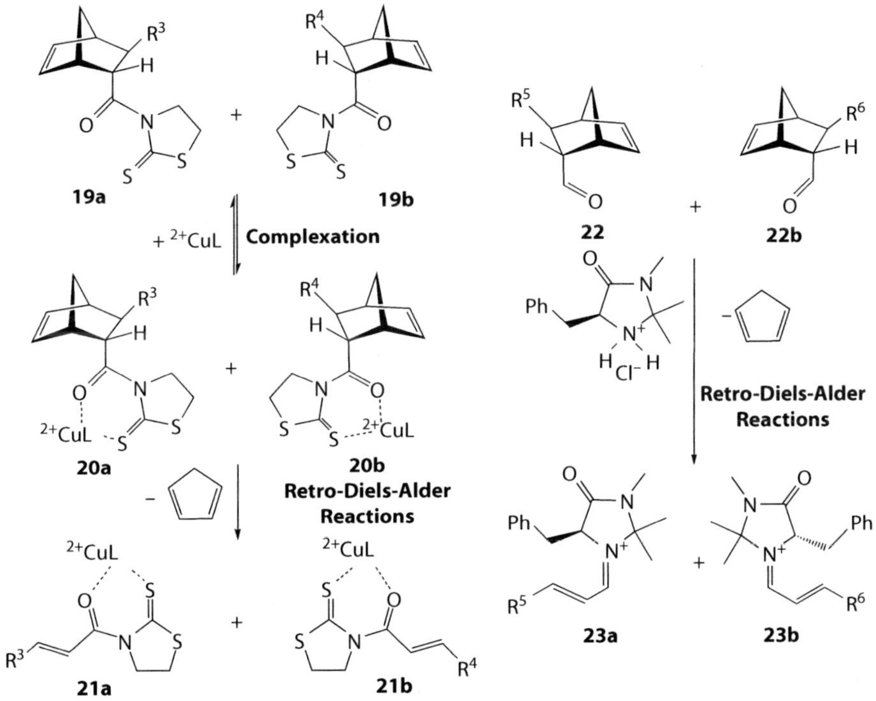


**SCHEME 9.6** ESI-MS screening of chiral $Cu^{II}$-bisoxazoline catalysts (L represents chiral bisoxazolineligand; $R^3$ and $R^4$ represent mass labels) and chiral organocatalysts for DA reactions by studying the retro-DA reactions of quasi-enantiomeric DA products (**19a/19b** and **22a/22b**, $R^5$ and $R^6$ represent mass labels). The ratios of reactive $Cu^{II}$ complexes **21a/21b** and the iminium ion intermediates **23a/23b** directly reflected the enantioselectivity of the chiral catalysts for the asymmetric DA reactions.

of organocatalysts could be screened simultaneously, which opens up new possibilities for the high-throughput screening of organocatalyst libraries.

## 9.3 SUMMARY AND OUTLOOK

The methods of MSEED and MSICD share two features: (1) a mass code makes it easy for ESI-MS to distinguish an individual pair of true enantiomeric reactants/products and catalytic intermediates/complexes; (2) both options evaluate chiral catalysts and asymmetric reactions without isolating, which forms the basis of a number of evaluating and screening techniques for asymmetric catalytic reactions and speeds up catalyst development considerably. Unlike MSEED, MSICD relies on the quantification of catalytic intermediates in the reaction solution rather than analysis of the products, which represents the intrinsic enantioselectivity of chiral catalysts. Moreover, the concept of back-reaction monitoring from products enhances the application range of ESI-MS-based screening in the studies of the most common asymmetric catalytic transformations, which use prochiral reactants. Although these concepts and approaches are attractive and promising, MSEED needs pure quasi-enantiomeric isomer substrates or derivatization regents; MSICD is only applicable to reactions proceeding via charged intermediates/complexes, and thus requires understanding the detailed knowledge about the reactive intermediates of the reaction. However, using the quasi-enantiomers instead of a racemic compound for catalyst screening provides a creative and bright new route to studies on the enantioselectivity of asymmetric catalytic reactions.

## ACKNOWLEDGMENTS

The authors gratefully acknowledge financial support from the National Natural Science Foundation of China (20875097, 20942002, and 20902104).

## REFERENCES

1. International Union of Pure and Applied Chemistry. asymmetric synthesis. *Compendium of chemical terminology*. Internet edition.
2. International Union of Pure and Applied Chemistry. Stereoselective synthesis. *Compendium of chemical terminology*. Internet edition.
3. Jacobsen, E. N., Pfaltz, A., Yamamoto, H., eds. 1999. *Comprehensive asymmetric catalysis*. Heidelberg: Springer.
4. Ojima, I., ed. 2000. *Catalytic asymmetric synthesis*. 2nd ed. New York: Wiley-VCH.
5. Mitchell, A. G. 1998. Racemic drugs: Racemic mixture, racemic compound, or pseudoracemate? *J. Pharm. Pharm. Sci.* 1:8–12.
6. Shin, Y. G., van Breemen, R. B. 2001. Analysis and screening of combinatorial libraries using mass spectrometry. *Biopharm. Drug Disposition* 22:353–72.
7. Pomerantz, S. C., McCloskey, J. A., Tarasow, T. M., Eaton, B. E. 1997. Deconvolution of combinatorial oligonucleotide libraries by electrospray ionization tandem mass spectrometry. *J. Am. Chem. Soc.* 119:3861–67.

8. Wu, Q. 1998. Multistage accurate mass spectrometry: A "basket in a basket" approach for structure elucidation and its application to a compound from combinatorial synthesis. *Anal. Chem.* 70:865–72.
9. Vestal, M. L. 2001. Methods of ion generation. *Chem. Rev.* 101:361–75.
10. McLuckey, S. A., Wells, J. M. 2001. Mass analysis at the advent of the 21st century. *Chem. Rev.* 101:571–606.
11. Munson, M. S. B., Field, F. H. 1966. Chemical ionization mass spectrometry. I. General introduction. *J. Am. Chem. Soc.* 88:2621–30.
12. Fales, H. M., Wright, G. J. 1977. Detection of chirality with the chemical ionization mass spectrometer"meso" ions in the gas phase. *J. Am. Chem. Soc.* 99:2339–40.
13. Barber, M., Bordoli, R. S., Sedgwick, R. D., Tyler, A. N. 1981. Fast atom bombardment of solids as an ion-source in mass-spectrometry. *Nature* 293:270–75.
14. Karas, M., Bachman, D., Bahr, U., Hillenkamp, F. 1987. Matrix-assisted ultraviolet-laser desorption of nonvolatile compounds. *Int. J. Mass Spectrom. Ion Processes* 78:53–68.
15. Tanaka, K., Waki, H., Ido, Y., Akita, S., Yoshida, Y., Yoshida, T. 1988. Protein and polymer analyses of up to m/z 100000 by laser ionization time-of-flight mass spectrometry. *Rapid Commun. Mass Spectrom.* 2:151–53.
16. Karas, M., Bachmann, D., Hillenkamp, F. 1988. Laser desorption ionization of proteins with molecular masses exceeding 10000 daltons. *Anal. Chem.* 60:2299–301.
17. Covey, T. R., Lee, E. D., Bruins, A. P., Henion, J. D. 1986. Liquid chromatography/mass spectrometry. *Anal. Chem.* 58:1451A–61A.
18. Fenn, J. B., Mann, M., Meng, C. K., Wong, S. F.,Whitehouse, C. M. 1989. Electrospray ionization for mass spectrometry of large biomolecules. *Science* 246:64–71.
19. Takáts, Z., Wiseman, J. M., Gologan, B., Cooks, R. G. 2004. Mass spectrometry sampling under ambient conditions with desorption electrospray ionization. *Science* 306:471–73.
20. Finn, M. G. 2002. Emerging methods for the rapid determination of enantiomeric excess. *Chirality* 14:534–40.
21. Finn, M. G. 2006. Emerging high-throughput screening methods for asymmetric induction. In *Chiral analysis*, ed. K. W. Busch and M. A. Busch, 79–94. Amsterdam: Elsevier.
22. Reetz, M. T. 2001. Combinatorial and evolution-based methods in the creation of enantioselective catalysts. *Angew. Chem. Int. Ed.* 40:284–310.
23. Baldwin, M. A., Howell, S. A., Welham, K. J., Winkler, F. J. 1988. Identification of chiral isomers by fast atom bombardment/mass spectrometry: Dialkyl tartrates. *Biomed. Environ. Mass Spectrom.* 16:357–60.
24. Sawada, M. 1997. Chiral recognition detected by fast atom bombardment mass spectrometry. *Mass Spectrom. Rev.* 16:73–90.
25. So, M. P., Wan, T. S. M., Chan, T.-W. D. 2000. Differentiation of enantiomers using matrix assisted laser desorption/ionization mass spectrometry. *Rapid Commun. Mass Spectrom.* 14:692–95.
26. Ramagiri, S., Gupte, R., Rakov, I., Yates, C. R., Miller, D. D. 2008. Quantitative chiral analysis of phthaloylglutamic acid and related analogs by a single ratio kinetic method using electrospray ionization and matrix-assisted laser desorption techniques. *Rapid Commun. Mass Spectrom.* 22:639–46.
27. Schug, K., Fryčák, P., Maier, N. M., Lindner, W. 2005. Measurement of solution-phase chiral molecular recognition in the gas phase using electrospray ionization-mass spectrometry. *Anal. Chem.* 77:3660–70.
28. Schug, K. A., Lindner, W. 2005. Chiral molecular recognition for the detection and analysis of enantiomers by mass spectrometric methods. *J. Sep. Sci.* 28:1932–55.
29. Nyadong, L., Hohenstein, E. G., Johnson, K., Sherrill, C. D., Green, M. D., Fernández, F. M. 2008. Desorption electrospray ionization reactions between host crown ethers and the influenza neuraminidase inhibitor oseltamivir for the rapid screening of Tamiflu®. *Analyst* 133:1513–22.

30. Tao, W. A., Zhang, D., Wang, F., Thomas, P. D., Cooks, R. G. 1999. Kinetic resolution of D,L-amino acids based on gas-phase dissociation of copper (II) complexes. *Anal. Chem.* 71:4427–29.
31. Lu, H. J., Guo, Y. L. 2003. Evaluation of chiral recognition characteristics of metal and proton complexes of di-o-benzoyl-tartaric acid dibutyl ester and L-tryptophan in the gas phase. *J. Am. Soc. Mass Spectrom.* 14:571–80.
32. Yu, C. T., Guo, Y. L., Chen, G. Q., Zhong, Y. W. 2004. Chiral recognition of zinc(II) ion complexes composed of bicyclo[3.3.0] octane-2,6-diol and S-naproxen probed by collisional-induced dissociation. *J. Am. Soc. Mass Spectrom.* 15:795–802.
33. Tao, W. A., Clark, R. L., Cooks, R. G. 2002. Quotient ratio method for quantitative enantiomeric determination by mass spectrometry. *Anal. Chem.* 74:3783–89.
34. Wu, L., Eduardo, M., Cooks, R. G. 2004. Chiral morphing and enantiomeric quantification in mixtures by mass spectrometry. *Anal. Chem.* 76:663–71.
35. Tao, W. A., Cooks, R. G. 2003. Chiral analysis by MS. *Anal. Chem.* 75:25A–31A.
36. Young, B. L., Wu, L.-M., Cooks, R. G. 2006. Mass spectral methods of chiral analysis. In *Chiral analysis*, ed. K. W. Busch and M. A. Busch, 595–659. Amsterdam: Elsevier.
37. Lebrilla, C. B. 2001. The gas-phase chemistry of cyclodextrin inclusion complexes. *Acc. Chem. Res.* 34:653–61.
38. Filippi, A., Giardini, A., Piccirillo, S., Speranza, M. 2000. Gas-phase enantioselectivity. *Int. J. Mass Spectrom.* 198:137–63.
39. Speranza, M. 2004. Enantioselectivity in gas-phase ion-molecule reactions. *Int. J. Mass Spectrom.* 232:277–317.
40. Schröder, D., Schwarz, H. 2003. Diastereoselective effects in gas-phase ion chemistry. *Top. Curr. Chem.* 225:133–52.
41. Zhang, Q., Curran, D. P. 2005. Quasienantiomers and quasiracemates: New tools for identification, analysis, separation, and synthesis of enantiomers. *Chem. Eur. J.* 11:4866–80.
42. Guo, J., Wu, J., Siuzdak, G., Finn, M. G. 1999. Measurement of enantiomeric excess by kinetic resolution and mass spectrometry. *Angew. Chem. Int. Ed.* 38:1755–58.
43. Schrader, W., Eipper, A., Pugh, D. J., Reetz, M. T. 2002. Second-generation MS-based high-throughput screening system for enantioselective catalysts and biocatalysts. *Can. J. Chem.* 80:626–32.
44. Pandiaraju, S., Chen, G., Lough, A., Yudin, A. K. 2001. Generation of highly enantioselective catalysts from the pseudoenatiomeric assembly of BINOL, $F_8$BINOL, and $Ti(OiPr)_4$. *J. Am. Chem. Soc.* 123:3850–51.
45. Koskinen, A. 1993. *Asymmetric synthesis of natural products*. Chichester, UK: John Wiley & Sons.
46. Vedejs, E., Chen, X. 1997. Parallel kinetic resolution. *J. Am. Chem. Soc.* 119:2584–85.
47. Kagan, H. B., Fiaud, J. C. 1988. Kinetic resolution. *Top. Stereochem.* 18:249–330.
48. Horeau, A., Nouaille, A. 1990. Micromethode de Determination de la Configuration des Alcools Secondaires par Dedoublement Cinetique. Emploi de la Spectrographie de Masse. *Tetrahedron Lett.* 31:2707–10.
49. Díaz, D. D., Yao, S.-L., Finn, M. G. 2001. Measurement of enantiomeric excess of amines by mass spectrometry following kinetic resolution with solid-phase chiral acylating agents. *Tetrahedron Lett.* 42:2617–19.
50. Yao, S., Meng, J.-C., Siuzdak, G., Finn, M. G. 2003. New catalysts for the asymmetric hydrosilylation of ketones discovered by mass spectrometry screening. *J. Org. Chem.* 68:2540–46.
51. Reetz, M. T., Becker, M. H., Klein, H. W., Stöckigt, D. 1999. A method for high-throughput screening of enantioselective catalysts. *Angew. Chem. Int. Ed.* 38:1758–61.
52. Eggert, T., Funke, S. A., Rao, N. M., Acharya, P., Krumm, H., Reetz, M. T., Jaeger, K.-E. 2005. Multiplex-PCR-based recombination as a novel high-fidelity method for directed evolution. *ChemBioChem* 6:1062–67.

53. Hinderling, C., Chen, P. 1999. Rapid screening of olefin polymerization catalyst libraries by electrospray ionization tandem mass spectrometry. *Angew. Chem. Int. Ed.* 38:2253–56.
54. Chen, P. 2003. Electrospray ionization tandem mass spectrometry in high-throughput screening of homogeneous catalysts. *Angew. Chem. Int. Ed.* 42:2832–47.
55. Martha, C. T., Elders, N., Krabbe, J. G., Kool, J., Niessen, W. M. A., Orru, R. V. A., Irth, H. 2008. Online screening of homogeneous catalyst performance using reaction detection mass spectrometry. *Anal. Chem.* 80:7121–27.
56. Santos, L. S., Knaack, L., Metzger, J. O. 2005. Investigation of chemical reactions in solution using API-MS. *Int. J. Mass. Spectrom.* 246:84–104.
57. Eberlin, M. N. 2007. Electrospray ionization mass spectrometry: A major tool to investigate reaction mechanisms both in solution and the gas phase. *Eur. J. Mass Spectrom.* 13:19–28.
58. Santos, L. S. 2008. Online mechanistic investigations of catalyzed reactions by electrospray ionization mass spectrometry: A tool to intercept transient species in solution. *Eur. J. Org. Chem.* 2:235–53.
59. Nugent, W. A., Licini, G., Bonchio, M., Bortolini, O., Finn, M. G., McCleland, B. W. 1998. Homogeneous catalysis as a tool for organic synthesis. *Pure & Appl. Chem.* 70:1041–46.
60. Marquez, C., Metzger, J. O. 2006. ESI-MS study on the aldol reaction catalyzed by L-proline. *Chem. Commun.* 14:1539–41.
61. Marquez, C. A., Fabbretti, F., Metzger, J. O. 2007. Electrospray ionization mass spectrometric study on the direct organocatalytic alpha-halogenation of aldehydes. *Angew. Chem. Int. Ed.* 46:6915–17.
62. Lou, S., Moquist, P. N., Schaus, S. E. 2007. Asymmetric allylboration of acyl imines catalyzed by chiral diols. *J. Am. Chem. Soc.* 129:15398–404.
63. Bao, H.-L., Zhou, J., Wang, Z., Guo, Y. L., You, T. P., Ding, K. L. 2008. Insight into the mechanism of the asymmetric ring-opening aminolysis of 4,4-dimethyl-3,5,8-trioxabicyclo[5.1.0]octane catalyzed by titanium/binolate/water system: Evidence for the Ti(binolate)(2)-bearing active catalyst entities and the role of water. *J. Am. Chem. Soc.* 130:10116–27.
64. Berkessel, A., Brandenburg, M., Schafer, M. 2008. Mass-spectrometric and kinetic studies on the mechanism and degradation pathways of titanium salalen catalysts for asymmetric epoxidation with aqueous hydrogen peroxide. *Adv. Synth. Catal.* 350:1287–94.
65. Markert, C., Pfaltz, A. 2004. Screening of chiral catalysts and catalyst mixtures by mass spectrometric monitoring of catalytic intermediates. *Angew. Chem. Int. Ed.* 43:2498–500.
66. Markert, C., Neuburger, M., Kulicke, K., Meuwly, M., Pfaltz, A. 2007. Palladium-catalyzed allylic substitution: Reversible formation of allyl-bridged dinuclear palladium(I) complexes. *Angew. Chem. Int. Ed.* 46:5892–95.
67. Muller, C. A., Pfaltz, A. 2008. Mass spectrometric screening of chiral catalysts by monitoring the back reaction of quasienantionneric products: Palladium-catalyzed allylic substitution. *Angew. Chem. Int. Ed.* 47:3363–66.
68. Markert, C., Rosel, P., Pfaltz, A. 2008. Combinatorial ligand development based on mass spectrometric screening and a double mass-labeling strategy. *J. Am. Chem. Soc.* 130:3234–35.
69. Teichert, A., Pfaltz, A. 2008. Mass spectrometric screening of enantioselective Diels-Alder reactions. *Angew. Chem. Int. Ed.* 47:3360–62.

# 10 Solution Phase vs. Gas Phase Chiral Recognition by ESI-MS
## *A Case Study of Two Chiral Selector Classes*

*Kevin A. Schug, Aruna B. Wijeratne, Bilal H. Bazzi, and Daniel W. Armstrong*

**CONTENTS**

10.1 Introduction .......... 181
10.2 Experimental .......... 184
10.3 Results and Discussion .......... 186
 10.3.1 Cinchona Alkaloid Carbamates .......... 186
 10.3.2 Antimony Tartrate .......... 190
10.4 Conclusion .......... 199
Acknowledgments .......... 199
References .......... 200

## 10.1 INTRODUCTION

The separation of enantiomers is a critical area of research, requiring continual development in order to keep pace with the development of new chiral entities designed to affect life processes, particularly in the pharmaceutical and agrochemical arenas. Since 1992, the U.S. Federal Drug Administration and the European Committee for Proprietary Medicinal Products have mandated that the properties of each enantiomer in a racemic mixture must be tested separately for therapeutic and toxic effects before the compound can be considered for marketing as a new drug. A move toward marketing drugs in single enantiomer forms has driven the need to either synthesize enantiomerically pure forms or make available separation strategies that can effectively resolve racemates. Furthermore, recent estimates of sales of chiral compounds in the United States (for all applications) place the value of this enterprise somewhere around \$20 billion annually (and approximately an order of magnitude higher globally[1]), providing a strong impetus for the continued development of such technologies.

Separation strategies for resolving racemic mixtures into pure enantiomers rely on the preferential association of one enantiomer form over another with a complementary chiral reference compound ("chiral selector") to form a diastereomeric complex. As depicted in Figure 10.1, if the free energy ($\Delta G$) for the association reaction between the chiral selector and each of the enantiomers is different, then $\Delta\Delta G \neq 0$ and the chiral selector is said to be enantioselective. In practice, even small enantioselectivity values can be sufficient to affect resolution and purification of enantiomers. Fundamentally, it is also important to have tools by which the magnitude of enantioselectivity for a given interaction system can be quantitatively assessed in an accelerated fashion. No chiral selector is currently available that can separate every racemic mixture, leaving an ever-present challenge to scientists involved in the development of such compounds.[2]

Enantioselective chromatographic and electrophoretic methods are currently the gold standard for assessing the purity of chiral compounds in both academic and industrial settings.[2–7] A variety of such techniques are available, some of which can boast determination of enantiomeric excess at the parts-per-million level.[8] For large-scale purification of enantiomers, preparative-scale[9,10] high-performance liquid chromatography (HPLC) and diastereomer crystallization[11] techniques are the dominant approaches. Although many of these approaches are well established, a wealth of new technologies related to chiral separations, such as those focusing on novel enantioselective nanomaterials,[12] continue to be developed.

While mass spectrometry (MS) has not reached the level of general (industrial) utility of many liquid phase separation techniques for enantiomeric excess

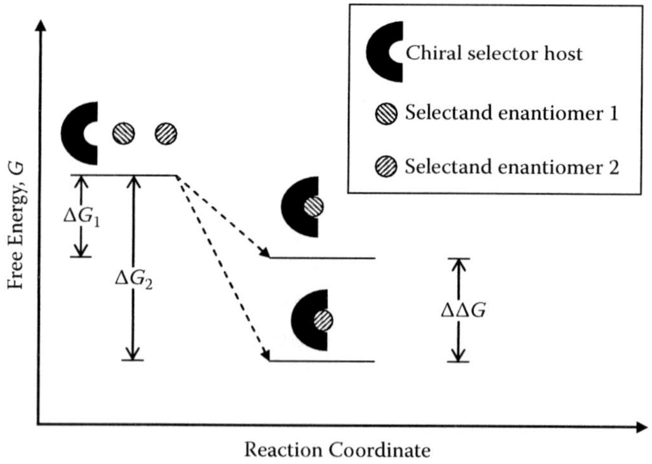


**FIGURE 10.1** Enantioselective discrimination is exhibited by a chiral selector host that can differentially bind ($\Delta\Delta G \neq 0$) enantiomers. In this example, the chiral selector shows a configurational preference for binding selectand enantiomer 2 (e.g., $\Delta G2 > \Delta G1$).

determinations, steady development of new methods and their application to a wide variety of chiral systems have emphasized the versatility of MS in providing mechanistic details for stereoselective interactions.[13] Fales and Wright were the first to report stereochemically controlled association phenomena using MS, focusing on self-association of dimethyl-*L*-tartrate enantiomeric isotopomers in the gas phase using chemical ionization (CI)–MS.[14] After this, a substantial number of publications were communicated, using a wide variety of mass spectrometric methods to study stereoselective phenomena in the solution/condensed phase (e.g., by fast atom bombardment [FAB],[15,16] matrix-assisted laser desorption/ionization [MALDI],[17] and electrospray ionization [ESI][13,18]) and in the gas phase (e.g., by ion-molecule or guest exchange reactions,[19–21] the kinetic method,[22] the direct dissociation of diastereomeric complexes,[23–25] and a variety of other experimental setups[26–28]). More recently, ESI-MS has become the dominant technique for generating and interrogating diastereomeric ion complexes by mass spectrometric methods.

The ability to glean information on both solvent-mediated solution phase and solvent-free gas phase molecular recognition from a qualitative and quantitative standpoint is a major advantage of ESI-MS. Many reviews on this subject are available, recounting research progress in a general sense,[29–31] as well as those focused more closely on biomacromolecular[32,33] and stereoselective[13,18] binding phenomena. General techniques used to directly obtain information relevant to the interactions forged by a system in solution include basic host-guest screening, as well as more rigorous competitive binding and titration experiments. A critical assumption that must be fulfilled for these approaches to provide meaningful information is that equilibrium has been established in solution and is not significantly perturbed during the ESI process. Thus, when an association reaction is kinetically stable on the timescale of the electrospray process (~0.1–1 ms),[34] ESI-MS can provide a snapshot of the distribution of species, and the shift in this distribution as a function of concentration can be probed to obtain quantitative binding information. The results of such experiments should be validated by data from complementary solution phase binding determination methods (e.g., UV/Vis, fluorescence, nuclear magnetic resonance [NMR], capillary electrophoresis, calorimetry, etc.[35–37]) when possible.

The isolation of complexes in the gas phase for interrogation by a variety of collisional, photonic, and electronic processes by tandem mass spectrometry provides additional information that can be useful for determination of enantiomeric excess and describing molecular recognition processes. Since isolated ionic complexes are not solvated, such experiments provide details about their relative strength and kinetic stability based solely on molecular contacts in a low dielectric environment. For example, collision threshold dissociation,[38,39] where the relative abundance of reactants and products is monitored as a function of activation energy, provides a convenient means for comparing relative gas phase stability. This and other tandem mass spectrometry experiments have both advantages and disadvantages. On one hand, probing the stability of complexes in the absence of solvation can provide new insights into molecular recognition mechanisms that might not be readily apparent when solvent is present. Such experimental data are particularly amenable for comparison with those from theoretical computations.

On the other hand, most molecular recognition processes occur, or are designed to occur, in a solution phase environment. In this case, data from gas phase experiments may be less relevant for studying the system of interest in the context where it will be used.

In general, the coherence between solution phase and gas phase relative affinity data is system dependent. Here, we present a contrast and comparison study between two enantioselective molecular recognition systems. The first system is the cinchona alkaloid carbamates, chiral selectors that have been used primarily for the separation of chiral acids in a variety of liquid phase separation techniques and are available in several commercial HPLC column formats.[40] The second is antimony(III)-*D*- and -*L*-tartrates, binuclear complexes that have been employed both as mobile phase additives and as immobilized stationary phases for the separation of inorganic and biologically relevant mixtures of enantiomers.[41] The data from solution and gas phase MS experiments provide evidence for significantly different enantioselective behavior in these two chiral selector systems. Both new and published data[42] emphasize assignment of cinchona alkaloid carbamates as ion exchange chiral selectors, requiring competing achiral species in solution to tune and modulate their interactions with the enantiomers they are used to separate. Stripped of solvent, these systems can actually show a reversal in enantioselective behavior toward the recognition of N-blocked amino acid enantiomers. In contrast, antimony(III)-*D*- and -*L*-tartrate more or less maintain their enantioselective preference toward a particular enantiomeric configuration in both the solution and the gas phase, indicating the role of the solvent or other solution modifiers to be less pronounced for achieving stereoselective recognition. This portion of the story is presented with new data focusing on the enantioselective discrimination of unblocked amino acids, but also describes some interesting nuances that require further investigation.

## 10.2 EXPERIMENTAL

All experiments were performed on a Thermo LCQ Deca XP quadrupole ion trap instrument equipped with a conventional ESI source (Thermo-Fisher Scientific, West Palm Beach, Florida). Samples were delivered by direct infusion to the source by an in-built syringe pump at 10 μl/min. All bulk solvents (water, methanol [MeOH], acetonitrile [ACN]) used were LC-MS grade and supplied by Mallincrodt-Baker (Phillipsburg, New Jersey).

Cinchona alkaloid carbamate chiral selector hosts, *tert*-butylcarbamoyl quinine (tBuCQN, **1a**), *tert*-butylcarbamoyl quinidine (tBuCQD, **1b**), diisopropylphenylcarbamoyl quinine (DIPPCQN, **2a**), diisopropylphenylcarbamoyl quidinine (DIPPCQD, **2b**), and enantiomerically pure *D*- and *L*-dinitrobenzoyl leucine (DNB-Leu) **3** selectand guests were synthesized, purified, and gifted by the Lindner group at the University of Vienna (Vienna, Austria).[43] The sulfate salt of quinine (QN) **4** was from Matheson, Coleman & Bell (Norwood, Ohio) and was used without further purification. The structures for these compounds are shown in Chart 10.1. Sample solutions were prepared by dissolving and mixing each host with each guest to create ten sample solutions with a final concentration of 10

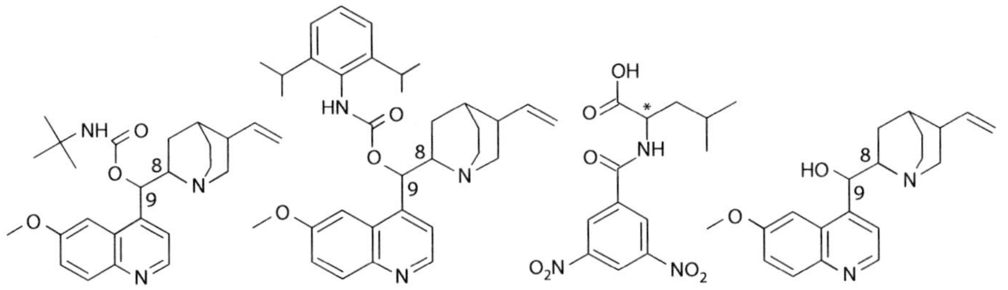

**1a** tBuCQN(8*S*, 9*R*)
**1b** tBuCQD(8*R*, 9*S*)

**2a** DIPPCQN (8*S*, 9*R*)
**2b** DIPPCQD (8*R*, 9*S*)

**3** DNB-Leu

**4** QN (8*S*, 9*R*)

**CHART 10.1**

μM host and 10 μM guest in 50/50 MeOH/water (v/v), including 100 μM sodium acetate (NaOAc; Sigma-Aldrich [St. Louis, Missouri]) For this system, collision threshold dissociation experiments were carried out. Prior to each measurement, the instrument was tuned to maximize the response of the 1:1 sodiated noncovalent complex in the positive ionization mode. The complex ion was then isolated in the ion trap and subjected to collision-activated dissociation using a systematic increase in activation voltage, from 0.500 V to 1.450 V endcap electrode potential in regular increments (activation time = 10 ms, q = 0.250). At each voltage, a minimum of fifty scans (where each scan is an average of three microscans) was recorded and averaged to obtain each data point. Each experiment was performed in triplicate.

The sodium salts of antimony(III)-*D*- and -*L*-tartrate as chiral selector hosts **5** were prepared according to a published procedure.[44] Antimony(III) oxide was purchased from Alfa Aesar (Ward Hill, Massachusetts) and *D*- and *L*-tartrate were from Sigma-Aldrich. Solution phase host-guest screening experiments were performed by analyzing the relative responses of ionic complexes formed between the hosts and each of the enantiomerically pure *D*- and *L*-forms of alanine (Ala; Sigma-Aldrich) **6**, valine (Val; Fluka Chemie [Buchs, Switzerland]) **7**, leucine (Leu; Fluka) **8**, and phenylalanine (Phe; Sigma-Aldrich) **9**. The achiral amino acid glycine (Gly; Sigma-Aldrich) **10** was used as an internal standard. The structures of these compounds are shown in Chart 10.2. Each solution was prepared with host and guest present, equimolar at 100 μ*M* in 75/25 ACN/water (v/v), including 100 m*M* formaldehyde (HCHO; Sigma-Aldrich). HCHO was added to the solution to facilitate better signal quality in the negative ionization mode according to a previous report in the literature.[45] Ionization and source parameters were optimized based on maximizing the ion abundance of the protonated 1:1 complex between antimony(III)-*L*-tartrate and glycine. Prepared solutions were directly infused and data were collected in triplicate, using a minimum of fifty scans for each replicate. For collision threshold experiments, the ion abundances of each 1:1 ionic complex of interest were individually optimized and subjected to collision-activated dissociation as described above for the cinchona alkaloid system (activation voltage = 0.150–0.900 V, activation time = 30 ms, q = 0.250).

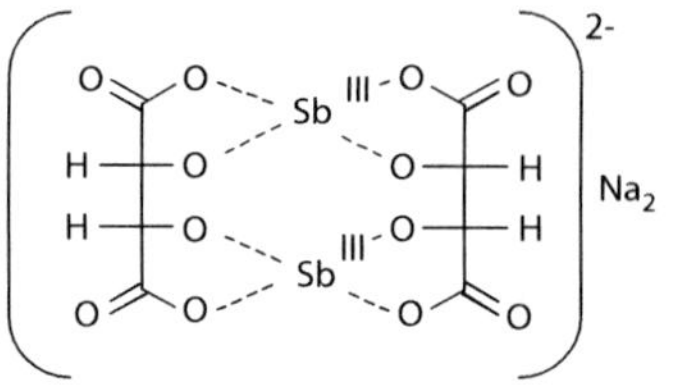
5

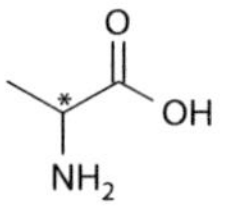
**6** Ala

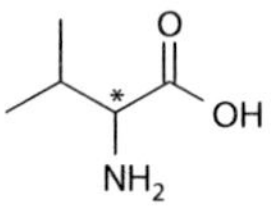
**7** Val

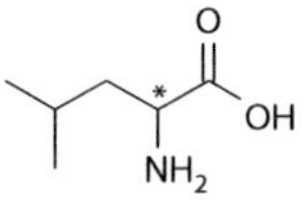
**8** Leu

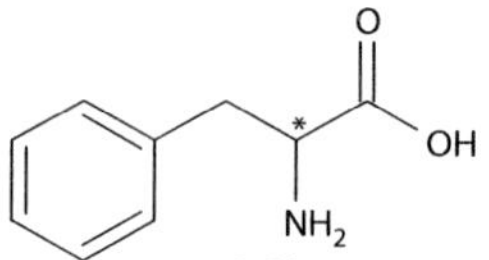
**9** Phe

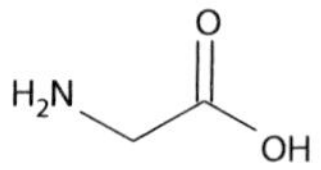
**10** Gly

**CHART 10.2**

## 10.3 RESULTS AND DISCUSSION

### 10.3.1 Cinchona Alkaloid Carbamates

Cinchona alkaloid chiral selector scaffolds have been employed for the separation of enantiomers in combination with a variety of liquid phase separation techniques.[43,46–51] An anion exchange mechanism has been established for the effective resolution of chiral acid enantiomers through these and a variety of other reports incorporating various experimental and theoretical methods, including NMR, calorimetry, IR, molecular modeling, and quantum mechanical calculations.[42,52–55] Stereochemically controlled 1:1 selector/select and association[42] is driven electrostatically by interaction between the protonated tertiary amine on the quinuclidene group of the chiral selector and a deprotonated carboxylic acid group on the chiral selectand. Simultaneous secondary interactions include hydrogen bonding, π-π, and van der Waals forces, the alignment of which differs between enantiomers of a selectand and the selector. These provide for differential association, as described by Figure 10.1. X-ray structural data taken for the interaction between *L*-**3** and chloro-tBuCQN, as shown in Figure 10.2, show the optimal alignment of these forces for a preferential selector-selectand binding pair (i.e., *D*-**3** does not readily crystallize with chloro-tBuCQN) in the solid phase.[43] In the solution phase, the presence of a competing achiral acid (e.g., acetate) is necessary to achieve optimal levels of enantiomer resolution, evidence that has long supported the designation of cinchona alkaloid carbamates as ion-exchange-type chiral selectors. For a large number of mechanistic studies, differentiation of the enantiomers of selectand **3** has served as the model analyte system most studied due to the high degree of enantioselectivity that can be achieved. It is also worthy to note that while quinine (**1a**, **2a**; 8*S*, 9*R*) and quinidine (**1b**, **2b**; 8*R*, 9*S*) derivatives are not enantiomers, they show pseudoenantiomeric behavior, displaying cross-chiral (opposing) configurational preferences for the association of *D*- vs. *L*-enantiomers of N-blocked amino acids, in general.[43]

It is further relevant in the context of this work to mention that a wide variety of mass spectrometric studies have been previously performed to study chiral recognition aspects and applications of cinchona alkaloid carbamates.[24,25,55–61]

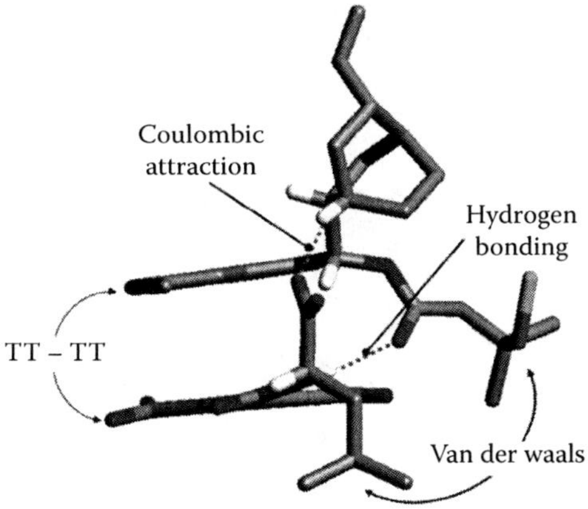


**FIGURE 10.2** (Color Figure 10.2 follows page 46.) Annotated X-ray crystal structure of L-DNB-Leu (bottom) in complex with chloro-tBuCQN (top). (Reprinted from Schug, K.A., Frycak, P., Maier, N.M., Lindner, W., Anal. Chem. 77 (2005): 3660–70. Copyright 2005, American Chemical Society. With permission.)

Overall, strong correlation has been shown for both the magnitude and the configurational preference of enantioselective binding between these chiral selectors and N-blocked amino acids when comparing results from liquid phase separation techniques and solution-phase-targeting mass spectrometry experiments. A corresponding study of the relative enantioselectivity observed from the direct dissociation of diastereomeric complexes between selectors **1**, **2**, and **4** with selectand $_3$ in the gas phase has not been reported. However, one prior study involving a closely related system,[25] where **1** was used to discriminate enantiomers of 3,5-dipropoxybenzoyl leucine, showed that enantioselectivity is largely diminished, if not abolished, in the absence of solvation. Such a finding provides further support for the necessary presence of competing species in solution to achieve an ion-exchange-type enantioselective discrimination mechanism. The studies reported here were done to (1) investigate whether this behavior holds for the model selectand **3**, and (2) check whether this behavior holds for a broader range of cinchona alkaloid carbamate chiral selector scaffolds.

Collision threshold dissociation (CTD) experiments in an ion trap provide a convenient means for assessing the relative binding affinity of noncovalent complexes in the gas phase. Figure 10.3 shows CTD plots where the percent relative precursor diastereomeric complex ion abundances, for all combinations of **1** and **2** in complex with **3** (observed following ESI-MS as a sodiated complex[59]), are followed through a systematic increase in applied activation voltage. The voltage required to dissociate 50% of the diastereomeric complex is conventionally denoted as $V_{50}$. Thus, given the same number of degrees of freedom in each of the diastereomeric complexes compared, absolute $V_{50}$ values can provide a relative measure of stability to delineate which enantiomer binds preferentially to each selector in the gas phase, and the difference between absolute $V_{50}$ values can be used to estimate the level of enantioselectivity present (i.e., $V_{50,2} - V_{50,1}$ is proportional to $\Delta G_2 - \Delta G_1 = \Delta\Delta G$). In each case,

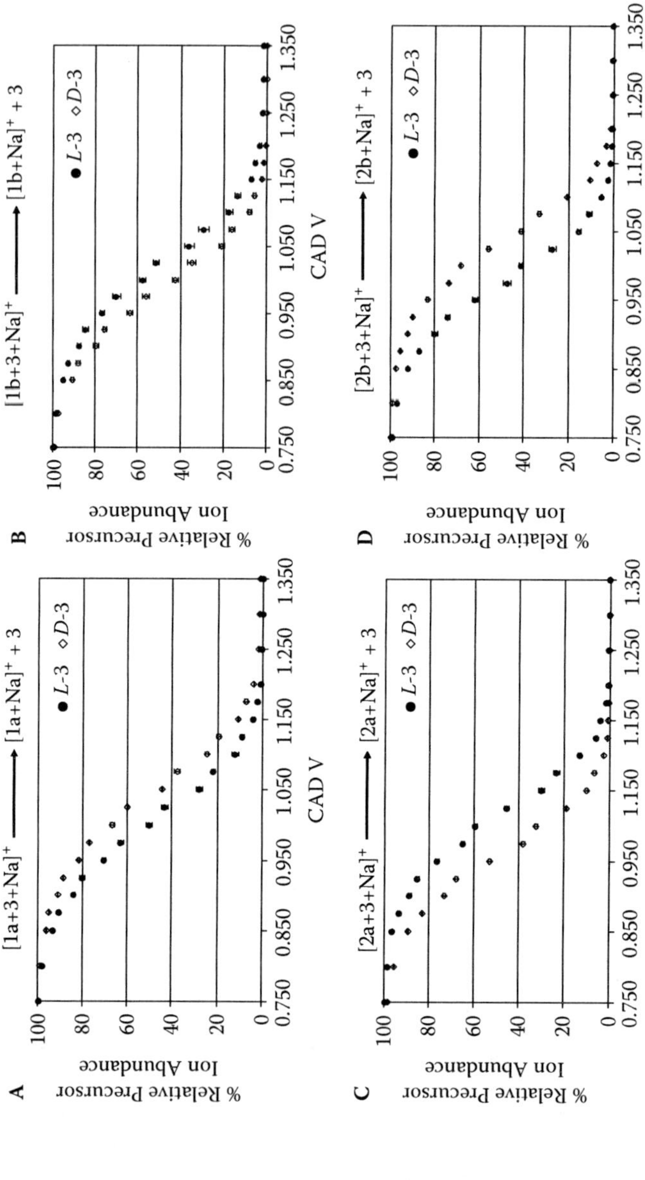


**FIGURE 10.3** Collision threshold dissociation of cinchona alkaloid diastereomeric complexes for chiral selectors (A) **1a** tBuCQN, (B) **1b** tBuCQD, (C) **2a** DIPPCQN, and (D) **2b** DIPPCQD binding enantiomers of *L*- and *D*-DNB-Leu **3**. Error bars are one standard deviation (n = 3).

**TABLE 10.1**
**$V_{50}$ and Enantioselectivity Values by CTD for Cinchona Alkaloid Carbamates**

| Selector | Selectand | $V_{50}$ ± SD [n = 3] (V) | $\alpha_{CTD}$ ± SD (V) [c.p.][a] | $\alpha_{HPLC}$ [c.p.][b] |
|---|---|---|---|---|
| 1a | *L*-**3** | 1.003 ± 0.003 | 0.040 ± 0.003 [*D*] | 15.8 [*L*] |
| | *D*-**3** | 1.043 ± 0.001 | | |
| 1b | *L*-**3** | 1.021 ± 0.004 | 0.03 ± 0.01 [*L*] | 12.5 [*D*] |
| | *D*-**3** | 0.99 ± 0.01 | | |
| 2a | *L*-**3** | 1.012 ± 0.002 | 0.04 ± 0.03 [*L*] | 3.5 [*L*] |
| | *D*-**3** | 0.97 ± 0.03 | | |
| 2b | *L*-**3** | 0.99 ± 0.02 | 0.04 ± 0.02 [*D*] | 2.7 [*D*] |
| | *D*-**3** | 1.034 ± 0.001 | | |
| 4 | *L*-**3** | 1.039 ± 0.003 | 0.012 ± 0.004 [*L*] | 1.2 [*D*] |
| | *D*-**3** | 1.027 ± 0.002 | | |

[a] c.p. denotes configuration preference, the most stable bound selectand in CTD and most highly retained enantiomer in HPLC. $\alpha_{CTD}$ denotes the absolute value of the difference in $V_{50}$ values presented with error propagation (SD denotes one standard deviation from the mean).

[b] HPLC values from Maier et al.[43] and Schug et al.,[57] measured in 80/20 MeOH/0.1 *M* ammonium acetate on an HPLC chiral stationary phase composed from the corresponding chiral selector.

unimolecular dissociation pathways were observed, where the sodiated free chiral selector is the only product.

Table 10.1 shows the results of the experiment, including the configurational preference and calculated enantioselectivity values (with error propagation) from the CTD experiments. Also given for comparison are the reported configurational preferences as a consensus from various HPLC and capillary electrophoresis (CE) determinations for these selectors in the solution phase, as well as the reported enantioselectivity values obtained by HPLC from the literature (in 80/20 MeOH/0.1 *M* ammonium acetate [v/v]; calculated as the ratio of HPLC capacity factors).[43,57] The comparison is intriguing. While chiral selector **2** maintains a consistent configurational preference in the gas phase CTD experiments relative to what is expected from solution phase experiments, chiral selectors **1** and **4** actually show a reversal of their expected configurational preferences, displaying significant enantioselectivity in the opposite direction in the absence of solvation. In each case, however, the pseudoenantiomeric reciprocal binding preferences are maintained (e.g., **1a** vs. **1b** and **2a** vs. **2b**).

An explanation for the reversal of observed enantioselectivity in the CTD data for chiral selectors **1** and **4** can be speculated based on the overexpression of electrostatic binding increments in the absence of solvent.[25] In solution, a delicate balance of noncovalent forces, tuned also by the presence of ion exchange factors, provides for a certain configurational preference. Stripped of solvent, a different picture is revealed. While electrostatic coulombic and hydrogen-bonding forces may be magnified in the gas phase, attractive contributions from solvophobic and Van der Waals interactions would be absent or diminished. For chiral selectors **1** and **4**, these changes in the relative magnitude of attractive and repulsive forces must create a different alignment of steric information in the diastereomeric complex, which, for DNB-Leu, reverses configurational preference. For chiral selector **2**, we can only speculate that the more rigid steric bulk provided by the diisopropylphenyl moiety resists some of these alterations to maintain the solution phase configurational preference in the gas phase. Finally, it is worthy to note that the magnitude of enantioselectivity values obtained by CTD in the gas phase is not comparable with the magnitude of those obtained by HPLC (and reproduced by solution-phase-targeting MS experiments[57]) in the solution phase. Again, the fact that recognition of chiral acid enantiomers by the cinchona alkaloid carbamate selectors is electrostatically driven means that such forces likely dominate in the absence of solvent. Contributions by other attenuating forces are obviously very important for the performance of this selector class in solution phase applications, and except for steric repulsions, these forces simply cannot contribute as appreciably in the gas phase.

### 10.3.2 Antimony Tartrate

Binuclear tartrato(4-)-metal-bridged complexes, of the general form as shown for antimony(III) tartrate **5**, but potentially incorporating a wide range of metal centers (e.g., also As(III), Bi(III), Cr(III), Cu(II), Fe(III), In(III), Ti(III), V(IV)O, W(VI)), have been of significant interest for their use in asymmetric catalysis,[62] medicinal chemistry,[63–66] and enantiomeric separations.[41,67–70] Antimony(III) tartrate, in particular, has enjoyed perhaps the most attention and has been known since medieval times for its use as an emetic drug (tartar emetic), during the 1600s as a therapeutic agent against widespread inflammatory conditions, in the twentieth century as an antifilarial compound, and more recently for its use in enantiomeric separations.[41,66] Use of tartar emetic as a medicine has fallen in and out of favor through the years, mainly due to its unpredictable and often severe toxic side effects. Even so, its promise as an analytical separations reagent has prompted a variety of fundamental studies to enumerate mechanistic details of its enantioselective performance, in a system-dependent fashion, by HPLC,[68–75] CE,[70,76–78] and MS.[79,80] However, a general understanding of these effects still remains elusive.

The goal of this study is to compare the solution phase and gas phase enantioselective behavior of disodium antimony(III) tartrate **5** toward several natural amino acids. Amino acids were chosen to provide evidence relevant not only to the use of tartar emetic in chiral separation applications, but also to the interaction of this compound with biologically relevant compounds. Our group has recently communicated interesting results along these lines focusing solely on the amino acid leucine.[81] Reported were the observation of discrete charge states exhibited by antimony(III) tartrate and an interesting disparity in their enantioselective recognition behavior.

Figure 10.4A shows a typical mass spectrum of disodium antimony(III)-*L*-tartrate (without selectand present), where the dominant signals correspond to dianionic $[M - 2Na]^{2-}$ ($m/z = 268.1$ Th), sodiated monoanionic $[M - Na]^{1-}$ ($m/z = 558.9$ Th), and protonated monoanionic $[M - 2Na + H]^{1-}$ ($m/z = 536.8$ Th) ion forms (the $m/z$ of the most abundant signal that corresponds to equivalent contributions from $^{121}$Sb [57.2 % natural abundance] and $^{123}$Sb [42.8 %] is given). Interestingly, upon addition of the selectand *D*-leucine, as shown in Figure 10.4B, significant complexation only forms with the dianionic and protonated monoanionic forms. Furthermore, comparing the degree

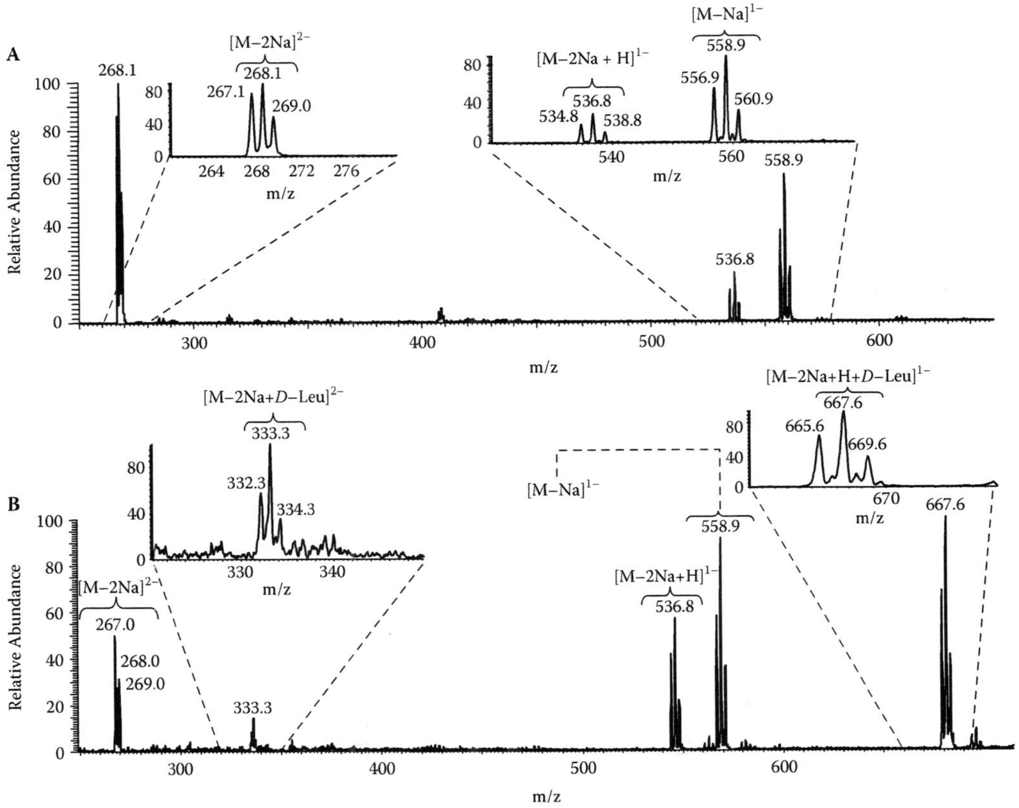


**FIGURE 10.4** Negative ionization mode ESI mass spectra for (A) antimony(III)-*L*-tartrate **5** and (B) antimony(III)-*L*-tartrate in complex with *D*-Leu **8**.

of complexation with different enantiomers of leucine, only the protonated monoanionic form showed appreciable enantioselectivity based on the relative abundance of ionic complexes, whereas the dianionic form provided no significant enantioselectivity (data not shown).[81] A consistent relationship, in terms of both enantioselective capacity and configurational preference, was also found in gas phase collision threshold dissociation data. It is these effects that were expanded in the current study with additional amino acids to help assess the generality of this phenomenon.

Solution phase host-guest screening experiments were performed by analyzing the relative complex formation by each enantiomerically pure antimony(III)-*D*- and -*L*-tartrate **5** with each enantiomerically pure form of the amino acids alanine **6**, valine **7**, leucine **8**, and phenylalanine **9**, in separate experiments. Normally, it would be possible to analyze the enantioselectivity provided for each amino acid by calculating the ratio of normalized complex ion abundances (normalized to the total ion signal) taken from subsequent runs;[13] however, the observed charge state dependence, and the desire to analyze multiple charge states from each spectrum separately, prompted a slightly different approach. Instead, an achiral competitive binding internal standard (glycine) was incorporated into each mixture. Thus, for each amino acid and charge state of interest, solution-phase-based enantioselectivity ($\alpha_{MS}$) was calculated, as shown in Equation 10.1 (using Phe as an example):

For the 2-charge state:

$$\left(\frac{int[\mathrm{Sb_2}L\mathrm{tar_2} \bullet D\text{-}\mathrm{Phe}]^{2-}}{int[\mathrm{Sb_2}L\mathrm{tar_2} \bullet \mathrm{Gly}]^{2-}}\right)_{\mathrm{run1}} \Bigg/ \left(\frac{int[\mathrm{Sb_2}L\mathrm{tar_2} \bullet L\text{-}\mathrm{Phe}]^{2-}}{int[\mathrm{Sb_2}L\mathrm{tar_2} \bullet \mathrm{Gly}]^{2-}}\right)_{\mathrm{run2}} = \alpha_{\mathrm{MS},2-} \tag{10.1a}$$

For the 1-charge state:

$$\left(\frac{int[\mathrm{Sb_2}L\mathrm{tar_2} \bullet D\text{-}\mathrm{Phe}+\mathrm{H}]^{1-}}{int[\mathrm{Sb_2}L\mathrm{tar_2} \bullet \mathrm{Gly}+\mathrm{H}]^{1-}}\right)_{\mathrm{run1}} \Bigg/ \left(\frac{int[\mathrm{Sb_2}L\mathrm{tar_2} \bullet L\text{-}\mathrm{Phe}+\mathrm{H}]^{1-}}{int[\mathrm{Sb_2}L\mathrm{tar_2} \bullet \mathrm{Gly}+\mathrm{H}]^{1-}}\right)_{\mathrm{run2}} = \alpha_{\mathrm{MS},1-} \tag{10.1b}$$

where *int* denotes ion abundance and • denotes a noncovalent complex between antimony(III)-*L*-tartrate ($Sb_2Ltar_2$) and the enantiomerically pure amino acid (or Gly). Figure 10.5 shows representative mass spectra for this system from which such values were calculated.

Table 10.2 summarizes the enantioselectivity data from the solution-phase-targeting selector-selectand screening experiments. Also included are the average relative amino acid/internal standard abundance ratios used to calculate the enantioselectivity values. With respect to the latter, it is apparent that there is a significant contribution to complex ion response from the identity of the amino acid. In other words, the much higher abundance ratios are likely not due to a significantly higher binding

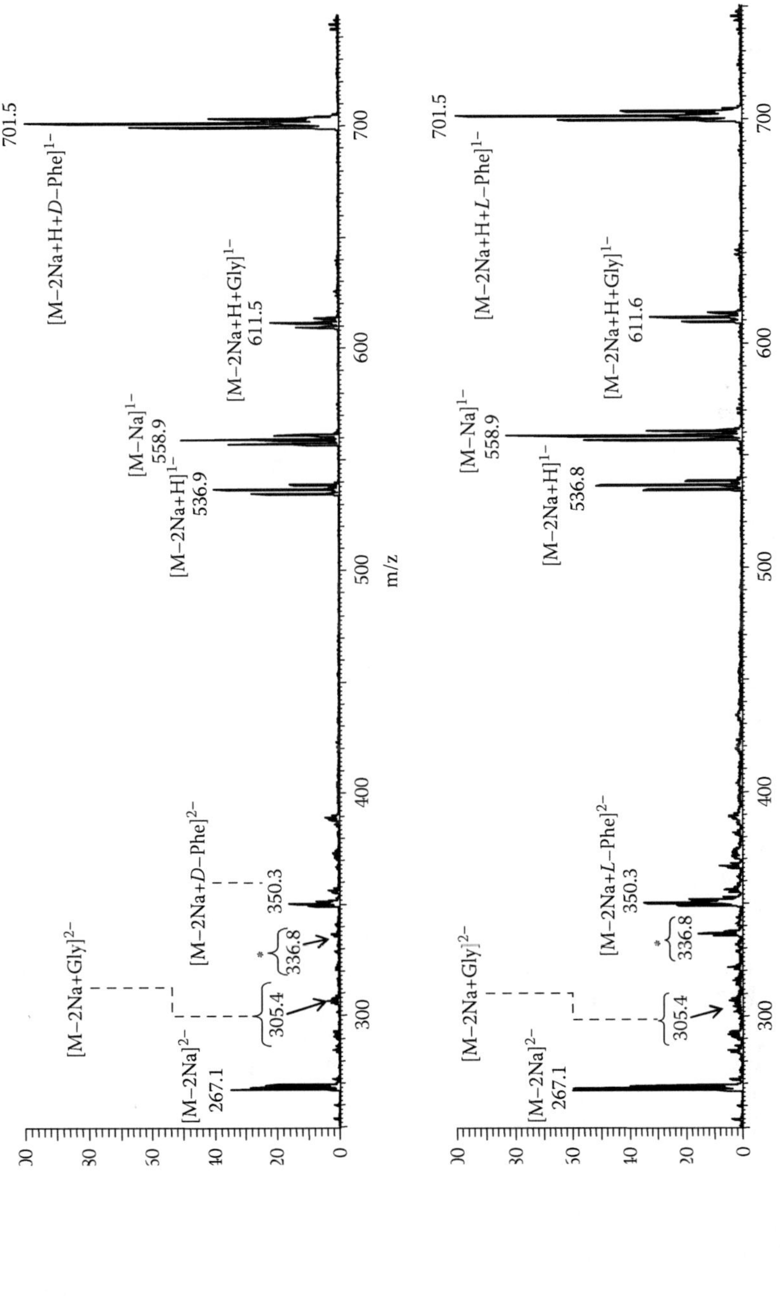


**FIGURE 10.5** Negative ionization mode ESI mass spectrum of antimony(III)-*L*-tartrate (denoted as M) in complex with (A) *D*-Phe and Gly (internal standard) and (B) *L*-Phe and Gly. Signals marked with an * could not be readily assigned.

**TABLE 10.2**
**Normalized Intensity Ratios[a] and Charge-State-Dependent Enantioselectivities from Solution-Phase-Targeting ESI-MS Selector-Selectand Screening Experiments**

| Antimony(III)-*X*-tartrate | Selectand | Intensity Ratio, 1– | $\alpha_{MS,1-}$ ± SD (c.p.)[b] | Intensity Ratio, 2– | $\alpha_{MS,2-}$ ± SD (c.p.)[b] |
|---|---|---|---|---|---|
| *L* | *L*-Ala | 1.00 ± 0.04 | 1.23 ± 0.06 (*D*) | 1.24 ± 0.24 | 1.2 ± 0.3 (np) |
| | *D*-Ala | 1.23 ± 0.04 | | 1.00 ± 0.04 | |
| | *L*-Val | 3.09 ± 0.10 | 1.07 ± 0.04 (*L*) | 3.05 ± 0.53 | 1.1 ± 0.2 (np) |
| | *D*-Val | 2.88 ± 0.02 | | 2.88 ± 0.02 | |
| | *L*-Leu | 14.12 ± 0.23 | 4.5 ± 0.2 (*D*) | 23.46 ± 5.22 | 1.3 ± 0.7 (np) |
| | *D*-Leu | 63.33 ± 2.48 | | 18.45 ± 8.78 | |
| | *L*-Phe | 3.17 ± 0.21 | 1.6 ± 0.1 (*D*) | 8.49 ± 1.05 | 1.3 ± 0.2 (*D*) |
| | *D*-Phe | 4.94 ± 0.01 | | 10.90 ± 0.85 | |
| *D* | *L*-Ala | 1.40 ± 0.02 | 1.44 ± 0.03 (*L*) | 1.15 ± 0.12 | 1.13 ± 0.07 (*L*) |
| | *D*-Ala | 0.97 ± 0.01 | | 1.02 ± 0.06 | |
| | *L*-Val | 7.70 ± 0.26 | 2.0 ± 0.2 (*L*) | 6.12 ± 0.45 | 1.3 ± 0.1 (*L*) |
| | *D*-Val | 3.79 ± 0.03 | | 4.57 ± 0.33 | |
| | *L*-Leu | 34.88 ± 2.18 | 1.5 ± 0.2 (*L*) | 15.98 ± 1.55 | 1.1 ± 0.2 (np) |
| | *D*-Leu | 23.63 ± 2.18 | | 17.05 ± 3.05 | |
| | *L*-Phe | 4.54 ± 0.90 | 1.0 ± 0.2 (np) | 9.86 ± 0.27 | 1.31 ± 0.05 (*D*) |
| | *D*-Phe | 4.53 ± 0.26 | | 12.96 ± 0.30 | |

[a] Normalized intensity ratios calculated from the relative ion abundance of each diastereomeric complex (charge state dependent) divided by the corresponding internal standard (glycine) complex (n = 3).
[b] Charge-state-dependent enantioselectivities calculated according to Equation 10.1 (n = 3, with error propagation); c.p. denotes configurational preference and np denotes no preference.

affinity for phenylalanine compared to the other amino acids, such as valine, which have smaller side chains. More likely is that larger, more hydrophobic side chains impart higher ionization efficiency for the complex due to increased surface activity.[82] Figure 10.6 shows a portion of a mass spectrum taken where the protonated monoanionic complexes between antimony(III)-*L*-tartrate and all five amino acids (in equimolar concentration) were evaluated simultaneously. The order of complex ion abundances approximately follows the relative hydrophobicity of the side chains: Leu > Phe > Val > Ala > Gly. What this means is that MS-based titration experiments,[18] were they pursued to obtain dissociation constants for these systems, would likely be subject to substantial systematic error due to disparities in response factors for the ionic complexes containing different amino acid selectands. Still, since enantioselectivity values are calculated only by comparing the relative abundances of diastereomeric complexes containing the same amino acids (in different enantiomeric configurations), such values are still expected to provide a good basis for comparison of enantioselective performance by the chiral selector.

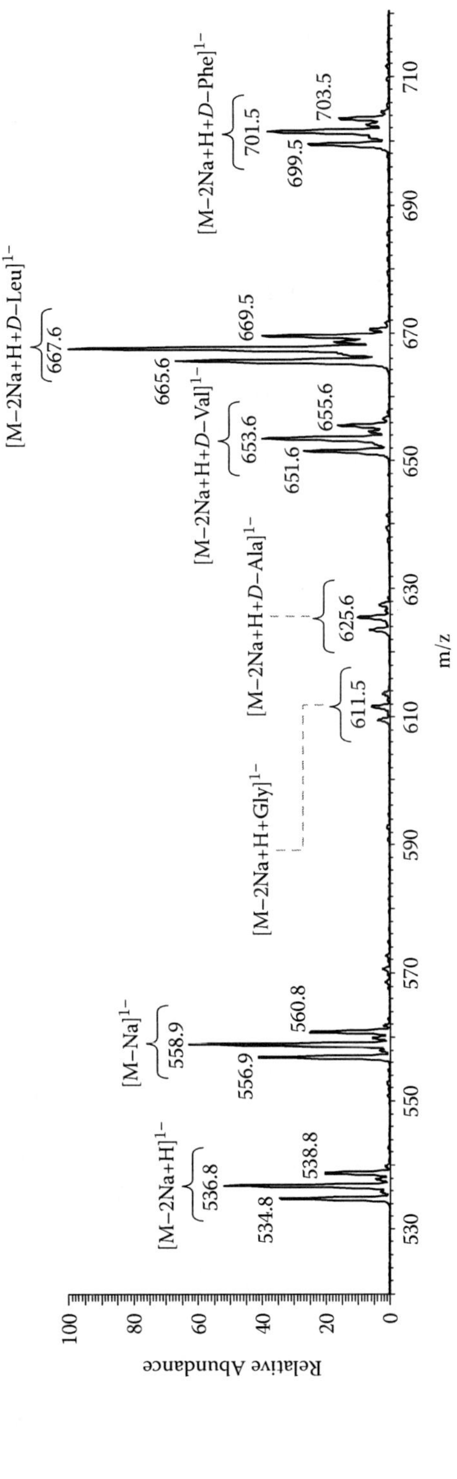


**FIGURE 10.6** Negative ionization mode ESI mass spectrum of antimony(III)-*L*-tartrate (denoted as M) in complex with each of five different amino acids (equimolar concentrations; all *D*-configuration) evaluated in a competitive binding experiment.

It is apparent that normalized intensity ratios and calculated enantioselectivity values are subject to a higher degree of variance, the greater the disparity between the ion intensity of the diastereomeric complex of interest and the corresponding internal standard complex. In this respect, while Gly was chosen because it is an achiral amino acid, the use of a different internal standard that provides a higher complex ion signal intensity may be preferred in future experiments. Even so, a consistent trend is largely revealed for the enantioselective recognition of amino acids by the protonated monoanionic complex (this form of antimony(III)-*L*-tartrate preferentially binds amino acids in the *D*-configuration, and vice versa), consistent with previous work focusing on leucine alone.[81] Two exceptions are noted: (1) for Val binding to antimony(III)-*L*-tartrate, which exhibits minimal enantioselectivity as well as a different configurational preference, and (2) for Phe binding to antimony(III)-*D*-tartrate, for which no enantioselectivity was observed. The data are less consistent for the dianionic form; no real trend emerged and a generally lower degree of enantioselectivity was recorded. A more rigorous evaluation of these effects could be obtained in a competitive binding format by incorporating deuterated enantiomer standards in a single run, avoiding run-to-run error, and allowing direct calculation of charge-state-dependent enantioselectivity. Unfortunately, such an approach is limited by the cost of enantiomerically pure deuterated standards. Furthermore, the preponderance of additional, unidentifiable signals in the low mass region (see, e.g., Figure 10.5) makes peak selection more difficult in this region, and subjects those values obtained for evaluating the enantioselectivity of the dianionic form to a higher degree of uncertainty.

To compare the enantioselective capacity of antimony(III) tartrate in the gas phase, and to check both the charge state dependence and configurational preference for a wider range of amino acids, CTD experiments were performed in a manner similar to that described above for the cinchona alkaloid carbamates. We have previously reported that, for leucine, configurational preferences and charge state dependencies are consistent with those observed in the solution phase.[81] Additionally, for antimony(III)-*L*-tartrate vs. antimony(III)-*D*-tartrate, a quantitative cross-chiral relationship was reported for enantioselectivity values, determined as the absolute magnitude of the difference between $V_{50}$ values from the CTD experiments. Figure 10.7 shows CTD plots for the dissociation of dianionic and protonated monoanionic diastereomeric complexes formed between antimony(III)-*L*-tartrate and each of the four amino acids investigated. Table 10.3 shows the $V_{50}$ values in each case, along with the calculated enantioselectivity values ($\alpha_{CTD}$). A consistent pattern was revealed. For CTD of the protonated monoanionic complexes, configurational preferences equivalent to the solution phase data were observed, with a cross-chiral relationship in going from antimony(III)-*L*-tartrate to antimony(III)-*D*-tartrate. Also consistent with previous results, dissociation of the dianionic form revealed no enantioselective discrimination. It is therefore possible that either a fundamental difference exists in the structural and functional arrangement of stereoselective recognition motifs between the dianionic and protonated monoanionic forms, or that some other unforeseen factor is contributing to the decreased stereoselective performance provided by the dianionic form. The fact that some of the CTD plots for the dissociation of the dianionic diastereomeric

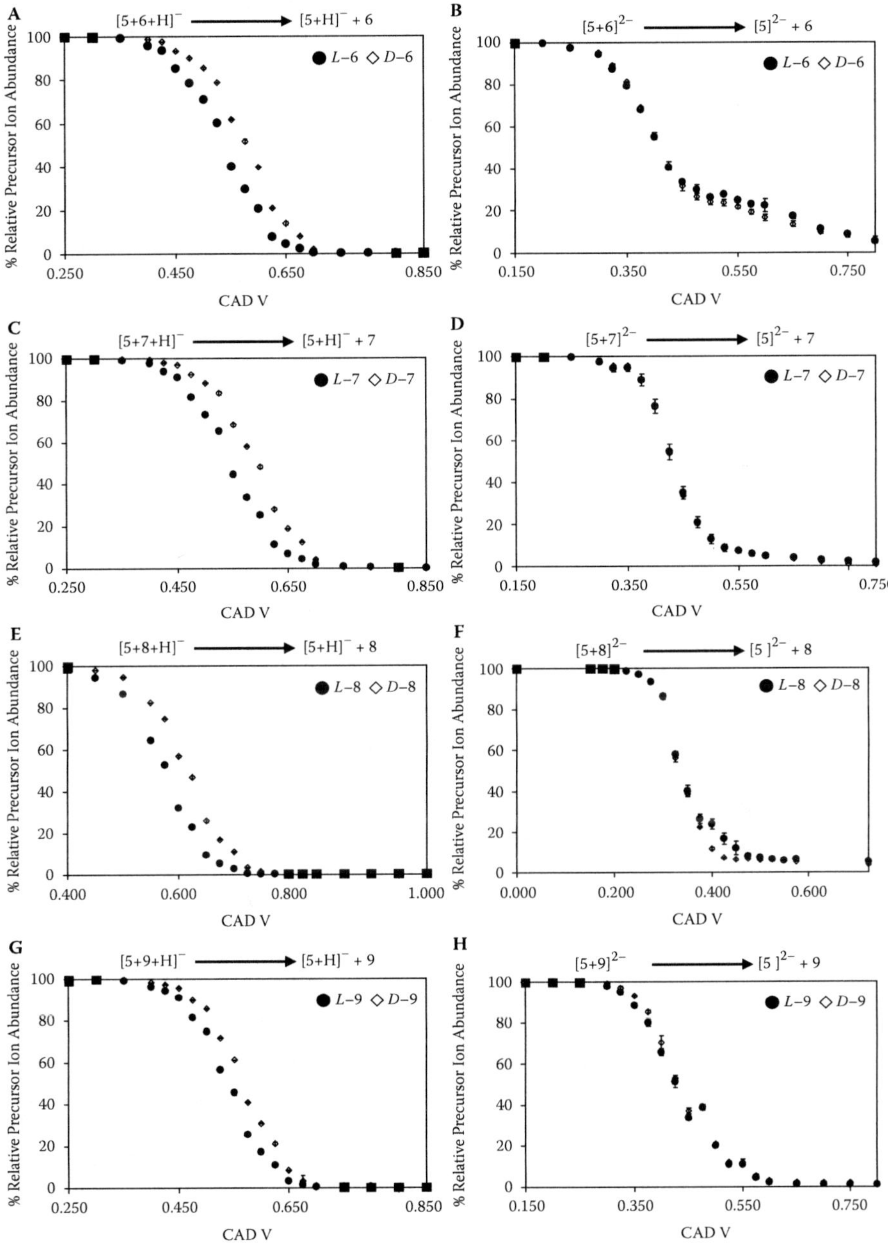


**FIGURE 10.7** Collision threshold dissociation of protonated monoanionic (A, C, E, G) and dianionic (B, D, F, H) diastereomeric complexes between antimony(III)-*L*-tartrate and alanine (A, B), valine (C, D), leucine (E, F), and phenylalanine (G, H). CAD V denotes the collision-activated dissociation voltage applied in the ion trap to dissociate the complexes.

**TABLE 10.3**
**$V_{50}$ and Enantioselectivity Values Recorded for Dissociation of Protonated Monoanionic Diastereomeric Complexes by Collision Threshold Dissociation**

| Selector | Selectand | $V_{50}$ ± SD [n = 3] (V) | $\alpha_{CTD}$ ± SD (V) [c.p.] |
|---|---|---|---|
| [$Sb_2$-*L*-$tar_2$][$H^+$] | *L*-Ala | 0.751 ± 0.002 | 0.037 ± 0.002 [*D*] |
| | *D*-Ala | 0.788 ± 0.001 | |
| | *L*-Val | 0.759 ± 0.002 | 0.043 ± 0.002 [*D*] |
| | *D*-Val | 0.802 ± 0.001 | |
| | *L*-Leu | 0.573 ± 0.000 | 0.041 ± 0.001 [*D*] |
| | *D*-Leu | 0.614 ± 0.001 | |
| | *L*-Phe | 0.735 ± 0.001 | 0.025 ± 0.001 [*D*] |
| | *D*-Phe | 0.760 ± 0.001 | |
| [$Sb_2$-*D*-$tar_2$][$H^+$] | *L*-Ala | 0.786 ± 0.006 | 0.035 ± 0.006 [*L*] |
| | *D*-Ala | 0.751 ± 0.002 | |
| | *L*-Val | 0.789 ± 0.002 | 0.039 ± 0.002 [*L*] |
| | *D*-Val | 0.750 ± 0.001 | |
| | *L*-Leu | 0.612 ± 0.001 | 0.040 ± 0.002 [*L*] |
| | *D*-Leu | 0.572 ± 0.001 | |
| | *L*-Phe | 0.771 ± 0.001 | 0.038 ± 0.002 [*L*] |
| | *D*-Phe | 0.733 ± 0.002 | |

complexes do not follow a clean sigmoidal shape (e.g., kinetic shifts were observed, as seen in Figure 10.7B for [**5** + **6**]$^{2-}$) may lend support to the latter notion. Further work is required to elucidate this aspect; however, these data provide some insight into where those efforts might be channeled.

From the results of both solution phase and gas phase data, it can be speculated that the protonated monoanionic form of antimony(III)-tartrate is mainly responsible for its observed enantioselective discrimination in various applications. Importantly, this information would be difficult to ascertain without the use of mass spectrometric experiments as described in this work. This hypothesis is supported also by the results of Martin et al.,[70] where the resolution of Ru(II) tris-diimine complexes was achieved at a pH of 2.5 using antimony(III) tartrate as a chiral selector in the run buffer. These conditions would favor the presence of the protonated monoanionic form. The mass spectrometric experiments also reveal the consistency of configurational preference in the presence and absence of solvent, contrary to the cinchona alkaloid carbamate experiments. Thus, for the protonated monoanionic form of antimony tartrate, the stereochemical arrangement of the associates that give rise to enantioselective binding appears to be less influenced by the presence of solvent. Whether the magnitude of the enantioselectivity can be tuned or optimized based on changing the percent organic, ionic strength, or pH of the solution remains to be further investigated. In addition to this, in future experiments we will also investigate the effects of different activation conditions in CTD experiments to gain more insight into the kinetics of the dissociation process. Through such experiments, it may be possible to learn more about apparent kinetic shifts observed at high activation voltages for the dianionic complexes in the present experiments. Theoretical computations and molecular modeling will also be pursued to provide a more comprehensive description of possible structural and functional differences between the different ion forms.

## 10.4 CONCLUSION

While experimental methods such as HPLC and CE remain the methods of choice for determining enantiomeric purity and separating chiral compounds into enantiomerically pure forms, MS still holds a special place for investigating distinct molecular level enantioselective mechanisms. Speed, sensitivity, and specificity are major advantages of MS compared to other techniques, such as IR and NMR, which can also be used for such purposes. Additionally, the ability to investigate both solution-phase-based interaction effects through full-scan ESI-MS and gas phase interaction effects in the absence of solvent through tandem mass spectrometry experiments provides an additional level of information that no other experimental technique can provide.

Two case studies were presented that show the disparities between solution and gas phase molecular recognition that can be encountered. Previous studies have provided a wealth of data supporting the classification of cinchona alkaloid carbamates as ion-exchange-type chiral selectors in the solution phase. However, stripped of solvent and the presence of competing anions, the relative orientation of the stereochemical information in different derivatives along with the increase in electrostatic forces shows a degree of unpredictability in terms of expected configurational preferences. For tBuCQN and tBuCQD **1**, the results show a classical example where solution phase and gas phase MS data are highly disparate. On the other hand, antimony(III) tartrate chiral recognition mechanisms are less well understood. While configurational preferences are maintained, providing good correlation between solution and gas phase data, the detailed intricacies in terms of the charge state dependence of enantioselectivities have also now been brought to light. More work is required to conclusively describe these effects; however, the data revealed by mass spectrometry indicate a compound class that may very well still be tuned to provide optimum performance both as an analytical separations reagent and as a useful therapeutic entity.

## ACKNOWLEDGMENTS

The authors acknowledge support of this work by the University of Texas at Arlington, the Robert A. Welch Foundation (Y-0026), and the National Science Foundation (CHE-0846310). Additional thanks is given to the Lindner group at the University of Vienna for providing some of the chemicals used in this study.

## REFERENCES


1. Darrow, J.J. 2007. *The patentability of enantiomers: Implications for the pharmaceutical industry*. Stan. Tech. L. Rev. 2. http://stlr.standford.edu/pdf/darrow-patentability.pdf (accessed March 29, 2009).
2. Maier, N.M., Franco, P., Lindner, W. 2001. Separation of enantiomers: Needs, challenges, perspectives. *J. Chromatogr. A* 906:3–33.
3. Ward, T.J. 2002. Chiral separations. *Anal. Chem.* 74:2863–872.
4. Davankov, V.A. 1997. Analytical chiral separation methods. *Pure Appl. Chem.* 69:1469–74.
5. Finn, M.G. 2002. Emerging methods for the rapid detection of enantiomeric excess. *Chirality* 14:534–40.
6. Armstrong, D.W. 1987. Optical isomer separation by liquid chromatography. *Anal. Chem.* 59:84A–91A.
7. Ward, T.J., Baker, B.A. 2008. Chiral separations. *Anal. Chem.* 80, 4363–72.
8. Armstrong, D.W., Duncan, J.D., Lee, S.H. 1991. Evaluation of D-amino acid levels in human urine and in commercial L-amino acid samples. *Amino Acids* 1:97–106.
9. Guiochon, G. 2002. Preparative liquid chromatography. *J. Chromatogr. A* 965:129–61.
10. Nelson, T.D., Welch, C.J., Rosen, J.D., Smitrovich, J.H., Huffman, M.A., McNamara, J.M., Mathre, D.J. 2004. Effective use of preparative chiral HPLC in a preclinical drug synthesis. *Chirality* 16:609–13.
11. Trung, T.Q., Kim, J.M., Kim, K.H. 2006. Preparative method of R-(–)-ibuprofen by diastereomer crystallization. *Arch. Pharm. Res.* 29:108–11.
12. Sancho, R., Minguillon, C. 2009. The chromatographic separation of enantiomers through nanoscale design. *Chem. Soc. Rev.* 38:797–805.
13. Schug, K.A., Lindner, W. 2005. Chiral molecular recognition for the detection and analysis of enantiomers by mass spectrometric methods. *J. Sep. Sci.* 28:1932–55.
14. Fales, H.M., Wright, G.J. 1977. Detection of chirality with the chemical ionization mass spectrometer. "Meso" ions in the gas phase. *J. Am. Chem. Soc.* 99:2339–40.
15. Sawada, M., Shizuma, M., Takai, Y., Yamada, H., Kaneda, T., Hanafusa, T. 1992. Enantioselectivity in FAB mass spectrometry. *J. Am. Chem. Soc.* 114:4405–4406.
16. Sawada, M. 1997. Chiral recognition detected by fast atom bombardment mass spectrometry. *Mass Spectrom. Rev.* 16:73–90.
17. So, M.P., Wan, T.S.M. 2000. Differentiation of enantiomers using matrix-assisted laser desorption/ionization mass spectrometry. *Rapid Commun. Mass Spectrom.* 14:692–95.
18. Schug, K.A. 2007. Solution phase enantioselective recognition and discrimination by electrospray ionization–mass spectrometry: State-of-the-art, methods, and an eye towards increased throughput measurements. *Combin. Chem. High Throughput Screen.* 10:301–16.
19. Speranza, M. 2004. Enantioselectivity in gas-phase ion-molecule reactions. *Int. J. Mass Spectrom.* 232:277–317.
20. Grigorean, G.S., Gronert, S., Lebrilla, C.B. 2002. Enantioselective gas-phase ion-molecule reactions in a quadrupole ion trap. *Int. J. Mass Spectrom.* 219:79–87.
21. Dearden, D.V., Liang, Y., Nicoll, J.B., Kellersberger, K. 2001. Study of gas-phase molecular recognition using Fourier transform ion cyclotron resonance mass spectrometry. *J. Mass Spectrom.* 36:989–97.
22. Tao, W.A., Cooks, R.G. 2003. Chiral analysis by MS. *Anal. Chem.* 75:25A–31A.
23. Gronert, S., Fagin, A.E., Okamoto, K. 2004. Stereoselectivity in the collision-activated reactions of gas phase salt complexes. *J. Am. Soc. Mass Spectrom.* 15:1509–16.
24. Czerwenka, C., Lindner, W. 2004. Enantiomer discrimination of peptides by tandem mass spectrometry: Influence of the peptide sequence on chiral recognition. *Rapid Commun. Mass Spectrom.* 18:2713–18.

25. Schug, K.A., Joshi, M.D., Frycak, P., Maier, N.M., Lindner, W. 2008. Investigation of monovalent and bivalent enantioselective molecular recognition by electrospray ionization–mass spectrometry and tandem mass spectrometry. *J. Am. Soc. Mass Spectrom.* 19:1629–42.
26. Filippi, A., Giardini, A., Piccirillo, S., Speranza, M. 2000. Gas-phase enantioselectivity. *Int. J. Mass Spectrom.* 198:137–63.
27. Speranza, M., Rondino, F., Satta, M., Paladini, A., Giardini, A., Catone, D., Piccirillo, S. 2009. Molecular and supramolecular chirality: R2PI spectroscopy as a tool for the gas-phase recognition of chiral systems of biological interest. *Chirality* 21:119–44.
28. Speranza, M., Satta, M., Piccirillo, S., Rondino, F., Paladini, A., Giardini, A., Filippi, A., Catone, D. 2005. Chiral recognition by mass-resolved laser spectroscopy. *Mass Spectrom. Rev.* 24:588–610.
29. Brodbelt, J.S. 2000. Probing molecular recognition by mass spectrometry. *Int. J. Mass Spectrom.* 200:57–69.
30. Schalley, C. A. 2001. Molecular recognition and supramolecular chemistry in the gas phase. *Mass Spectrom. Rev.* 20:253–309.
31. Daniel, J.M., Friess, S.D., Rajagopalan, S., Wendt, S., Zenobi, R. 2002. Quantitative determination of noncovalent binding interactions using soft ionization mass spectrometry. *Int. J. Mass Spectrom.* 216:1–27.
32. Nesatyy, V. J. 2002. Mass spectrometry evaluation of the solution and gas-phase properties of noncovalent protein complexes. *Int. J. Mass Spectrom.* 221:147–61.
33. Heck, A. J. R., van den Heuvel, R. H. H. 2004. Investigation of intact protein complexes by mass spectrometry. *Mass Spectrom. Rev.* 23:368–89.
34. Wortmann, A., Kistler-Momotova, A., Zenobi, R., Heine, M.C., Wilhelm, O., Pratsinis, S.E. 2007. Shrinking droplets in electrospray ionization and their influence on chemical equilibria. *J. Am. Soc. Mass Spectrom.* 18:385–393.
35. Jelesarov, I., Bosshard, H.R. 1999. Isothermal titration calorimetry and differential scanning calorimetry as complementary tools to investigate the energetics of biomolecular recognition. *J. Mol. Recog.* 12:3–18.
36. Fielding, L. 2000. Determination of association constants (Ka) from solution NMR data. *Tetrahedron* 56:6151–70.
37. Hirose, K. 2001. A practical guide for the determination of binding constants. *J. Incl. Phenom. Macrocyc. Chem.* 39:193–209.
38. Wan, K. X., Gross, M. L., Shibue, T. 2000. Gas-phase stability of double-stranded oligodeoxynucleotides and their noncovalent complexes with DNA-binding drugs as revealed by collision activation in an ion trap. *J. Am. Soc. Mass Spectrom.* 11:450–57.
39. Colorado, A., Brodbelt, J. 1996. An empirical approach to estimation of critical energies by using a quadrupole ion trap. *J. Am. Soc. Mass Spectrom.* 7:1116–25.
40. Gasparrini, F., Misiti, D., Villani, C. 2001. High-performance liquid chromatography chiral stationary phases based on low-molecular-mass selectors. *J. Chromatogr. A* 906:35–50.
41. Wijeratne, A.B., Schug, K.A. 2009. Molecular recognition properties of tartrates and metal-tartrates in solution and gas phase. *J. Sep. Sci.* 32:1537–47.
42. Maier, N.M., Schefzick, S., Lombardo, G.M., Feliz, M., Rissanen, K., Lindner, W., Lipkowitz, K.B. 2002. Elucidation of the chiral recognition mechanism of cinchona alkaloid carbamate-type receptors for 3,5-dinitrobenzoyl amino acids. *J. Am. Chem. Soc.* 124:8611–29.
43. Maier, N.M., Nicoletti, L., Laemmerhofer, M., Lindner, W. 1999. Enantioselective ion exchangers based on cinchona alkaloid-derived carbamates: Influence of C8/C9 stereochemistry on chiral recognition. *Chirality* 11:522–28.

44. Marcovich, D., Tapscott, R.E. 1980. Carbon-13 NMR studies on arsenic(III) and antimony(III) dihydroxydicarboxylate complexes. *J. Am. Chem. Soc.* 102:5712–17.
45. Wu, Z., Gao, W., Phelps, M.A., Wu, D., Miller, D.D., Dalton, T.J. 2004. Favorable effects of weak acids on negative-ion electrospray ionization mass spectrometry. *Anal. Chem.* 76:839–47.
46. Laemmerhofer, M., Lindner, W. 1996. Quinine and quinidine derivatives as chiral selectors. I. Brush type chiral stationary phases for high-performance liquid chromatography based on cinchonan carbamates and their application as chiral anion exchangers. *J. Chromatogr. A* 741:33–48.
47. Mandl, A., Nicoletti, L., Laemmerhofer, M., Lindner, W. 1999. Quinine versus carbamoylated quinine-based chiral anion exchangers. A comparison regarding enantioselectivity for N-protected amino acids and other chiral acids. *J. Chromatogr. A* 858:1–11.
48. Bicker, W., Laemmerhofer, M., Lindner, W. 2004. Direct high-performance liquid chromatographic method for enantioselective and diastereoselective determination of selected pyrethroic acids. *J. Chromatogr. A* 1035:37–46.
49. Czerwenka, C., Laemmerhofer, M., Maier, N.M., Rissanen, K., Lindner, W. 2002. Direct high performance liquid chromatographic separation of peptide enantiomers: Study on chiral recognition by systematic evaluation of the influence of structural features of the chiral selectors on enantioselectivity. *Anal. Chem.* 74:5658–66.
50. Laemmerhofer, M., Tobler, E., Zarbl, E., Lindner, W., Svec, F., Fréchet, J.M. 2003. Macroporous monolithic chiral stationary phases for capillary electrochromatography: New chiral monomer derived from cinchona alkaloid with enhanced enantioselectivity. *Electrophoresis* 24:2986–99.
51. Laemmerhofer, M., Lindner, W. 2004. Chiral separations by capillary electrophoresis using cinchona alkaloid derivatives as chiral counter-ions. *Methods Mol. Biol.* 243:323–42.
52. Schefzick, S., Laemmerhofer, M., Lindner, W., Lipkowitz, K., Jalaie, M. 2000. Comparative molecular field analysis of quinine derivatives used as chiral selectors in liquid chromatography: 3D QSAR for the purposes of molecular design of chiral stationary phases. *Chirality* 12:742–50.
53. Lah, J., Maier, N.M., Lindner, W., Vesnaver, G. 2001. Thermodynamics of binding of (R)- and (S)-dinitrobenzoyl leucine to cinchona alkaloids and their tert-butylcarbamate derivatives in methanol: Evaluation of enantioselectivity by spectroscopic (CD, UV) and microcalorimetric (ITC) titrations. *J. Phys. Chem. B* 105:1670–78.
54. Wirz, R., Buergi, T., Lindner, W., Baiker, A. 2004. Absolute configuration modulation attenuated total reflection IR spectroscopy: An in situ method for probing chiral recognition in liquid chromatography. *Anal. Chem.* 76:5319–30.
55. Czerwenka, C., Zhang, M.M., Kaehlig, H.P., Maier, N.M., Lipkowitz, K.B., Lindner, W. 2003. Chiral recognition of peptide enantiomers by cinchona alkaloid derived chiral selectors: Mechanistic investigations by liquid chromatography, NMR spectroscopy, and molecular modeling. *J. Org. Chem.* 68:8315–27.
56. Czerwenka, C., Maier, N.M., Lindner, W. 2004. Enantiomer discrimination by mass spectrometry: Noncovalent interactions of an N-derivatized dipeptide with various cinchona alkaloid derivatives and comparison with enantioselective liquid-phase separations. *Anal. Bioanal. Chem.* 379:1039–44.
57. Schug, K.A., Frycak, P., Maier, N.M., Lindner, W. 2005. Measurement of solution-phase chiral molecular recognition in the gas phase using electrospray ionization–mass spectrometry. *Anal. Chem.* 77:3660–70.
58. Schug, K.A., Maier, N.M., Lindner, W. 2006. Deuterium isotope effects observed during competitive binding chiral recognition electrospray ionization–mass spectrometry of cinchona alkaloid-based systems. *J. Mass Spectrom.* 41:157–61.

59. Schug, K.A., Maier, N.M., Lindner, W. 2006. Chiral recognition mass spectrometry: Remarkable effects observed from the relative ion abundances of ternary diastereomeric complexes using electrospray ionization. *Chem. Commun.* 414–16.
60. Frycak, P., Schug, K.A. 2007. On-line dynamic titration: Determination of dissociation constants for noncovalent complexes using Gaussian concentration profiles by electrospray ionization mass spectrometry. *Anal. Chem.* 79:5407–13.
61. Frycak, P., Schug, K.A. 2009. High throughput multiplexed method for evaluation of enantioselective performance of chiral selectors by HPLC-ESI-MS and dynamic titration: Cinchona alkaloid carbamates discriminating N-blocked amino acids. *Chirality* DOI: 10.1002/chir.20691.
62. Williams, I.D., Pedersen, S.F., Sharpless, K.B., Lippard, S.J. 1984. Crystal structures of two titanium tartrate asymmetric epoxidation catalysts. *J. Am. Chem. Soc.* 106:6340–41.
63. Plimmer, H.G., Bateman, H.R. 1908. Further results on the experimental treatment of trypanosomiasis: Being a progress report to a committee of the Royal Society. *Proc. R. Soc. London B* 80:477–487.
64. Tracy, J.W., Webster, L.T. 1996. In *Goodman & Gilman's the pharmacological basis of therapeutics*, ed. J.G. Hardman, L.E. Limbird, P.B. Molinoff, R.W. Ruddon, A.G. Gilman, 987–1008. 9th ed. New York: McGraw-Hill Book Co.
65. Mansour, T.E. 1979. Chemotherapy of parasitic worms: New biochemical strategies. *Science* 205:462–69.
66. Duffin, J., René, P. 1991. "Anti-moine; anti-biotique": The public fortunes of the secret properties of antimony potassium tartrate (tartar emetic). *J. Hist. Med. Allied Sci.* 46:440–56.
67. Tapscott, E.R., Belford, R.L., Paul, I.C. 1969. Stereochemistry of tartrato(4-)-bridged binuclear complexes. *Coord. Chem. Rev.* 4:323–59.
68. Sun, P., Krishnan, A., Yadav, A., Singh, S., MacDonnell, F. M., Armstrong, D. W. 2007. Enantiomeric separations of ruthenium(II) polypyridyl complexes using high-performance liquid chromatography (HPLC) with cyclodextrin chiral stationary phases (CSPs). *Inorg. Chem.* 46:10312–20.
69. Nakazawa, H., Yoneda, H. 1978. Chromatographic study of optical resolution. II. Separation of optically active cobalt(III) complexes using potassium antimony D-tartrate as eluent. *J. Chromatogr.* 160:89–99.
70. Martin, S.E., Connatser, R.M., Kane-Maguire, N.A.P., Wheeler, J.F. 2001. Capillary electrophoresis with laser-induced fluorescence detection for chiral analysis and DNA binding studies of ruthenium(II) Tris-diimine complexes. *Anal. Chim. Acta* 445:21–27.
71. Yoneda, H., Nakatani, K., Yamazaki, S. 1985. Chromatographic study of optical resolution. XI. Optimum conditions for optical resolution of uncharged complexes. *J. Chromatogr.* 319:35–41.
72. Miyoshi, K., Izumoto, S., Nakai, K., Yoneda, H. 1986. Mechanism of chiral recognition of octahedral metal complexes effected by bis(μ-d-tartrato)diantimonate(III) anion in solution. II. Cage complexes of the type [Co(N)6]3+. *Inorg. Chem.* 25:4654–57.
73. Izumoto, S., Miyoshi, K., Yoneda, H. 1987. Chromatographic study of optical resolution. XIV. Optical resolution of amine complex cations of the type [Co(N)6]3+ by reversed-phase ion-pair chromatography using bis(μ-d-tartrato)diantimonate(III) and an arsenic analog as an ion-pairing reagent. *Bull. Chem. Soc. Jpn.* 60:3199–205.
74. Oi, N., Kitahara, H., Aoki, F. 1995. Direct separation of carboxylic acid and amine enantiomers by high-performance liquid chromatography on reversed-phase silica gels coated with chiral copper(II) complexes. *J. Chromatogr. A* 707:380–83.

75. Mizuta, T., Sasaki, K., Yamane, H., Miyoshi, K., Yoneda, H. 1998. Chromatographic resolution of a diastereomeric mixture of tris(trans-1,2-diaminocyclohexane)cobalt(III) by bis(μ-tartrato)diantimonate(III) anion and association modes found in their diastereomeric salts. *Bull. Chem. Soc. Jpn.* 71:1055–64.
76. Shelton, C.M., Seaver, K.E., Wheeler, J.F., Kane-Maguire, N.A.P. 1997. Application of capillary electrophoresis for the assessment of enantiomeric purity of α-diimine transition metal complexes. *Inorg. Chem.* 36:1532–33.
77. Harris, J.E., Desai, N., Seaver, K.E., Watson, R.T., Kane-Maguire, N.A., Wheeler, J.F. 2001. Chiral separations of transition metal complexes using capillary zone electrophoresis. *J. Chromatogr. A* 919:427–36.
78. Kodama, S., Yamamoto, A., Matsunaga, A., Soga, T., Hayakawa, K. 2001. Direct chiral resolution of malic acid in apple juice by ligand-exchange capillary electrophoresis using copper(II)-L-tartaric acid as a chiral selector. *Electrophoresis* 22:3286–90.
79. Arakawa, R., Kobayashi, M., Ama, T. 2000. Chiral recognition in association between antimony potassium tartrate and bis(L-alaninate)ethylenediamine cobalt(III) complexes using electrospray ionization mass spectrometry. *J. Am. Soc. Mass Spectrom.* 11:804–8.
80. Arakawa, R., Kobayashi, M., Fukuo, T., Shiraiwa, T. 2001. Studies on the association of 2-thiazolidinecarboxylic acid and antimony potassium tartrate: Chiral recognition and prediction of absolute configuration by electrospray ionization mass spectrometry. *Rapid Commun. Mass Spectrom.* 15:685–89.
81. Wijeratne, A.B., Spencer, S.E., Gracia, J., Armstrong, D.W., Schug, K.A. 2009. Antimony(III)-D,L-tartrates exhibit proton-assisted enantioselective binding in solution and in the gas phase. *J. Am. Soc. Mass Spectrom.* 20:2100–05.
82. Cech, N.B., Enke, C.G. 2000. Relating electrospray ionization response to nonpolar character of small peptides. *Anal. Chem.* 72:2717–23.

# 11 Recognition of Amino Acid Chirality in Polypeptide Ions by MS/MS

*Hongqian Yang and Roman A. Zubarev*

**CONTENTS**


11.1 Introduction .......... 205
11.2 MS/MS Approaches for Chiral Recognition in Peptides .......... 206
11.2.1 Gas-Phase Structures of Polypeptides .......... 206
11.2.2 Mechanisms of Backbone Bond Cleavages in Gas-Phase Peptide Ions .......... 207
11.2.3 Quantitative Aspect of Chiral Recognition by MS/MS .......... 209
11.2.4 Examples of Chiral Recognition with CAD .......... 210
11.2.5 Examples of Chiral Recognition with ECD .......... 214
11.2 Conclusion .......... 216
References .......... 216


## 11.1 INTRODUCTION

In terms of structural change, L → D chirality inversion of an amino acid residue side chain in polypeptides is the most subtle posttranslational modification.[1] It presents a significant challenge for detection by tandem mass spectrometry (MS/MS), as it does not result in either molecular mass deviation or appearance of specific fragment masses. When all amino acids in the polypeptide chain invert their chirality, direct MS/MS distinguishing becomes impossible. However, an MS/MS solution exists if one or several residues in the polypeptide chain are epimerized (epimers are diastereomers that differ in configuration of only one stereogenic center), while the rest preserve their chiral orientation. As early as 1985, Tabet et al. have differentiated diastereoisomers of acetylated diphenylananine peptide by tandem mass spectrometry.[2] This and all subsequent chiral differentiation efforts have been based exclusively on fragment abundances. In this article we describe direct MS/MS approaches for chiral recognition in peptides. Differentiation by tandem MS of peptide-molecule clusters is discussed in other chapters of this book.

## 11.2 MS/MS APPROACHES FOR CHIRAL RECOGNITION IN PEPTIDES

### 11.2.1 Gas-Phase Structures of Polypeptides

Linear polypeptide chains are very floppy molecules, and at typical experimental conditions in the gas phase their ions do not possess a defined structure. Even polypeptide ions trapped at very low temperatures (ca. 6 K) in a cryogenic trap exhibit a great variety of conformations.[3] Yet molecular dynamics simulations (MDSs) revealed that at any given moment, the structure of a polypeptide ion is not completely random but only moderately deviating from one or several "average trajectories."[4] These trajectories represent typical conformations that have different probabilities of occurrence. They are mostly defined by the location and the solvation pattern of ionizing protons, because protonation and proton sharing are the most energetic intramolecular interactions. Protonation sites are mostly determined by intrinsic basicities of amino acid residues and the N terminus, which are parameters not affected by the L → D inversion. Proton solvation (sharing) is usually afforded by proximal carbonyls and other polar functionalities, e.g., side chains of other residues of the same polypeptide chain, and is strongly affected by the backbone structure, including the chirality of amino acid residues. As an example, Figure 11.1 shows average gas-phase trajectories of Trp-cage (NLYIQWLKDGGPSSGRPPPS) dications simulated at 305 K.[4] The two all-L isomers have protonations at $Arg^{16}$ and $Lys^{8}$, and at $Arg^{16}$ and $Gln^{5}$,

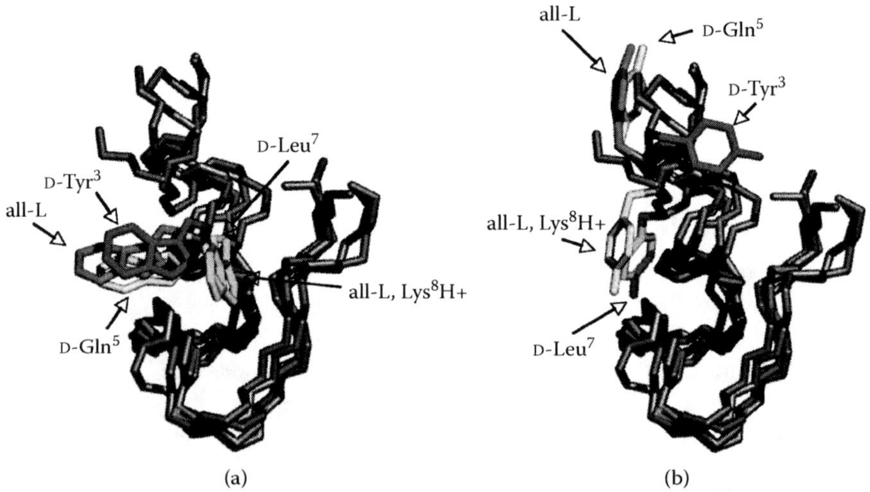


**FIGURE 11.1** (Color Figure 11.1 follows page 46.) Gas-phase structures of Trp-cage (NLYIQWLKDGGPSSGRPPPS) dications calculated as an average trajectory at 305 K shown in comparison (best fit) with each other. The positions of the critical Tyr3 and Trp6 side chains are shown in panels (a) and (b), respectively. Gln5 residue was protonated if not indicated otherwise. The second proton is located on Arg16 in all cases. (Reproduced from Patriksson, A., Adams, C. M., Kjeldsen, F., Raber, J., van der Spoel, D., Zubarev, R. A., Int. J. Mass Spectrom. 248 (2005): 124–35. With permission.)

**TABLE 11.1**
**MDS Results for Trp-Cage 2+ Stereoisomers (the Other Protonated Residue Is $Arg^{16}$): The Number of ps-Long Snapshots at 305 K When a Backbone Carbonyl Establishes a Hydrogen Bond with the Protonated $Gln^5$ Side Chain**

| | $Asn^1$ | $Leu^2$ | $Tyr^3$ | $Ile^4$ | $Gln^5$ | $Pro^{19}$ |
|---|---|---|---|---|---|---|
| D-$Tyr^3$ | 326 | 23 | 0 | 36 | 194 | 134,356 |
| D-$Gln^5$ | 7,965 | 4,261 | 1,022 | 407 | 1,462 | 58,890 |
| D-$Leu^7$ | 10,573 | 1,100 | 1 | 4,519 | 71 | 44,278 |
| $Lys^8H^+$ | — | 8 | 689 | 987 | 138,847 | 90,300 |

*Source:* Patriksson, A., Adams, C. M., Kjeldsen, F., Raber, J., van der Spoel, D., Zubarev, R. A., *Int. J. Mass Spectrom.* 248 (2005): 124–35.

**TABLE 11.2**
**MDS Results for the $Gln^5$-Protonated Form of Stereoisomers of Trp-Cage 2+ Ions (the Other Protonated Residue is $Arg^{16}$): Percentage of Time at 305 K When a Backbone Carbonyl Establishes a Neutral Hydrogen Bond with Any Other Part of the Molecule**

| | $Asn^1$ | $Leu^2$ | $Tyr^3$ | $Ile^4$ | $Gln^5$ | $Pro^{19}$ |
|---|---|---|---|---|---|---|
| D-$Tyr^3$ | 57.6 | 87.0 | 85.9 | 65.1 | 13.0 | N/A |
| D-$Gln^5$ | 72.5 | 58.8 | 39.8 | 54.5 | 29.9 | N/A |
| D-$Leu^7$ | 68.8 | 71.9 | 53.3 | 49.7 | 30.4 | 52.1 |
| $Lys^8H^+$ | 65.3 | 49.8 | 74.7 | 44.1 | 73.4 | N/A |

*Source:* Patriksson, A., Adams, C. M., Kjeldsen, F., Raber, J., van der Spoel, D., Zubarev, R. A., *Int. J. Mass Spectrom.* 248 (2005): 124–35.
*Note:* N/A = data were not calculated.

respectively. The second charge isotopomer is present in all-L, D-$Tyr^3$, D-$Leu^7$, and D-$Gln^5$ isoforms. All isoforms give similar yet distinct average trajectories. The differences are not only in the atom positions and bond angles, but also in the pattern of hydrogen bonds, both charged as well as neutral bonds (Tables 11.1 and 11.2[4]).

### 11.2.2 Mechanisms of Backbone Bond Cleavages in Gas-Phase Peptide Ions

In collision-activated dissociation (CAD), peptide fragmentation produces backbone cleavages (mostly *b* and *y* ions originating from peptide bond rupture; Scheme 11.1), small immonium ions, and small neutral losses (mostly water and ammonia) from both precursor molecular ions and the backbone fragments. In electron capture

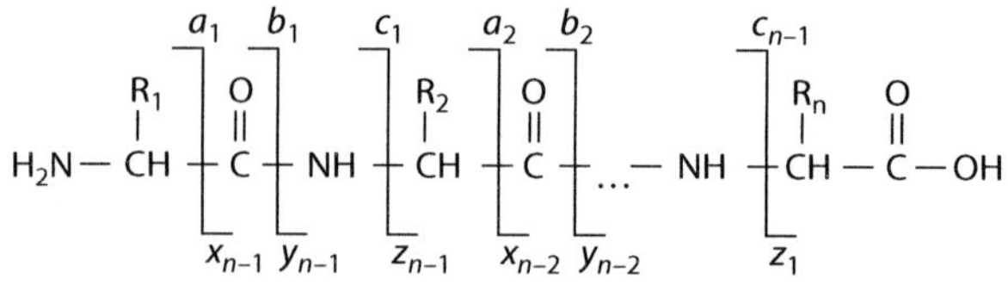


**SCHEME 11.1**

dissociation (ECD), the backbone fragments are typically *c* and *z* ions, and there are abundant charge-reduced species as well as small neutral losses from the latter species. All the above fragments can in principle be used for chiral differentiation. However, since the abundance measurements are most accurate for intense peaks, backbone fragments are the species most commonly used for this purpose.

Backbone fragmentation in CAD is a proton-induced process.[5,6] In most situations, collisional activation releases a mobile proton that moves between backbone carbonyls until it eventually attaches to the backbone -NH- group, whose attachment is followed by rapid C-N bond dissociation. In relatively short sequences, e.g., tryptic peptides, in which the number of ionizing protons exceeds the number of strongly basic groups (mainly arginine side chains), the carbonyl-carbonyl proton transfer is usually facile, and the cleavage probability of a particular peptide bond depends mostly upon the intrinsic carbonyl basicity.[7] The cleavage after the second N-terminal residue ($b_2$, $y_{N-2}$ cleavage) deviates from this general picture for reasons that are still debated in literature.[8] Another deviation is due to acidic amino acids that can donate a proton from their side chain, i.e., Asp and Glu. To a lesser extent, the frequencies of cleavages after amino acids His, Trp, Gln, Glu, Asp, and Asn also deviate from the general trend.[7] The reason for this deviation is the propensity of side chains of these amino acids to create hydrogen bonds with backbone carbonyls, including their own carbonyls. These hydrogen bonds can pose a hindrance for proton transfer, which can reduce the cleavage frequency on one side of the residue while increasing it on the other side. Other hydrogen bonds, most importantly those involved in the formation of helical and sheet-like structures, also influence the cleavage abundances, usually reducing them at the site of extensive hydrogen bonding.[9] Since L → D substitution alters the pattern of hydrogen bonding (see above), fragment abundances in CAD are influenced by chiral changes.

The systematic nature of the CAD fragment abundance changes upon chiral substitution is not yet described in literature. In our recent experiments, we monitored the abundances of the $b_n^+$, $y_{N-n}^{2+}$ fragments of 3+ Trp-cage precursors as a function of L → D substitution of the first five residues. The results are shown in Figure 11.2. As a rule of thumb, the fragment abundances "upstream" of the substitution (i.e., toward the C terminus) increase, especially for the cleavage site one residue away from the substituted residue (e.g., $b_3$+ for the substituted second residue, $b_4$+ for the third residue, etc.). This is consistent with the disruption of the N-terminal alpha-helical structure in Trp-cage by L → D inversion.[4] A notable exception to this rule is the effect of the conversion of the first and second residues on $b_2$+ ions, which is negative. This exception agrees well with the distinctly different formation mechanism of $b_2$+ ions compared to larger *b* ions.[8]

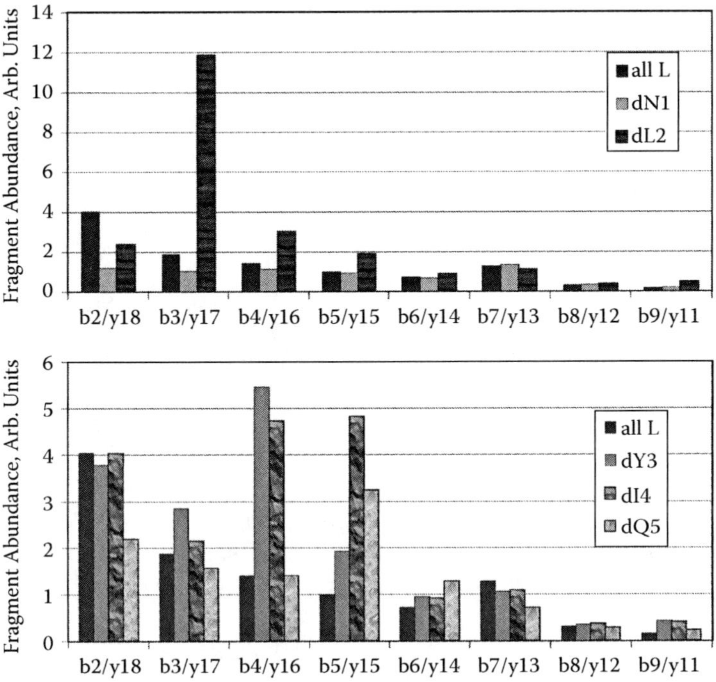


**FIGURE 11.2** Experimentally measured abundances of the $b_n^+$, $y_{N-n}^{2+}$ fragments of 3+ Trp-cage precursors as a function of L → D substitution of the first five residues. The abundance values were normalized by the abundances of the fragments $y_{13}^{2+}$ and $y_{14}^{2+}$.

In ECD, the sensitivity to hydrogen bonding is even stronger than in CAD.[10] One recently proposed ECD mechanism (Scheme 11.2) directly postulates that the sites of N-$C_\alpha$ bond cleavages are involved in neutral hydrogen bonding.[4]

In both CAD and ECD, the temperature of ions plays an important role. At higher temperatures the molecules undergo more frequent interconversions between typical conformations, and the difference between them becomes blurrier.[4] As a result, the pattern of hydrogen bonding becomes less specific, and the chiral recognition is reduced.[10]

Among other factors influencing the chiral recognition is the nature of the ionizing adduct. It appears that the protons are optimal charge carriers: cationization of peptides with singly charged metal ions (e.g., Na+) in MALDI gives poor fragmentation spectra,[11] while adducts with doubly charged cations ($Co^{2+}$ and $Mn^{2+}$) in ESI produce poor chiral recognition.[12]

### 11.2.3 Quantitative Aspect of Chiral Recognition by MS/MS

The fragment ion branching ratios (e.g., the ratio of abundances of certain fragment ions) are $R_D$ and $R_L$ for the species with D- and L-chiral forms, respectively.[13] The quantitative measure of chiral recognition is the chiral recognition factor, $R_{chiral}$:

$$R_{chiral} = R_D/R_L \tag{11.1}$$

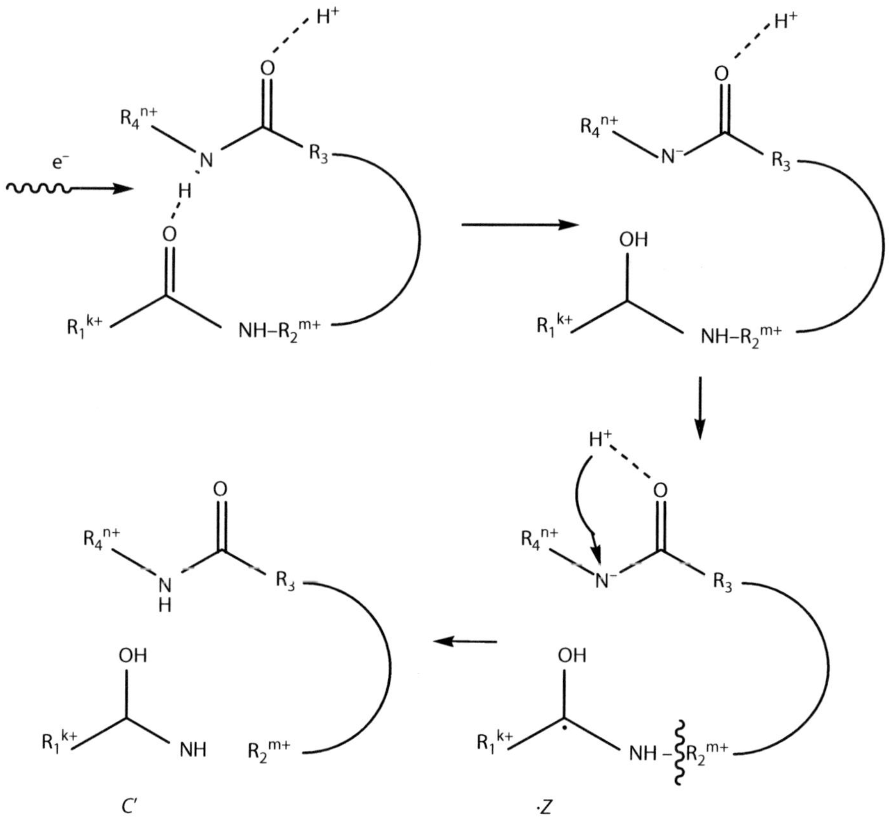


**SCHEME 11.2**

with $R_{chiral}$ = 1 meaning no chiral differentiation and $R_{chiral}$ > 1 meaning positive differentiation. The typical values are $R_{chiral}$ = 2–3 in CAD and $R_{chiral}$ = 5–10 in ECD.[14] If the molecule represents a mixture containing both D- and L-forms with the molar fraction α of the D-form, the experimentally measured fragment ion branching ratio for the mixture is $R_M$:[14]

$$\ln R_M = \ln(R_L) + \ln(R_{chiral})\,\alpha \tag{11.2}$$

Equation 11.2 can be used to obtain a linear calibration plot and then quantitatively measure the D-form peptide content in mixtures.

### 11.2.4 Examples of Chiral Recognition with CAD

The opioid peptides dermorphin (YAFGYPS) and deltorphin (YYAFEVVG) are biologically active in the D-Ala$_2$ form as highly specific ligands to human μ- and δ-opioid receptors, respectively. Since the all-L forms of these peptides are inactive, there is a need for their differentiation from the active forms. Figure 11.3a shows CAD mass spectra of 2+ ions of all-L and D-Ala$_2$ forms of dermorphin.[14] The ratio of

$b_5$ and $y_5$ fragments was used for chiral determination, giving $R_L$ ~ 2.3 and $R_D$ ~ 4.8, which corresponds to $R_{chiral}$ ~ 2.1 (Figure 11.3b). Figure 11.3c shows the calibration plot quantifying the admixture of D-$Ala_2$ form to all-L form.

Another example of chiral recognition by CAD of 2+ ions of the peptide LVFFAEDVGSNK (fragments 17–28 of amyloid β-protein) is shown in Figure 11.4.[12] Here five ion fragments (four for D-Asp and one for D-Ser) showed significant differences compared to the all-L form, with the maximum chiral recognition factor of 2.6. It is quite typical for more bulky amino acid residues to produce higher $R_{chiral}$ than smaller side chains.

Singly charged ions, including those produced by MALDI, also afford chiral recognition.[11] Figure 11.5 demonstrates an example of a MALDI TOF-TOF mass spectrum of the acetylated form of dermorphin produced by CAD. The gray areas represent the regions of mass spectra where the fragment ion abundances vary with chiral substitution. The obtained chiral recognition with $b_5/y_5$ ratios was higher for CAD ($R_{chiral}$ ~ 1.9) than for metastable decomposition ($R_{chiral}$ ~ 1.6). However, even metastable decomposition can provide large chiral recognition factors when a bulky residue is substituted: in the same work, the peptide ASAWTNDEC gave $R_{chiral}$ ~ 7.1 for chiral inversion of the tryptophan residue (Figure 11.6).

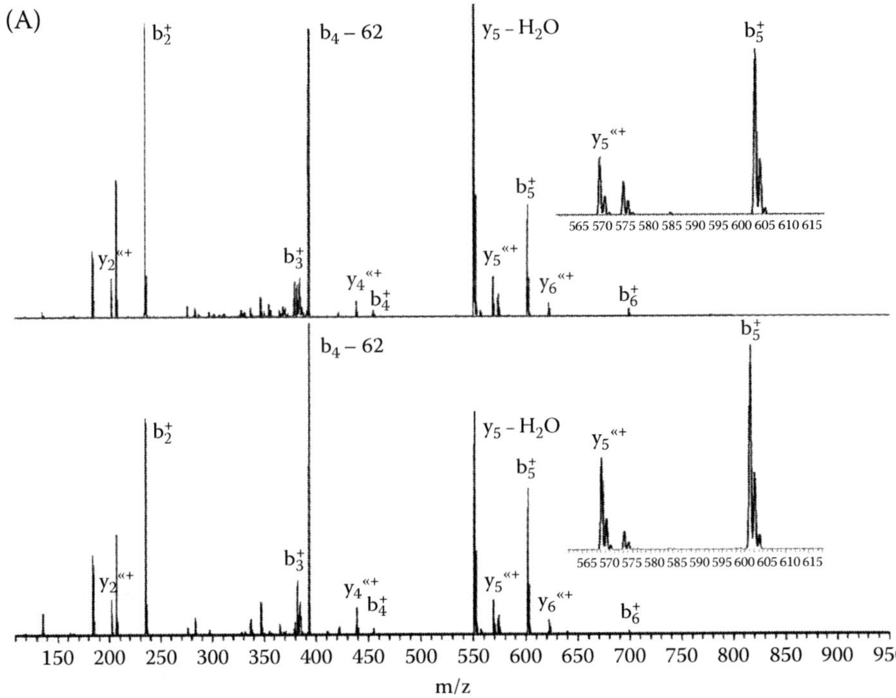


**FIGURE 11.3** (A) CAD spectra of the opioid peptide dermorphin, all L-form (top) and D-$Ala_2$ (bottom). The $b_5$ and $y_5$ fragments shown in the upper right panels were used for chiral determination, with $R_L$ ~ 2.3 and $R_D$ ~ 4.8 at an excitation of 50 arb. Units. (Reproduced from Adams, C., Zubarev, R. A., *Anal. Chem.* 77 (2005): 4571–80. With permission.) (*continued*)

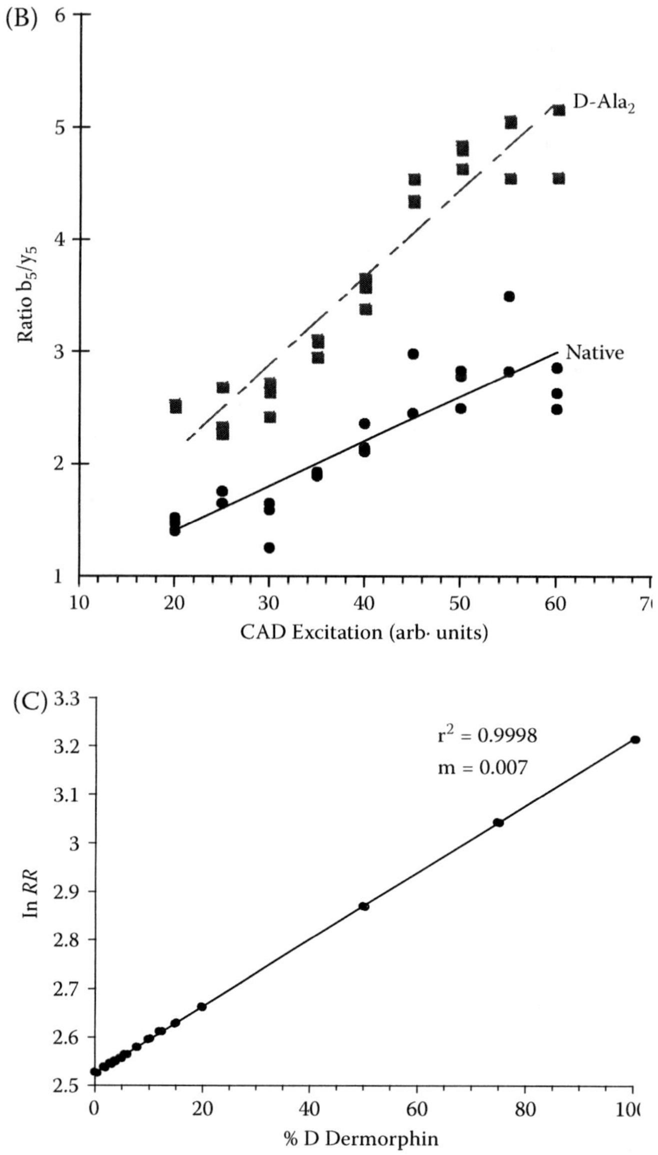


**FIGURE 11.3 (continued)** (B) The effects of CAD excitation on fragment ratios; the $R_{chiral}$ value remained relatively constant. (C) Calibration plot quantifying the admixture of D-Ala$_2$ form to all-L form of the dermorphin peptide. (Reproduced from Adams, C., Zubarev, R. A., *Anal. Chem.* 77 (2005): 4571–80. With permission.)

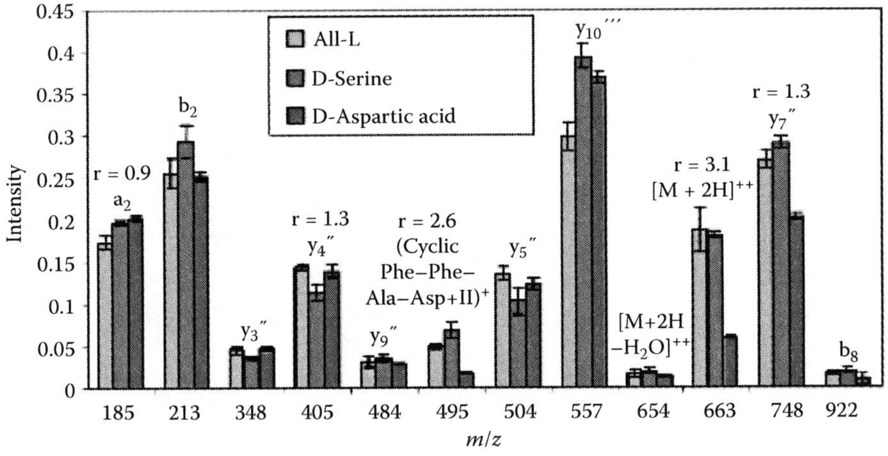


**FIGURE 11.4** Chiral recognition by CAD in 2+ ions of the peptide LVFFAEDVGSNK (fragments 17–28 of amyloid β-protein). Ion ratios show significant differences ($p < 0.01$) compared to the all-L form. (Reproduced from Serafin, S. V., Maranan, R., Zhang, K., Morton, T. H., *Anal. Chem.*, 77 (2005): 5480–87. With permission.)

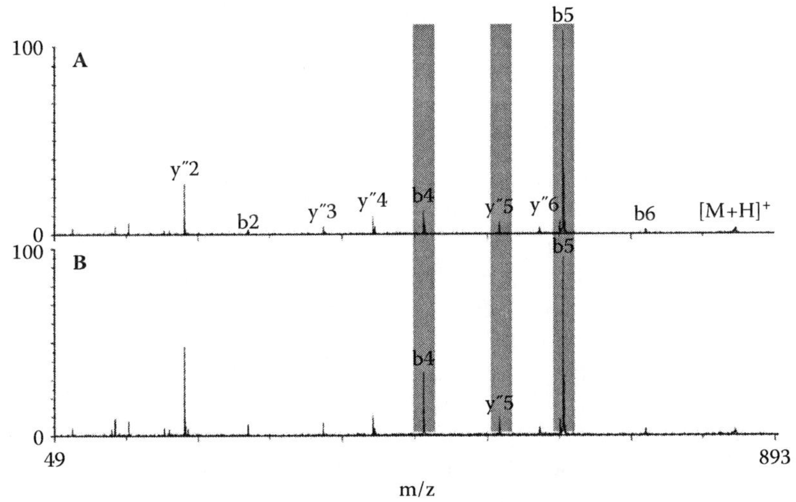


**FIGURE 11.5** MALDI TOF-TOF mass spectrum (CAD) of 1+ ions of the acetylated form of dermorphin. A) Ac-dermorphin; B) Ac-(LA)$_2$- dermorphin. (Adopted from Sachon, E., Clodic, G., Galanth, C., Amiche, M., Ollivaux, C., Soyez, D., Bolbach, G., *Anal. Chem.* 81 (2009): 4389–96.)

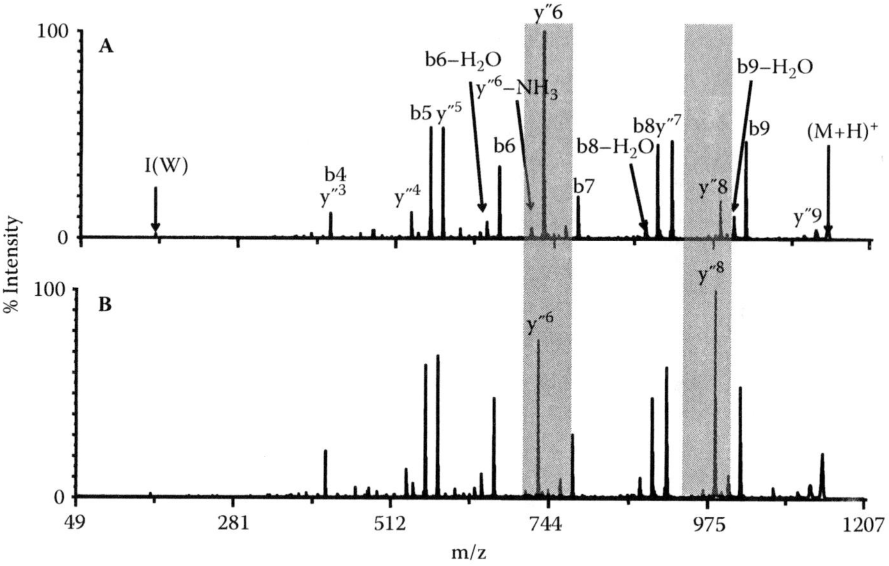


**FIGURE 11.6** MALDI TOF-TOF mass spectrum (metastable decomposition) of 1+ ions of the peptide ASAWTNDEC. A) all-L; B) $(DW)_4$. (Reproduced from Sachon, E., Clodic, G., Galanth, C., Amiche, M., Ollivaux, C., Soyez, D., Bolbach, G., *Anal. Chem.* 81 (2009): 4389–96. With permission.)

## 11.2.5 Examples of Chiral Recognition with ECD

Systematic investigations of the effect of chiral substitution on ECD results were performed in Patriksson et al.,[4] Adams et al.,[10,14,16] and Polfer et al.[15] Figure 11.7a shows a series ECD mass spectra of Trp-cage dications.[4] Not surprisingly, L → D substitution of the bulky tyrosine residue produced a significant effect: compare the abundances of $z_{18}$ and $z_{19}$ ions. This pair of fragments was used for quantification of the D-Tyr[3] admixture to the all-L form with a detection threshold of 1%.[14] In general, the position of the most significant changes in ECD fragment abundances is indicative of the location of the chiral substitution.[15]

Another feature of ECD mass spectra that can be used for chiral recognition is the stability of the charge-reduced species, which are the precursor dications that captured an electron but remained quasi-stable.[16] Their apparent stability depends upon the amount of released recombination energy, which in turn is a function of the distance between the two ionizing protons. Consistent with this picture, Figure 11.8 shows linear dependence of the parameter Lambda (chiral recognition factor calculated for the reduced species) of the intercharge distances determined by MDS.

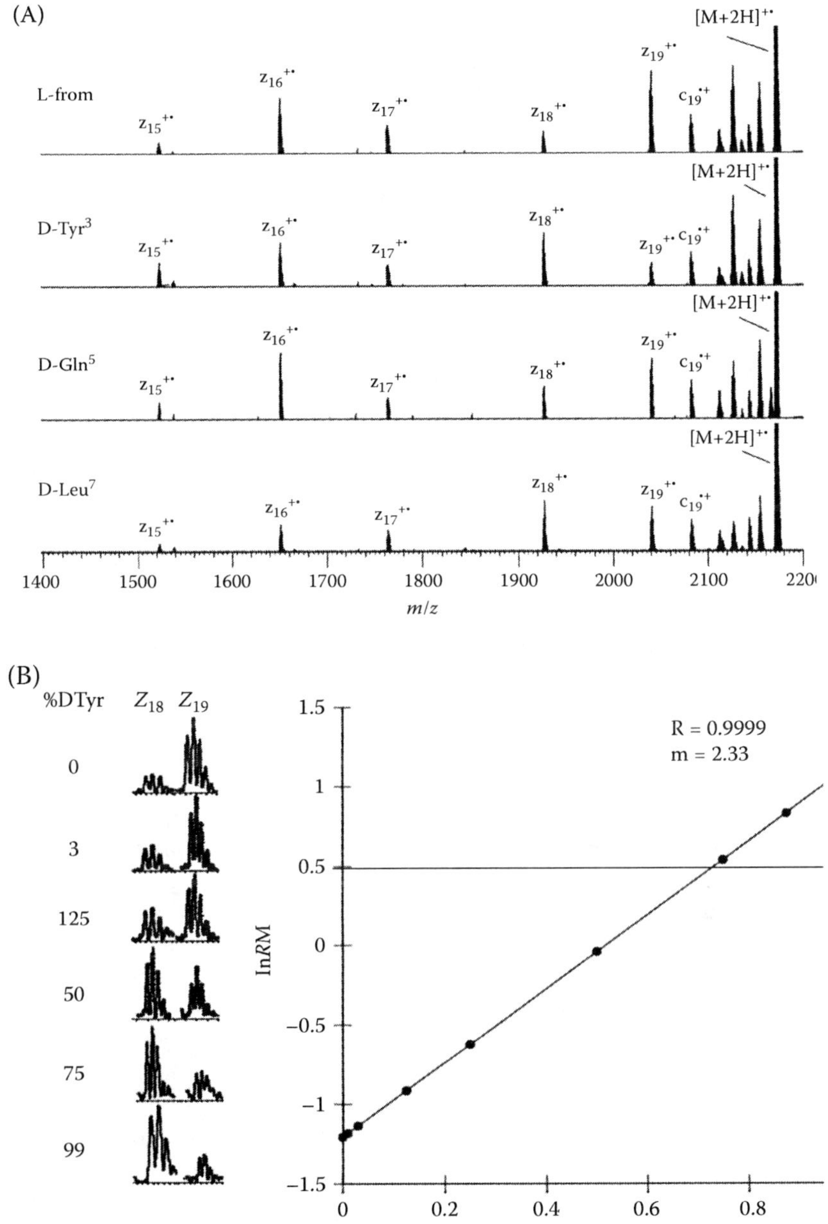


**FIGURE 11.7** (A) ECD fragment abundances of 2+ molecular ions for Trp-cage stereoisomers. Note the large abundance variation between $z_{18}$ and $z_{19}$ fragments for D-Tyr[3], which thereby generated a $R_{\text{chiral}}$ of 10. (Reproduced from Adams, C., Zubarev, R. A., *Anal. Chem.* 77 (2005): 4571–80. With permission.) (B) Variation in $z_{18}$ and $z_{19}$ fragment abundances as a function of D-amino acid content, a property that is quantifiable as seen in the plot to the right. (Reproduced from Adams, C., Zubarev, R. A., *Anal. Chem.* 77 (2005): 4571–80. With permission.)

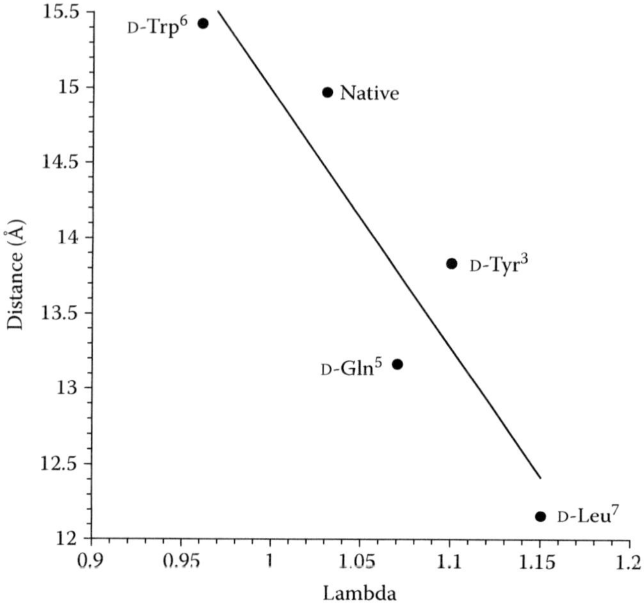


**FIGURE 11.8** Dependence of the parameter Lambda (chiral recognition factor calculated for the reduced species) of the intercharge distances determined by MDS. (From Adams, C. M., Kjeldsen, F., Patriksson, A., van der Speel, D., Gräslund, A., Papadopoulos, E., Zubarev, R. A. *Int. J. Mass Spectrom.* 253 (2006): 263–73. With permission.)

## 11.2 CONCLUSION

Both collisional and electron capture mechanisms of ion excitation produce fragments whose abundances can be used for chiral differentiation in peptide stereoisomers. The mechanism of differentiation is acting through the sensitivity of fragmentation frequencies to the hydrogen-bonding pattern. Since ECD is more sensitive to hydrogen bonding than CAD and acts at lower internal ion temperatures, this technique affords higher chiral recognition than CAD.

## REFERENCES

1. Kreil, G. 1994. Conversion of L- to D-amino acids: A posttranslational reaction. *Science* 266:996–97.
2. Tabet, J. C., Kagan, H. B., Poulin, J. C., Meyer, D., Fraisse, D. 1985. Differentiation of a diastereoisomeric dipeptide by mass spectrometry–mass spectrometry. *Spectroscopy* 4:81–90.
3. Rizzo, T. R., Stearns, J. A., Boyarkin, O. V. 2009. Spectroscopic studies of cold, gas-phase biomolecular ions. *Int. Rev. Phys. Chem.* 28(3).
4. Patriksson, A., Adams, C. M., Kjeldsen, F., Raber, J., van der Spoel, D., Zubarev, R. A. 2005. Prediction of N-$C_\alpha$ bond cleavage frequencies in electron capture dissociation of Trp-cage dications by force-field molecular dynamics simulations. *Int. J. Mass Spectrom.* 248:124–35.

5. Dongre, A. R., Somogyi, A., Wysocki, V. H. 1996. Surface-induced dissociation: An effective tool to probe structure, energetics and fragmentation mechanisms of protonated peptides. *J. Mass Spectrom.* 31:339–50.
6. Burlet, O., Yang, C. Y., Gaskell, S. J. 1992. Influence of cysteine to cysteic acid oxidation on the collision-activated decomposition of protonated peptides: Evidence for intraionic interactions. *J. Am. Soc. Mass Spectrom.* 3:337–44.
7. Savitski, M. M., Kjeldsen, F., Nielsen, M. L., Garbuzynskiy, S. O., Galzitskaya, O. V., Surin, A. K., Zubarev, R. A. 2007. Backbone carbonyl basicities relate gas-phase fragmentation of peptides and protein folding. *Angew. Chem. Int. Ed.* 46:1481–84.
8. Savitski, M. M., Fälth, M., Fung, E. Y. M., Adams, C. M., Zubarev, R. A. 2008. Bifurcating fragmentation behavior of gas-phase tryptic peptide dications in collisional activation. *J. Am. Soc. Mass Spectrom.* 19:1755–63.
9. Zhang, Z. Q. 2005. Prediction of low-energy collision-induced dissociation spectra of peptides with three or more charges. *Anal. Chem.* 77:6364–73.
10. Adams, C., Budnik, B. A., Haselmann, K. F., Kjeldsen, F., Zubarev, R. A. 2004. Electron capture dissociation distinguishes a single D-amino acid in a protein and probes the tertiary structure. *J. Am. Soc. Mass Spectrom.* 15:1087–98.
11. Sachon, E., Clodic, G., Galanth, C., Amiche, M., Ollivaux, C., Soyez, D., Bolbach, G. 2009. D-amino acid detection in peptides by MALDI-TOF-TOF. *Anal. Chem.* 81:4389–96.
12. Serafin, S. V., Maranan, R., Zhang, K., Morton, T. H. 2005. Mass spectrometric differentiation of linear peptides composed of L-amino acids from isomers containing one D-amino acid residue. *Anal. Chem.* 77:5480–87.
13. Tao, W. A., Cooks, R. G. 2003. Chiral analysis by MS. *Anal. Chem.* 75:25A–31A.
14. Adams, C., Zubarev, R. A. 2005. Distinguishing and quantification of peptides and proteins containing D-amino acids by tandem mass spectrometry. *Anal. Chem.* 77:4571–80.
15. Polfer, N. C., Haselmann, K. F., Langridge-Smith, P. R. R., Barran, P. E. 2005. Structural investigation of naturally occurring peptides by electron capture dissociation and AMBER force field modelling. *Mol. Phys.* 15:1481–89.
16. Adams, C. M., Kjeldsen, F., Patriksson, A., van der Spoel, D., Gräslund, A., Papadopoulos, E., Zubarev, R. A. 2006. Probing solution-phase and gas-phase structures of Trp-cage cations by chiral substitution and spectroscopic techniques. *Int. J. Mass Spectrom.* 253:263–73.

# Index


## A

Addition structure, 51
AI/MeLac (cis-1-amino-2-indanol/methyl lactate), 56–57
AMBER
  equilibrium methods, 136, 139
  high-res MS, contact pairs, 30
Amino acids
  amino acid, co-selector, 154–158
  chiral auxiliaries, 148–154, 159–163
  chiral recognition ratio method, 147–148
  co-selector, 154–158
  discrimination, 144–147
  DNA quartets, 159–163
  enantioselectivity, 147–148
  fundamentals, *xv,* 143–144, 164
  mass spectrometry, 144–147
  relative GCA binding order, 158–159
Angle-resolved photoelectroniphotoion coincidence (ARPEPICO), 23
Antimony tartrate, 190–199
Anti-structures, 36
APCI, asymmetric catalytic reactions, 168
Argon, 42–43
Aromatic chromophores, 50–57
ARPEPICO, *see* Angle-resolved photoelectroniphotoion coincidence (ARPEPICO)
Associated conformers, 37
Asymmetric catalytic reactions
  fundamentals, *xiv, xv–xvi,* 167–168, 176
  MSEED method, 168–170
  MSICD method, 171–175
Attachment, 42–44
*ax-ax* conformations, 124, 126–127

## B

Backbone bond cleavages, 207––209
Basis set superposition error (BSSE), 30, 49
Biexponential kinetics, 121, 123, 126
BIRD, *see* Blackbody infrared radiative dissociation (BIRD)
Blackbody infrared radiative dissociation (BIRD), 90
B3LYP/631G/B3LYP/6-31(d,p) theory, 30, 50
Boltzmann characteristics, 16
Born–Oppenheimer approximation, 39
B spline methods, 5, 11
BSSE, *see* Basis set superposition error (BSSE)

## C

CAD, *see* Collision-activated dissociation (CAD)
Calorimetry, 183
Camphor, 7–11
Capillary electrophoresis, 183, 189–190
Caraway seeds example, 28
Case study, selector classes
  antimony tartrate, 190–199
  cinchona alkaloid carbamates, 186–190
  discussion, 189–199
  experimental approach, 184–185
  fundamentals, 181–184, 199
  results, 189–199
Catalysts, screening
  fundamentals, *xiv, xv–xvi,* 167–168, 176
  MSEED method, 168–170
  MSICD method, 171–175
Challenges, PECD, 22
Characteristic structures, 51
Characterization, *xiv,* 133–140
Chemical structure, 15–16
Chiral auxiliaries
  deoxy oligonucleotides, 148–154
  DNA quartets, 159–163
Chiral molecule attachment, 44
Chiral recognition ratio method, 147–148
Chiral scattering, 11–15
Chiral stationary phase (CSP), 144
CID, *see* Collision-induced dissocation (CID)
CIMS (isobutane chemical ionization), 90
Cinchona alkaloid carbamates, 186–190
Circularly polarized radiation (CPL), 2–4, 24
Circumstellar/interstellar ice photochemistry, 24
Cis-1-amino-2-indanol/methyl lactate, 56–57
CMS-Xα, *see* Continuum multiple scattering (CMS-Xα)
Collision-activated dissociation (CAD)
  backbone bond cleavages, 207–209
  recognition examples, 210–214
Collisionally activated metal-bound complexes, 91–95
Collision-induced dissocation (CID)
  deoxy oligonucleotides, 150–153
  DNA quartets, 159, 161
  enantioselectivity, 89–94
  GCA binding order, 158
  host-guest chemistry, 62
Collisions, argon, 42
Collision threshold (CTD), 185, 187–190, 196–198
Compact G conformations, 36–37

Computational methods, 136
Cone conformation, 106
Conformational sensitivity, 13
Contact pairs, *see* Molecular contact pairs
Continuum multiple scattering (CMS-X$\alpha$)
  fenchone *vs.* camphor, 16
  PECD study methods, 5
Cook's kinetic method, 89, 115–116, *see also* Kinetics
Co-selector, amino acid as, 154–158
Coulomb behavior, 6, 104
CPL, *see* Circularly polarized radiation (CPL)
Cumin example, 28
Cycle structure, 51–52
Cytochrome *c*, 104

## D

Deformation energy (Edef), diastereoisomers formation
  addition structure, 51
  characteristic structures, 51
  cis-1-amino-2-indanol/methyl lactate (AI/MeLac), 56–57
  cycle structure, 51–52
  experimental approach, 48–49
  fundamentals, *xiii–xiv*, 47–48, 57–58
  head-to-head structure, 51
  insertion structure, 51
  methyl mandelate/methyl lactate (MeMan/MeLac), 54–55
  $\alpha$-methyl-2-naphthalenemethanol/amino alcohols, 52–54
  results, 50–57
  theoretical approach, 49–50
DELICIOUS/DELICIOUS-VMI
  camphor, 7–8
  fenchone *vs.* camphor, 15–16
  glycidol conformers, 16
  PECD, 4, 23
Density Functional Theory (DFT), 6
Deoxy oligonucleotides
  amino acid, co-selector, 154–158
  chiral auxiliaries, 148–154, 159–163
  chiral recognition ratio method, 147–148
  discrimination, 144–147
  DNA quartets, 159–163
  enantioselectivity, 147–148
  fundamentals, *xv*, 143–144, 164
  mass spectrometry, 144–147
  relative GCA binding order, 158–159
Dependency values, 67–69, 71–72
DEs, *see* Dissocation energies (DEs)
DESI, asymmetric catalytic reactions, 168
Developments, photoelectron circular dichroism, 22–23
DFT, *see* Density Functional Theory (DFT)
Diastereoisomers formation
  addition structure, 51
  characteristic structures, 51
  cis-1-amino-2-indanol/methyl lactate (AI/MeLac), 56–57
  cycle structure, 51–52
  experimental approach, 48–49
  fundamentals, *xiii–xiv*, 47–48, 57–58
  head-to-head structure, 51
  insertion structure, 51
  methyl mandelate/methyl lactate (MeMan/MeLac), 54–55
  $\alpha$-methyl-2-naphthalenemethanol/amino alcohols, 52–54
  results, 50–57
  theoretical approach, 49–50
Diastereoisomization, 31
Diastereomer crystallization technique, 182
Diastereomeric adduct ions, 145
Differential measurements, 4
Dimerization, 43–44
Direct infrared absorption, 44–45
Discrimination, 144–147
Dissocation energies (DEs), 49
DNA quartets, 159–163
DNA triplets, 150–151
Double resonance, *see* Infrared and ultraviolet (IR/UV) double resonance

## E

ECCE package, 136
ECD, *see* Electron capture dissociation (ECD)
Edef, *see* Deformation energy (Edef), diastereoisomers formation
Electron capture dissociation (ECD), 90, 207–209, 214–216
Electron transfer dissociation (ETD), 90
Electrospray ionization, *see* ESI
Enantiodiscrimination, 134–135, 139
Enantiomeric discrimination, amino acids
  chiral auxiliaries, 148–154, 159–163
  chiral recognition ratio method, 147–148
  co-selector, 154–158
  discrimination, 144–147
  DNA quartets, 159–163
  enantioselectivity, 147–148
  fundamentals, *xv*, 143–144, 164
  mass spectrometry, 144–147
  relative GCA binding order, 158–159
Enantioselectivity
  chiral recognition ratio method, 147–148
  collisionally activated metal-bound complexes, 91–95
  cytochrome *c*, 104
  defined, 87
  experimental methodologies, 89–90

fundamentals, *xiv–xv,* 87–89, 127–128
ion recognition, 90–127
kinetic, 88
MSEED method, 168–170
MSICD method, 171–175
oligosaccharides, 95–104
resorcinarenes, 104, 106–120
tetra-amide macrocycles, 120–127
thermally activated proton-bound complexes, 95–127
thermodynamic, 88
Eppink and Parker VMI design, 4
*eq-eq* conformations, 124, 126–127
Equilibrium methods
computational methods, 136
experimental approach, 135
fundamentals, *xiv,* 133–135
results, 136–140
ESI and MALDI MS/enantiomer-labeled method, 81
ESI (electrospray ionization)
chiral selector classes, 183
deoxy oligonucleotides, 148
DNA quartets, 159
enantioselectivity, 89
oligosaccharides, 95
ESI-FT-ICR-MS technique, 104, 108–109, 121–128
ESI-MS
antimony tartrate, 190–199
cinchona alkaloid carbamates, 186–190
discussion, 189–199
experimental approach, 184–185
fundamentals, 181–184, 199
results, 189–199
ESI-MS, enantioselectivity evaluation, 187, 189–190
fundamentals, *xiv, xv–xvi,* 167–168, 176
MSEED method, 168–170
MSICD method, 171–175
ESI-QHQ-CID, 91–95
ETD, *see* Electron transfer dissociation (ETD)
Ethanol, transient recognition
argon, 42–43
attachment, 42–43
chiral molecule attachment, 44
collisions, 42
dimerization, 43–44
experimental techniques, 44–45
fluorination effects, 41
fundamentals, 39–40, 45
parity violating forces, 40
torsional chirality, 40–41
water attachment, 43
EtOH dimers, 31, 36–37
Evaluation, enantioselectivity
fundamentals, *xiv, xv–xvi,* 167–168, 176
MSEED method, 168–170
MSICD method, 171–175
Experimental approaches
chiral molecular contact pairs, 28–30
diastereoisomers formation, 48–49
enantioselectivity, 89–90
equilibrium methods, 135
photoelectron circular dichroism, 4–5
solution phase *vs.* gas phase, 184–185
transient recognition, ethanol, 44–45

## F

FAB
asymmetric catalytic reactions, 168
chiral selector classes, 183
enantioselectivity, 89
FABMS/enantiomer-labeled guest method
dependency values, 67–69, 71–72
fundamentals, 63–64, 66
initial concentrations, 67–68
isotope effects, 66–67
matrix, 72
screening new hosts, 72–73
temperature, 69, 71
FABMS/enantiomer-labeled host method, 75–77
FABMS/relative peak intensity method, 63
Fast atom bombardment, *see* FABMS
Features, photoelectron circular dichroism, 7–11
FE dimers, 31, 36–37
Fenchone *vs.* camphor, 15–16
Fluorescence, 183
Fluorination effects, 41
Formalism, 2–4
Forward-backward asymmetry, 3
Fourier-transform infrared (FTIR) spectrometry, 28, *see also* High-resolution MS, contact pairs
Fourier-transform ion cyclotron resonance (FTICR/MS)
enantiodiscrimination, 135
enantioselectivity, 90, 128
equilibrium methods, 138
host-guest methods, 63
Fourier-transform MW (FTMW) spectrometry, 28, *see also* High-resolution MS, contact pairs
FTICR, *see* Fourier-transform ion cyclotron resonance (FTICR/MS)

## G

Gas phase *vs.* solution phase
antimony tartrate, 190–199
cinchona alkaloid carbamates, 186–190
discussion, 189–199
experimental approach, 184–185

fundamentals, 181–184, 199
results, 189–199
Gaussian 98 engine, 136
GCA binding order, 158–159
Glycidol, 16–20, 31–32
Guests
dependency values, 67–69, 71–72
FABMS/enantiomer-labeled host method, 76–77
fundamentals, 63–64, 66
initial concentrations, 67–68
isotope effects, 66–67
matrix, 72
screening new hosts, 72–73
solution phase screening, 192
temperature, 69, 71

## H

Hairpin structure, 154
Handshakes, dimerization, 43
Head-to-head structure, 51
Helicity, 3, 24
Heterochiral conformers, 31–32
HF/3-21G theory, 30
Highest occupied molecular orbital (HOMO)
camphor, 8–11
fenchone *vs.* camphor, 16
glycidol conformers, 16, 19
High-performance liquid chromatography (HPLC), 182, 189–190
High-resolution MS, contact pairs
experimental details, 28–30
fundamentals, *xiii,* 27–28, 31, 37
induction, 33, 35–36
permanent-permanent interactions, 31–32
permanent-transient interactions, 33, 35–36
recognition, 31–32
synchronization, 36–37
theoretical approaches, 30
transient-transient interactions, 36–37
High-throughput screening, *see* Asymmetric catalytic reactions
HOMO, *see* Highest occupied molecular orbital (HOMO)
Homochirality, 23–24, 90
Hosts
cytochrome *c,* 104
FABMS/enantiomer-labeled guest method, 72–73
FABMS/enantiomer-labeled host method, 75–77
guest method relationship, 76–77
mass spectrometry, 72–73, 75–77
oligosaccharides, 95–104
recognition, mass spectrometry, 72–73, 75–77
resorcinarenes, 104, 106–120
screening, 72–73
screening new hosts, 72–73
solution phase screening, 192
tetra-amide macrocycles, 120–127
HPLC, *see* High-performance liquid chromatography (HPLC)

## I

Induction, 33, 35–36
Infrared and Raman detection, transient recognition
argon, 42–43
attachment, 42–43
chiral molecule attachment, 44
collisions, 42
dimerization, 43–44
experimental techniques, 44–45
fluorination effects, 41
fundamentals, 39–40, 45
parity violating forces, 40
torsional chirality, 40–41
water attachment, 43
Infrared and ultraviolet (IR/UV) double resonance, 48–49
Initial concentrations, 67–68
Inserted conformers, 37
Insertion structure, 51
Integral measurements, 4
Interactions, chiral molecular contact pairs
permanent-permanent, 31–32
permanent-transient interactions, 33, 35–36
transient-transient interactions, 36–37
Interstellar/circumstellar ice photochemistry, 24
Ionized orbitals, 10–11
Ion recognition
collisionally activated metal-bound complexes, 91–95
cytochrome *c,* 104
fundamentals, 90–91
oligosaccharides, 95–104
resorcinarenes, 104, 106–120
tetra-amide macrocycles, 120–127
thermally activated proton-bound complexes, 95–127
Ion trap techniques, 90, 184
IR/UV, *see* Infrared and ultraviolet (IR/UV) double resonance
Isotope effects, 66–67

## K

Kinetics, *see also* Cook's kinetic method
asymmetric catalytic reactions, 170
biexponential, 121, 123, 126
chiral ion recognition, 91
chiral recognition, 145

diastereomeric adduct ions, 145
enantioselectivity, 88, 92
oligosaccharides, 95–96

**L**

LCAO, *see* B spline methods
Legendre properties, 3–5
Lyman α energy, 24

**M**

Macromodel package, 136
Magic angle, 5
MALDI
asymmetric catalytic reactions, 168
backbone bond cleavages, 209
chiral recognition, CAD, 211
chiral selector classes, 183
enantioselectivity, 89
MALDI MS/enantiomer-labeled method, 81
Mass spectrometry (MS)
antimony tartrate, 190
chiral discrimination, 144–147
cinchona alkaloid carbamates, 186
dependency values, 67–69, 71–72
ESI and MALDI MS/enantiomer-labeled method, 81
FABMS/enantiomer-labeled guest method, 63–73
FABMS/enantiomer-labeled host method, 75–77
FABMS/relative peak intensity method, 63
fundamentals, *xiv*, 61–62, 82
guest method relationship, 76–77
initial concentrations, 67–68
isotope effects, 66–67
matrix, 72
screening new hosts, 72–73
tandem, 147
temperature, 69, 71
Mass spectrometry (MS), deoxy oligonucleotides
amino acid, co-selector, 154–158
chiral auxiliaries, 148–154, 159–163
chiral recognition ratio method, 147–148
discrimination, 144–147
DNA quartets, 159–163
enantioselectivity, 147–148
fundamentals, *xv*, 143–144, 164
mass spectrometry, 144–147
relative GCA binding order, 158–159
Matrix, 72
Matrix-assisted laser desorption ionization, *see* MALDI
MeMan/MeLac (methyl mandelate/methyl lactate), 54–55
Methyl functional groups, 35
Methyl mandelate/methyl lactate (MeMan/MeLac), 54–55
α-methyl-2-naphthalenemethanol/amino alcohols, 52–54
Mint example, 28
MM, *see* Molecular modeling (MM)
MMFF94, 136, 139
Molecular beam clustering techniques, 23
Molecular contact pairs
experimental details, 28–30
fundamentals, *xiii*, 27–28, 31, 37
induction, 33, 35–36
permanent-permanent interactions, 31–32
permanent-transient interactions, 33, 35–36
recognition, 31–32
synchronization, 36–37
theoretical approaches, 30
transient-transient interactions, 36–37
Molecular modeling (MM), 97, 100, 128
Møller–Plesset (MP-2) perturbation theory, 30
Monte Carlo techniques, 136
MP-2, *see* Møller–Plesset (MP-2) perturbation theory
MSEED method, 168–170
MSICD method, 171–175
MS/MS approaches, polypeptides, 206–214
Multidimensional nature, PECD spectroscopy, 2
Multiplexed (MUX) sprayer system, 170
Multipoint interactions, DNA quartets, 162
MUX (multiplexed sprayer system), 170

**N**

Nuclear magnetic resonance (NMR) spectroscopy
chiral selector classes, 183
host-guest chemistry, 62, 75
NWChem engine, 136

**O**

OH functional groups
AI/MeLac, 56
deformation energy, 51–52
permanent-permanent interactions, 31–32
Oligonucleotides, *see* Deoxy oligonucleotides
Oligosaccharides, 95–104
Ox molecule, 35

**P**

Parity violating forces, 40
Permanent-permanent interactions, 31–32
Permanent-transient interactions, 33, 35–36
Perturbation exchange theory, 49
PES, *see* Potential energy surface (PES)
Photoelectron circular dichroism (PECD)

applications, 22–23
camphor, 7–11
chemical structure, 15–16
chiral scattering, 11–15
developments, 22–23
experimental approaches, 4–5
fenchone *vs.* camphor, 15–16
formalism, 2–4
fundamentals, *xi–xii,* xvi, 1–2
glycidol conformers, 16–20
homochirality, 23–24
main features, 7–11
structure probe, 11–20
study methods, 4–6
theoretical approaches, 5–6

PO dimers, 31–32

Polypeptide ions, amino acid chirality recognition
backbone bond cleavages, 207––209
CAD examples, 210–214
ECD examples, 214–216
fundamentals, 205, 214, 216
gas-phase structures, 206–207
MS/MS approaches, 206–214
quantitative aspect, 209–210

Potential energy surface (PES), 49

## Q

Quantitative aspect, polypeptides, 209–210

## R

Racemic mixtures/racemization, 41–42, 90

Raman detection, *see* Infrared also and Raman detection, transient recognition
experimental techniques, 44–45
torsional chirality, 41
water attachment, 43

Recognition
chiral molecular contact pairs, 31–32
defined, 27, 40
equilibrium methods, *xiv,* 133–140

Recognition, amino acid chirality
backbone bond cleavages, 207––209
CAD examples, 210–214
ECD examples, 214–216
fundamentals, 205, 214, 216
gas-phase structures, 206–207
MS/MS approaches, 206–214
quantitative aspect, 209–210

Recognition, mass spectrometry
dependency values, 67–69, 71–72
ESI and MALDI MS/enantiomer-labeled method, 81
FABMS/enantiomer-labeled guest method, 63–73
FABMS/enantiomer-labeled host method, 75–77
FABMS/relative peak intensity method, 63
fundamentals, *XIV,* 61–62, 82
guest method relationship, 76–77
initial concentrations, 67–68
isotope effects, 66–67
matrix, 72
screening new hosts, 72–73
temperature, 69, 71

Relative GCA binding order, 158–159

Resorcinarenes, 104, 106–120

Results
diastereoisomers formation, 50–57
equilibrium methods, 136–140
solution phase *vs.* gas phase, 189–199

RR/RSastereoisomer, 54–55, 56

## S

Screening, chiral catalysts
fundamentals, *xiv, xv xvi,* 167 168, 176
MSEED method, 168–170
MSICD method, 171–175

Screening new hosts, 72–73

"Seeding" Earth, 25

Selector classes, case study
antimony tartrate, 190–199
cinchona alkaloid carbamates, 186–190
discussion, 189–199
experimental approach, 184–185
fundamentals, 181–184, 199
results, 189–199

Solvents, 50–57

Spearmint example, 28

SR, *see* Synchrotron radiation (SR)

Stereomutation, *see* High-resolution MS, contact pairs
argon attachment, 42–43
fluorination effects, 41
fundamentals, 40
microwave spectroscopy, 45
torsional chirality, 41

Structure probe, PECD
chemical structure, 15–16
chiral scattering, 11–15
fenchone *vs.* camphor, 15–16
glycidol conformers, 16–20

Structures, polypeptides, 206–207

Synchronization
chiral molecular contact pairs, 36–37
dimerization, 44

Synchrotron radiation (SR), 4, 22

Syn-structures, 36

## T

Temperature, 69, 71
Tetra-amide macrocycles, 120–127
Theoretical approaches
  chiral molecular contact pairs, 30
  diastereoisomers formation, 49–50
  photoelectron circular dichroism, 5–6
Thermally activated proton-bound complexes
  cytochrome *c*, 104
  ion recognition, 95–127
  oligosaccharides, 95–104
  resorcinarenes, 104, 106–120
  tetra-amide macrocycles, 120–127
Time-of-flight (TOF) spectrometry, 23, 170, 211
Torsional chirality, 40–41
Transient recognition, ethanol
  argon, 42–43
  attachment, 42–43
  chiral molecule attachment, 44
  collisions, 42
  dimerization, 43–44
  experimental techniques, 44–45
  fluorination effects, 41
  fundamentals, 39–40, 45
  parity violating forces, 40
  torsional chirality, 40–41
  water attachment, 43
Transient-transient interactions, 36–37

## U

UV/Vis, 183

## V

van der Waals complexes, 42
Velocity map imaging (VIM), 4
VMI, *see* Velocity map imaging (VIM)
VUV-PES method, 19
VUV radiation, 16, 24
VUV SR photon flux improvements, 22

## W

Water attachment, 43
Watson–Crick technique, 144

## X

X-ray photoelectron spectrum (XPS), 19

## Z

Zero-point energy (ZPE), 30, 49